精通 TensorFlow 1.x(影印版)

Mastering TensorFlow 1.x

Armando Fandango 著

南京　东南大学出版社

图书在版编目(CIP)数据

精通 TensorFlow 1.x:英文/(美)阿曼多·范丹戈(Armando Fandango)著. —影印本. —南京:东南大学出版社,2019.3

书名原文:Mastering TensorFlow 1.x

ISBN 978-7-5641-8292-2

Ⅰ.①精… Ⅱ.①阿… Ⅲ.①人工智能-算法-英文 Ⅳ.①TP18

中国版本图书馆 CIP 数据核字(2019)第 025336 号

图字:10-2018-493 号

精通 TensorFlow 1.x(影印版)

出版发行:东南大学出版社
地　　址:南京四牌楼 2 号　　邮编:210096
出 版 人:江建中
网　　址:http://www.seupress.com
电子邮件:press@seupress.com
印　　刷:常州市武进第三印刷有限公司
开　　本:787 毫米×980 毫米　16 开本
印　　张:29.75
字　　数:583 千字
版　　次:2019 年 3 月第 1 版
印　　次:2019 年 3 月第 1 次印刷
书　　号:ISBN 978-7-5641-8292-2
定　　价:108.00 元

本社图书若有印装质量问题,请直接与营销部联系。电话(传真):025-83791830

Mapt

Foreword

TensorFlow and Keras are a key part of the "Data Science for Internet of Things" course, which I teach at the University of Oxford. My TensorFlow journey started with Keras. Over time, in our course, we increasingly gravitated towards core TensorFlow in addition to Keras. I believe many people's 'TensorFlow journey' will follow this trajectory.

Armando Fandango's book "Mastering TensorFlow 1.x" provides a road map for this journey. The book is an ambitious undertaking, interweaving Keras and core TensorFlow libraries. It delves into complex themes and libraries such as Sonnet, distributed TensorFlow with TF Clusters, deploying production models with TensorFlow Serving, TensorFlow mobile, and TensorFlow for embedded devices.

In that sense, this is an advanced book. But the author covers deep learning models such as RNN, CNN, autoencoders, generative adversarial models, and deep reinforcement learning through Keras. Armando has clearly drawn upon his experience to make this complex journey easier for readers.

I look forward to increased adoption of this book and learning from it.

Ajit Jaokar

Data Science for IoT Course Creator and Lead Tutor at the University of Oxford / Principal Data Scientist.

Contributors

About the author

Armando Fandango creates AI-empowered products by leveraging his expertise in deep learning, computational methods, and distributed computing. He advises Owen.ai Inc on AI product strategy. He founded NeuraSights Inc. with the goal of creating insights using neural networks. He is the founder of Vets2Data Inc., a non-profit organization assisting US military veterans in building AI skills.

Armando has authored books titled Python Data Analysis - 2nd Edition and Mastering TensorFlow and published research in international journals and conferences.

I would like to thank Dr. Paul Wiegand (UCF), Dr. Brian Goldiez (UCF), Tejas Limkar (Packt), and Tushar Gupta (Packt) for being able to complete this book. This work would not be possible without their inspiration.

About the reviewer

Nick McClure is currently a senior data scientist at PayScale Inc in Seattle, Washington, USA. Previously, he worked at Zillow and Caesar's Entertainment. He has degrees in applied mathematics from the University of Montana and the College of Saint Benedict and Saint John's University. He has also authored TensorFlow Machine Learning Cookbook by Packt.

He has a passion for learning and advocating for analytics, machine learning, and artificial intelligence. he occasionally puts his thoughts and musings on his blog, `fromdata.org`, or through his Twitter account at `@nfmcclure`.

Packt is searching for authors like you

If you're interested in becoming an author for Packt, please visit `authors.packtpub.com` and apply today. We have worked with thousands of developers and tech professionals, just like you, to help them share their insight with the global tech community. You can make a general application, apply for a specific hot topic that we are recruiting an author for, or submit your own idea.

Table of Contents

Preface

Google's TensorFlow has become a major player and a go-to tool for developers to bring smart processing within an application. TensorFlow has become a major research and engineering tool in every organization. Thus, there is a need to learn advanced use cases of TensorFlow that can be implemented in all kinds of software and devices to build intelligent systems. TensorFlow is one of its kind, with lots of new updates and bug fixes to bring smart automation into your projects. So in today's world, it becomes a necessity to master TensorFlow in order to create advanced machine learning and deep learning applications. *Mastering TensorFlow* will help you learn all the advanced features TensorFlow has to offer. This book funnels down the key information to provide the required expertise to the readers to enter the world of artificial intelligence, thus extending the knowledge of intermediate TensorFlow users to the next level. From implementing advanced computations to trending real-world research areas, this book covers it all. Get to the grips with this highly comprehensive guide to make yourself well established in the developer community, and you'll have a platform to contribute to research works or projects.

Who this book is for

This book is for anyone who wants to build or upgrade their skills in applying TensorFlow to deep learning problems. Those who are looking for an easy-to-follow guide that underlines the intricacies and complex use cases of deep learning will find this book useful. A basic understanding of TensorFlow and Python is required to get the most out of the book.

What this book covers

`Chapter 1`, *TensorFlow 101*, recaps the basics of TensorFlow, such as how to create tensors, constants, variables, placeholders, and operations. We learn about computation graphs and how to place computation graph nodes on various devices such as GPU. We also learn how to use TensorBoard to visualize various intermediate and final output values.

`Chapter 2`, *High-Level Libraries for TensorFLow*, covers several high-level libraries such as TF Contrib Learn, TF Slim, TFLearn, Sonnet, and Pretty Tensor.

`Chapter 3`, *Keras 101*, gives a detailed overview of the high-level library Keras, which is now part of the TensorFlow core.

`Chapter 4`, *Classical Machine Learning with TensorFlow*, teaches us to use TensorFlow to implement classical machine learning algorithms, such as linear regression and classification with logistic regression.

`Chapter 5`, *Neural Networks and MLP with TensorFlow and Keras*, introduces the concept of neural networks and shows how to build simple neural network models. We also cover how to build deep neural network models known as MultiLayer Perceptrons.

`Chapter 6`, *RNNs with TensorFlow and Keras*, covers how to build Recurrent Neural Networks with TensorFlow and Keras. We cover the internal architecture of RNN, Long Short-Term Networks (LSTM), and Gated Recurrent Units (GRU). We provide a brief overview of the API functions and classes provided by TensorFlow and Keras to implement RNN models.

`Chapter 7`, *RNN for Time Series Data with TensorFlow and Keras*, shows how to build and train RNN models for time series data and provide examples in TensorFlow and Keras libraries.

`Chapter 8`, *RNN for Text Data with TensorFlow and Keras*, teaches us how to build and train RNN models for text data and provides examples in TensorFlow and Keras libraries. We learn to build word vectors and embeddings with TensorFlow and Keras, followed by LSTM models for using embeddings to generate text from sample text data.

`Chapter 9`, *CNN with TensorFlow and Keras*, covers CNN models for image data and provides examples in TensorFlow and Keras libraries. We implement the LeNet architecture pattern for our example.

`Chapter 10`, *Autoencoder with TensorFlow and Keras*, illustrates the Autoencoder models for image data and again provides examples in TensorFlow and Keras libraries. We show the implementation of Simple Autoencoder, Denoising Autoencoder, and Variational Autoencoders.

`Chapter 11`, *TensorFlow Models in Production with TF Serving*, teaches us to deploy the models with TensorFlow Serving. We learn how to deploy using TF Serving in Docker containers and Kubernetes clusters.

`Chapter 12`, *Transfer Learning and Pre-Trained Models*, shows the use of pretrained models for predictions. We learn how to retrain the models on a different dataset. We provide examples to apply the VGG16 and Inception V3 models, pretrained on the ImageNet dataset, to predict images in the COCO dataset. We also show examples of retraining only the last layer of the models with the COCO dataset to improve the predictions.

`Chapter 13`, *Deep Reinforcement Learning*, covers reinforcement learning and the OpenAI gym. We build and train several models using various reinforcement learning strategies, including deep Q networks.

`Chapter 14`, *Generative Adversarial Networks*, shows how to build and train generative adversarial models in TensorFLow and Keras. We provide examples of SimpleGAN and DCGAN.

`Chapter 15`, *Distributed Models with TensorFlow Clusters*, covers distributed training for TensorFLow models using TensorFLow clusters. We provide examples of asynchronous and synchronous update methods for training models in data-parallel fashion.

`Chapter 16`, *TensorFlow Models on Mobile and Embedded Platforms*, shows how to deploy TensorFlow models on mobile devices running on iOS and Android platforms. We cover both TF Mobile and TF Lite APIs of the TensorFlow Library.

`Chapter 17`, *TensorFlow and Keras in R*, covers how to build and train TensorFlow models in R statistical software. We learn about the three packages provided by R Studio that implement the TF Core, TF Estimators, and Keras API in R.

`Chapter 18`, *Debugging TensorFlow Models*, tells us strategies and techniques to find problem hotspots when the models do not work as expected. We cover TensorFlow debugger, along with other methods.

`Appendix`, *Tensor Processing Units*, gives a brief overview of Tensor Processing Units. TPUs are futuristic platforms optimized to train and run TensorFlow models. Although not widely available yet, they are available on the Google Cloud Platform and slated to be available soon outside the GCP.

To get the most out of this book

1. We assume that you are familiar with coding in Python and the basics of TensorFlow and Keras.
2. If you haven't done already, then install Jupyter Notebooks, TensorFlow, and Keras.
3. Download the code bundle for this book that contains the Python, R, and notebook code files.

4. Practice with the code as you read along the text and try exploring by modifying the provided sample code.
5. To practice the Android chapter, you will need Android Studio and an Andrioid device.
6. To practice the iOS chapter, you will need an Apple computer with Xcode and an Apple device.
7. To practice the TensorFlow chapter, you will need Docker and Kubernetes installed. Instruction for installing Kubernetes and Docker on Ubuntu are provided in the book.

Download the example code files

You can download the example code files for this book from your account at `www.packtpub.com`. If you purchased this book elsewhere, you can visit `www.packtpub.com/support` and register to have the files emailed directly to you.

You can download the code files by following these steps:

1. Log in or register at `www.packtpub.com`.
2. Select the **SUPPORT** tab.
3. Click on **Code Downloads & Errata**.
4. Enter the name of the book in the **Search** box and follow the onscreen instructions.

Once the file is downloaded, please make sure that you unzip or extract the folder using the latest version of:

- WinRAR/7-Zip for Windows
- Zipeg/iZip/UnRarX for Mac
- 7-Zip/PeaZip for Linux

The code bundle for the book is also hosted on GitHub at `https://github.com/PacktPublishing/Mastering-TensorFlow-1x`. We also have other code bundles from our rich catalog of books and videos available at `https://github.com/PacktPublishing/`. Check them out!

Conventions used

There are a number of text conventions used throughout this book.

`CodeInText`: Indicates code words in text, database table names, folder names, filenames, file extensions, pathnames, dummy URLs, user input, and Twitter handles. Here is an example: "The most common practice is to use the `with</kbd> block, which will be shown later in this chapter."`

A block of code is set as follows:

```
from datasetslib.ptb import PTBSimple
ptb = PTBSimple()
ptb.load_data()
print('Train :',ptb.part['train'][0:5])
print('Test: ',ptb.part['test'][0:5])
print('Valid: ',ptb.part['valid'][0:5])
print('Vocabulary Length = ',ptb.vocab_len)
```

Bold: Indicates a new term, an important word, or words that you see onscreen. For example, words in menus or dialog boxes appear in the text like this. Here is an example: "Select **System info** from the **Administration** panel."

Warnings or important notes appear like this.

Tips and tricks appear like this.

Get in touch

Feedback from our readers is always welcome.

General feedback: Email `feedback@packtpub.com` and mention the book title in the subject of your message. If you have questions about any aspect of this book, please email us at `questions@packtpub.com`.

Errata: Although we have taken every care to ensure the accuracy of our content, mistakes do happen. If you have found a mistake in this book, we would be grateful if you would report this to us. Please visit `www.packtpub.com/submit-errata`, selecting your book, clicking on the Errata Submission Form link, and entering the details.

Piracy: If you come across any illegal copies of our works in any form on the Internet, we would be grateful if you would provide us with the location address or website name. Please contact us at `copyright@packtpub.com` with a link to the material.

If you are interested in becoming an author: If there is a topic that you have expertise in and you are interested in either writing or contributing to a book, please visit `authors.packtpub.com`.

Reviews

Please leave a review. Once you have read and used this book, why not leave a review on the site that you purchased it from? Potential readers can then see and use your unbiased opinion to make purchase decisions, we at Packt can understand what you think about our products, and our authors can see your feedback on their book. Thank you!

For more information about Packt, please visit `packtpub.com`.

1
TensorFlow 101

TensorFlow is one of the popular libraries for solving problems with machine learning and deep learning. After being developed for internal use by Google, it was released for public use and development as open source. Let us understand the three models of the TensorFlow: **data model**, **programming model**, and **execution model**.

TensorFlow data model consists of tensors, and the programming model consists of data flow graphs or computation graphs. TensorFlow execution model consists of firing the nodes in a sequence based on the dependence conditions, starting from the initial nodes that depend on inputs.

In this chapter, we will review the elements of TensorFlow that make up these three models, also known as the core TensorFlow.

We will cover the following topics in this chapter:

- TensorFlow core
    - Tensors
    - Constants
    - Placeholders
    - Operations
    - Creating tensors from Python objects
    - Variables
    - Tensors generated from library functions

- Data flow graph or computation graph
    - Order of execution and lazy loading
    - Executing graphs across compute devices - CPU and GPGPU
    - Multiple graphs
- TensorBoard overview

This book is written with a practical focus in mind, hence you can clone the code from the book's GitHub repository or download it from Packt Publishing. You can follow the code examples in this chapter with the Jupyter Notebook `ch-01_TensorFlow_101` included in the code bundle.

What is TensorFlow?

According to the TensorFlow website (`www.tensorflow.org`):

> *TensorFlow is an open source library for numerical computation using data flow graphs.*

Initially developed by Google for its internal consumption, it was released as open source on November 9, 2015. Since then, TensorFlow has been extensively used to develop machine learning and deep neural network models in various domains and continues to be used within Google for research and product development. TensorFlow 1.0 was released on February 15, 2017. Makes one wonder if it was a Valentine's Day gift from Google to machine learning engineers!

TensorFlow can be described with a data model, a programming model, and an execution model:

- **Data model** comprises of tensors, that are the basic data units created, manipulated, and saved in a TensorFlow program.
- **Programming model** comprises of data flow graphs or computation graphs. Creating a program in TensorFlow means building one or more TensorFlow computation graphs.
- **Execution** model consists of firing the nodes of a computation graph in a sequence of dependence. The execution starts by running the nodes that are directly connected to inputs and only depend on inputs being present.

To use TensorFlow in your projects, you need to learn how to program using the TensorFlow API. TensorFlow has multiple APIs that can be used to interact with the library. The TF APIs or libraries are divided into two levels:

- **Lower-level library**: The lower level library, also known as TensorFlow core, provides very fine-grained lower level functionality, thereby offering complete control on how to use and implement the library in the models. We will cover TensorFlow core in this chapter.
- **Higher-level libraries**: These libraries provide high-level functionalities and are comparatively easier to learn and implement in the models. Some of the libraries include TF Estimators, TFLearn, TFSlim, Sonnet, and Keras. We will cover some of these libraries in the next chapter.

TensorFlow core

TensorFlow core is the lower level library on which the higher level TensorFlow modules are built. The concepts of the lower level library are very important to learn before we go deeper into learning the advanced TensorFlow. In this section, we will have a quick recap of all those core concepts.

Code warm-up - Hello TensorFlow

As a customary tradition when learning any new programming language, library, or platform, let's write the simple Hello TensorFlow code as a warm-up exercise before we dive deeper.

We assume that you have already installed TensorFlow. If you have not, refer to the TensorFlow installation guide at `https://www.tensorflow.org/install/` for detailed instructions to install TensorFlow.

Open the file `ch-01_TensorFlow_101.ipynb` in Jupyter Notebook to follow and run the code as you study the text.

1. Import the TensorFlow Library with the following code:

```
import tensorflow as tf
```

2. Get a TensorFlow session. TensorFlow offers two kinds of sessions: `Session()` and `InteractiveSession()`. We will create an interactive session with the following code:

```
tfs = tf.InteractiveSession()
```

The only difference between `Session()` and `InteractiveSession()` is that the session created with `InteractiveSession()` becomes the default session. Thus, we do not need to specify the session context to execute the session-related command later. For example, say that we have a session object, `tfs`, and a constant object, `hello`. If `tfs` is an `InteractiveSession()` object, then we can evaluate `hello` with the code `hello.eval()`. If `tfs` is a `Session()` object, then we have to use either `tfs.hello.eval()` or a `with` block. The most common practice is to use the `with` block, which will be shown later in this chapter.

3. Define a TensorFlow constant, `hello`:

```
hello = tf.constant("Hello TensorFlow !!")
```

4. Execute the constant in a TensorFlow session and print the output:

```
print(tfs.run(hello))
```

5. You will get the following output:

```
'Hello TensorFlow !!'
```

Now that you have written and executed the first two lines of code with TensorFlow, let's look at the basic ingredients of TensorFlow.

Tensors

Tensors are the basic elements of computation and a fundamental data structure in TensorFlow. Probably the only data structure that you need to learn to use TensorFlow. A tensor is an n-dimensional collection of data, identified by rank, shape, and type.

Rank is the number of dimensions of a tensor, and **shape** is the list denoting the size in each dimension. A tensor can have any number of dimensions. You may be already familiar with quantities that are a zero-dimensional collection (scalar), a one-dimensional collection (vector), a two-dimensional collection (matrix), and a multidimensional collection.

A scalar value is a tensor of rank 0 and thus has a shape of [1]. A vector or a one-dimensional array is a tensor of rank 1 and has a shape of [columns] or [rows]. A matrix or a two-dimensional array is a tensor of rank 2 and has a shape of [rows, columns]. A three-dimensional array would be a tensor of rank 3, and in the same manner, an n-dimensional array would be a tensor of rank *n*.

Refer to the following resources to learn more about tensors and their mathematical underpinnings:

- Tensors page on Wikipedia, at `https://en.wikipedia.org/wiki/Tensor`
- Introduction to Tensors guide from NASA, at `https://www.grc.nasa.gov/www/k-12/Numbers/Math/documents/Tensors_TM2002211716.pdf`

A tensor can store data of one type in all its dimensions, and the data type of its elements is known as the data type of the tensor.

You can also check the data types defined in the latest version of the TensorFlow library at `https://www.tensorflow.org/api_docs/python/tf/DType`.

At the time of writing this book, the TensorFlow had the following data types defined:

TensorFlow Python API data type	Description
`tf.float16`	16-bit half-precision floating point
`tf.float32`	32-bit single-precision floating point
`tf.float64`	64-bit double-precision floating point
`tf.bfloat16`	16-bit truncated floating point
`tf.complex64`	64-bit single-precision complex
`tf.complex128`	128-bit double-precision complex
`tf.int8`	8-bit signed integer
`tf.uint8`	8-bit unsigned integer
`tf.uint16`	16-bit unsigned integer
`tf.int16`	16-bit signed integer

tf.int32	32-bit signed integer
tf.int64	64-bit signed integer
tf.bool	Boolean
tf.string	String
tf.qint8	Quantized 8-bit signed integer
tf.quint8	Quantized 8-bit unsigned integer
tf.qint16	Quantized 16-bit signed integer
tf.quint16	Quantized 16-bit unsigned integer
tf.qint32	Quantized 32-bit signed integer
tf.resource	Handle to a mutable resource

We recommend that you should avoid using the Python native data types. Instead of the Python native data types, use TensorFlow data types for defining tensors.

Tensors can be created in the following ways:

- By defining constants, operations, and variables, and passing the values to their constructor.
- By defining placeholders and passing the values to `session.run()`.
- By converting Python objects such as scalar values, lists, and NumPy arrays with the `tf.convert_to_tensor()` function.

Let's examine different ways of creating Tensors.

Constants

The constant valued tensors are created using the `tf.constant()` function that has the following signature:

```
tf.constant(
  value,
  dtype=None,
  shape=None,
  name='Const',
  verify_shape=False
```

```
)
```

Let's look at the example code provided in the Jupyter Notebook with this book:

```
c1=tf.constant(5,name='x')
c2=tf.constant(6.0,name='y')
c3=tf.constant(7.0,tf.float32,name='z')
```

Let's look into the code in detail:

- The first line defines a constant tensor `c1`, gives it value 5, and names it x.
- The second line defines a constant tensor `c2`, stores value 6.0, and names it y.
- When we print these tensors, we see that the data types of `c1` and `c2` are automatically deduced by TensorFlow.
- To specifically define a data type, we can use the `dtype` parameter or place the data type as the second argument. In the preceding code example, we define the data type as `tf.float32` for `c3`.

Let's print the constants `c1`, `c2`, and `c3`:

```
print('c1 (x): ',c1)
print('c2 (y): ',c2)
print('c3 (z): ',c3)
```

When we print these constants, we get the following output:

```
c1 (x):  Tensor("x:0", shape=(), dtype=int32)
c2 (y):  Tensor("y:0", shape=(), dtype=float32)
c3 (z):  Tensor("z:0", shape=(), dtype=float32)
```

In order to print the values of these constants, we have to execute them in a TensorFlow session with the `tfs.run()` command:

```
print('run([c1,c2,c3]) : ',tfs.run([c1,c2,c3]))
```

We see the following output:

```
run([c1,c2,c3]) :  [5, 6.0, 7.0]
```

Operations

TensorFlow provides us with many operations that can be applied on Tensors. An operation is defined by passing values and assigning the output to another tensor. For example, in the provided Jupyter Notebook file, we define two operations, `op1` and `op2`:

```
op1 = tf.add(c2,c3)
op2 = tf.multiply(c2,c3)
```

When we print `op1` and `op2`, we find that they are defined as Tensors:

```
print('op1 : ', op1)
print('op2 : ', op2)
```

The output is as follows:

```
op1 :  Tensor("Add:0", shape=(), dtype=float32)
op2 :  Tensor("Mul:0", shape=(), dtype=float32)
```

To print the value of these operations, we have to run them in our TensorFlow session:

```
print('run(op1) : ', tfs.run(op1))
print('run(op2) : ', tfs.run(op2))
```

The output is as follows:

```
run(op1) :  13.0
run(op2) :  42.0
```

The following table lists some of the built-in operations:

Operation types	Operations
Arithmetic operations	`tf.add`, `tf.subtract`, `tf.multiply`, `tf.scalar_mul`, `tf.div`, `tf.divide`, `tf.truediv`, `tf.floordiv`, `tf.realdiv`, `tf.truncatediv`, `tf.floor_div`, `tf.truncatemod`, `tf.floormod`, `tf.mod`, `tf.cross`
Basic math operations	`tf.add_n`, `tf.abs`, `tf.negative`, `tf.sign`, `tf.reciprocal`, `tf.square`, `tf.round`, `tf.sqrt`, `tf.rsqrt`, `tf.pow`, `tf.exp`, `tf.expm1`, `tf.log`, `tf.log1p`, `tf.ceil`, `tf.floor`, `tf.maximum`, `tf.minimum`, `tf.cos`, `tf.sin`, `tf.lbeta`, `tf.tan`, `tf.acos`, `tf.asin`, `tf.atan`, `tf.lgamma`, `tf.digamma`, `tf.erf`, `tf.erfc`, `tf.igamma`, `tf.squared_difference`, `tf.igammac`, `tf.zeta`, `tf.polygamma`, `tf.betainc`, `tf.rint`

Matrix math operations	tf.diag, tf.diag_part, tf.trace, tf.transpose, tf.eye, tf.matrix_diag, tf.matrix_diag_part, tf.matrix_band_part, tf.matrix_set_diag, tf.matrix_transpose, tf.matmul, tf.norm, tf.matrix_determinant, tf.matrix_inverse, tf.cholesky, tf.cholesky_solve, tf.matrix_solve, tf.matrix_triangular_solve, tf.matrix_solve_ls, tf.qr, tf.self_adjoint_eig, tf.self_adjoint_eigvals, tf.svd
Tensor math operations	tf.tensordot
Complex number operations	tf.complex, tf.conj, tf.imag, tf.real
String operations	tf.string_to_hash_bucket_fast, tf.string_to_hash_bucket_strong, tf.as_string, tf.encode_base64, tf.decode_base64, tf.reduce_join, tf.string_join, tf.string_split, tf.substr, tf.string_to_hash_bucket

Placeholders

While constants allow us to provide a value at the time of defining the tensor, the placeholders allow us to create tensors whose values can be provided at runtime. TensorFlow provides the `tf.placeholder()` function with the following signature to create placeholders:

```
tf.placeholder(
  dtype,
  shape=None,
  name=None
  )
```

As an example, let's create two placeholders and print them:

```
p1 = tf.placeholder(tf.float32)
p2 = tf.placeholder(tf.float32)
print('p1 : ', p1)
print('p2 : ', p2)
```

We see the following output:

```
p1 :  Tensor("Placeholder:0", dtype=float32)
p2 :  Tensor("Placeholder_1:0", dtype=float32)
```

Now let's define an operation using these placeholders:

```
op4 = p1 * p2
```

TensorFlow allows using shorthand symbols for various operations. In the earlier example, `p1 * p2` is shorthand for `tf.multiply(p1,p2)`:

```
print('run(op4,{p1:2.0, p2:3.0}) : ',tfs.run(op4,{p1:2.0, p2:3.0}))
```

The preceding command runs the `op4` in the TensorFlow Session, feeding the Python dictionary (the second argument to the `run()` operation) with values for `p1` and `p2`.

The output is as follows:

```
run(op4,{p1:2.0, p2:3.0}) :  6.0
```

We can also specify the dictionary using the `feed_dict` parameter in the `run()` operation:

```
print('run(op4,feed_dict = {p1:3.0, p2:4.0}) : ',
      tfs.run(op4, feed_dict={p1: 3.0, p2: 4.0}))
```

The output is as follows:

```
run(op4,feed_dict = {p1:3.0, p2:4.0}) :  12.0
```

Let's look at one last example, with a vector being fed to the same operation:

```
print('run(op4,feed_dict = {p1:[2.0,3.0,4.0], p2:[3.0,4.0,5.0]}) : ',
    tfs.run(op4,feed_dict = {p1:[2.0,3.0,4.0], p2:[3.0,4.0,5.0]}))
```

The output is as follows:

```
run(op4,feed_dict={p1:[2.0,3.0,4.0],p2:[3.0,4.0,5.0]}):[  6.  12.  20.]
```

The elements of the two input vectors are multiplied in an element-wise fashion.

Creating tensors from Python objects

We can create tensors from Python objects such as lists and NumPy arrays, using the `tf.convert_to_tensor()` operation with the following signature:

```
tf.convert_to_tensor(
  value,
  dtype=None,
  name=None,
  preferred_dtype=None
  )
```

Let's create some tensors and print them for practice:

1. Create and print a 0-D Tensor:

```
tf_t=tf.convert_to_tensor(5.0,dtype=tf.float64)

print('tf_t : ',tf_t)
print('run(tf_t) : ',tfs.run(tf_t))
```

The output is as follows:

```
tf_t :  Tensor("Const_1:0", shape=(), dtype=float64)
run(tf_t) : 5.0
```

2. Create and print a 1-D Tensor:

```
a1dim = np.array([1,2,3,4,5.99])
print("a1dim Shape : ",a1dim.shape)

tf_t=tf.convert_to_tensor(a1dim,dtype=tf.float64)

print('tf_t : ',tf_t)
print('tf_t[0] : ',tf_t[0])
print('tf_t[0] : ',tf_t[2])
print('run(tf_t) : \n',tfs.run(tf_t))
```

The output is as follows:

```
a1dim Shape :  (5,)
tf_t :  Tensor("Const_2:0", shape=(5,), dtype=float64)
tf_t[0] :  Tensor("strided_slice:0", shape=(), dtype=float64)
tf_t[0] :  Tensor("strided_slice_1:0", shape=(), dtype=float64)
run(tf_t) :
 [ 1.    2.    3.    4.    5.99]
```

3. Create and print a 2-D Tensor:

```
a2dim = np.array([(1,2,3,4,5.99),
                  (2,3,4,5,6.99),
                  (3,4,5,6,7.99)
                 ])
print("a2dim Shape : ",a2dim.shape)

tf_t=tf.convert_to_tensor(a2dim,dtype=tf.float64)

print('tf_t : ',tf_t)
print('tf_t[0][0] : ',tf_t[0][0])
print('tf_t[1][2] : ',tf_t[1][2])
print('run(tf_t) : \n',tfs.run(tf_t))
```

The output is as follows:

```
a2dim Shape :  (3, 5)
tf_t :  Tensor("Const_3:0", shape=(3, 5), dtype=float64)
tf_t[0][0] :  Tensor("strided_slice_3:0", shape=(), dtype=float64)
tf_t[1][2] :  Tensor("strided_slice_5:0", shape=(), dtype=float64)
run(tf_t) :
 [[ 1.    2.    3.    4.    5.99]
  [ 2.    3.    4.    5.    6.99]
  [ 3.    4.    5.    6.    7.99]]
```

4. Create and print a 3-D Tensor:

```
a3dim = np.array([[[1,2],[3,4]],
                  [[5,6],[7,8]]
                 ])
print("a3dim Shape : ",a3dim.shape)

tf_t=tf.convert_to_tensor(a3dim,dtype=tf.float64)

print('tf_t : ',tf_t)
print('tf_t[0][0][0] : ',tf_t[0][0][0])
print('tf_t[1][1][1] : ',tf_t[1][1][1])
print('run(tf_t) : \n',tfs.run(tf_t))
```

The output is as follows:

```
a3dim Shape :  (2, 2, 2)
tf_t :  Tensor("Const_4:0", shape=(2, 2, 2), dtype=float64)
tf_t[0][0][0] :  Tensor("strided_slice_8:0", shape=(),
dtype=float64)
tf_t[1][1][1] :  Tensor("strided_slice_11:0", shape=(),
```

```
dtype=float64)
run(tf_t) :
 [[[ 1.  2.][ 3.  4.]]
  [[ 5.  6.][ 7.  8.]]]
```

TensorFlow can seamlessly convert NumPy `ndarray` to TensorFlow tensor and vice-versa.

Variables

So far, we have seen how to create tensor objects of various kinds: constants, operations, and placeholders. While working with TensorFlow to build and train models, you will often need to hold the values of parameters in a memory location that can be updated at runtime. That memory location is identified by variables in TensorFlow.

In TensorFlow, variables are tensor objects that hold values that can be modified during the execution of the program.

While `tf.Variable` appears similar to `tf.placeholder`, there are subtle differences between the two:

tf.placeholder	**tf.Variable**
`tf.placeholder` defines input data that does not change over time	`tf.Variable` defines variable values that are modified over time
`tf.placeholder` does not need an initial value at the time of definition	`tf.Variable` needs an initial value at the time of definition

In TensorFlow, a variable can be created with `tf.Variable()`. Let's see an example of placeholders and variables with a linear model:

$$y = W \times x + b$$

1. We define the model parameters `w` and `b` as variables with initial values of `[.3]` and `[-0.3]`, respectively:

```
w = tf.Variable([.3], tf.float32)
b = tf.Variable([-.3], tf.float32)
```

2. The input `x` is defined as a placeholder and the output `y` is defined as an operation:

```
x = tf.placeholder(tf.float32)
y = w * x + b
```

3. Let's print `w`, `v`, `x`, and `y` and see what we get:

```
print("w:",w)
print("x:",x)
print("b:",b)
print("y:",y)
```

We get the following output:

```
w: <tf.Variable 'Variable:0' shape=(1,) dtype=float32_ref>
x: Tensor("Placeholder_2:0", dtype=float32)
b: <tf.Variable 'Variable_1:0' shape=(1,) dtype=float32_ref>
y: Tensor("add:0", dtype=float32)
```

The output shows that `x` is a placeholder tensor and `y` is an operation tensor, while `w` and `b` are variables with shape `(1,)` and data type `float32`.

Before you can use the variables in a TensorFlow session, they have to be initialized. You can initialize a single variable by running its initializer operation.

For example, let's initialize the variable `w`:

```
tfs.run(w.initializer)
```

However, in practice, we use a convenience function provided by the TensorFlow to initialize all the variables:

```
tfs.run(tf.global_variables_initializer())
```

You can also use the `tf.variables_initializer()` function to initialize only a set of variables.

The global initializer convenience function can also be invoked in the following manner, instead of being invoked inside the `run()` function of a session object:

```
tf.global_variables_initializer().run()
```

After initializing the variables, let's run our model to give the output for values of x = [1,2,3,4]:

```
print('run(y,{x:[1,2,3,4]}) : ',tfs.run(y,{x:[1,2,3,4]}))
```

We get the following output:

```
run(y,{x:[1,2,3,4]}) :  [ 0.          0.30000001  0.60000002  0.90000004]
```

Tensors generated from library functions

Tensors can also be generated from various TensorFlow functions. These generated tensors can either be assigned to a constant or a variable, or provided to their constructor at the time of initialization.

As an example, the following code generates a vector of 100 zeroes and prints it:

```
a=tf.zeros((100,))
print(tfs.run(a))
```

TensorFlow provides different types of functions to populate the tensors at the time of their definition:

- Populating all elements with the same values
- Populating elements with sequences
- Populating elements with a random probability distribution, such as the normal distribution or the uniform distribution

Populating tensor elements with the same values

The following table lists some of the tensor generating library functions to populate all the elements of the tensor with the same values:

Tensor generating function	Description
`zeros( shape, dtype=tf.float32, name=None )`	**Creates a tensor of the provided shape, with all elements set to zero**

Tensor generating function	Description

`zeros_like(`
` tensor,`
` dtype=None,`
` name=None,`
` optimize=True`
`)`
	Creates a tensor of the same shape as the argument, with all elements set to zero

`ones(`
` shape,`
` dtype=tf.float32,`
` name=None`
`)`
	Creates a tensor of the provided shape, with all elements set to one

`ones_like(`
` tensor,`
` dtype=None,`
` name=None,`
` optimize=True`
`)`
	Creates a tensor of the same shape as the argument, with all elements set to one

`fill(`
` dims,`
` value,`
` name=None`
`)`
	Creates a tensor of the shape as the `dims` argument, with all elements set to `value`; for example, `a = tf.fill([100],0)`

Populating tensor elements with sequences

The following table lists some of the tensor generating functions to populate elements of the tensor with sequences:

Tensor generating function	Description

`lin_space(`
` start,`
` stop,`
` num,`
` name=None`
`)`
	Generates a 1-D tensor from a sequence of `num` numbers within the range `[start, stop]`. The tensor has the same data type as the `start` argument.
	For example, `a = tf.lin_space(1,100,10)` generates a tensor with values `[1,12,23,34,45,56,67,78,89,100]`.

`range(`
` limit,`
` delta=1,`
` dtype=None,`
` name='range'`
`)`
 	
`range(`
` start,`
` limit,`
` delta=1,`
` dtype=None,`
` name='range'`
`)`
	Generates a 1-D tensor from a sequence of numbers within the range [`start, limit`], with the increments of `delta`. If the `dtype` argument is not specified, then the tensor has the same data type as the `start` argument. This function comes in two versions. In the second version, if the `start` argument is omitted, then `start` becomes number 0.
	For example, `a = tf.range(1,91,10)` generates a tensor with values `[1,11,21,31,41,51,61,71,81]`. Note that the value of the `limit` argument, that is 91, is not included in the final generated sequence.

Populating tensor elements with a random distribution

TensorFlow provides us with the functions to generate tensors filled with random valued distributions.

The distributions generated are affected by the graph-level or the operation-level seed. The graph-level seed is set using `tf.set_random_seed`, while the operation-level seed is given as the argument `seed` in all of the random distribution functions. If no seed is specified, then a random seed is used.

More details on random seeds in TensorFlow can be found at the following link: `https://www.tensorflow.org/api_docs/python/tf/set_random_seed`.

The following table lists some of the tensor generating functions to populate elements of the tensor with random valued distributions:

Tensor generating function	**Description**

`random_normal(`
` shape,`
` mean=0.0,`
` stddev=1.0,`
` dtype=tf.float32,`
` seed=None,`
` name=None`
`)`
	Generates a tensor of the specified shape, filled with values from a normal distribution: `normal(mean, stddev)`.

truncated_normal(shape, mean=0.0, stddev=1.0, dtype=tf.float32, seed=None, name=None)	Generates a tensor of the specified shape, filled with values from a truncated normal distribution: `normal(mean, stddev)`. Truncated means that the values returned are always at a distance less than two standard deviations from the mean.
random_uniform(shape, minval=0, maxval=None, dtype=tf.float32, seed=None, name=None)	Generates a tensor of the specified shape, filled with values from a uniform distribution: `uniform([minval, maxval))`.
random_gamma(shape, alpha, beta=None, dtype=tf.float32, seed=None, name=None)	Generates tensors of the specified shape, filled with values from gamma distributions: `gamma(alpha,beta)`. More details on the `random_gamma` function can be found at the following link: `https://www.tensorflow.org/api_docs/python/tf/random_gamma`.

Getting Variables with tf.get_variable()

If you define a variable with a name that has been defined before, then TensorFlow throws an exception. Hence, it is convenient to use the `tf.get_variable()` function instead of `tf.Variable()`. The function `tf.get_variable()` returns the existing variable with the same name if it exists, and creates the variable with the specified shape and initializer if it does not exist. For example:

```
w = tf.get_variable(name='w',shape=[1],dtype=tf.float32,initializer=[.3])
b = tf.get_variable(name='b',shape=[1],dtype=tf.float32,initializer=[-.3])
```

The initializer can be a tensor or list of values as shown in above examples or one of the inbuilt initializers:

- `tf.constant_initializer`
- `tf.random_normal_initializer`
- `tf.truncated_normal_initializer`
- `tf.random_uniform_initializer`
- `tf.uniform_unit_scaling_initializer`
- `tf.zeros_initializer`
- `tf.ones_initializer`
- `tf.orthogonal_initializer`

In distributed TensorFlow where we can run the code across machines, the `tf.get_variable()` gives us global variables. To get the local variables TensorFlow has a function with similar signature: `tf.get_local_variable()`.

Sharing or Reusing Variables: Getting already-defined variables promotes reuse. However, an exception will be thrown if the reuse flags are not set by using `tf.variable_scope.reuse_variable()` or `tf.variable.scope(reuse=True)`.

Now that you have learned how to define tensors, constants, operations, placeholders, and variables, let's learn about the next level of abstraction in TensorFlow, that combines these basic elements together to form a basic unit of computation, the data flow graph or computational graph.

Data flow graph or computation graph

A **data flow graph** or **computation graph** is the basic unit of computation in TensorFlow. We will refer to them as the **computation graph** from now on. A computation graph is made up of nodes and edges. Each node represents an operation (`tf.Operation`) and each edge represents a tensor (`tf.Tensor`) that gets transferred between the nodes.

A program in TensorFlow is basically a computation graph. You create the graph with nodes representing variables, constants, placeholders, and operations and feed it to TensorFlow. TensorFlow finds the first nodes that it can fire or execute. The firing of these nodes results in the firing of other nodes, and so on.

Thus, TensorFlow programs are made up of two kinds of operations on computation graphs:

- Building the computation graph
- Running the computation graph

The TensorFlow comes with a default graph. Unless another graph is explicitly specified, a new node gets implicitly added to the default graph. We can get explicit access to the default graph using the following command:

```
graph = tf.get_default_graph()
```

For example, if we want to define three inputs and add them to produce output $y = x_1 + x_2 + x_3$, we can represent it using the following computation graph:

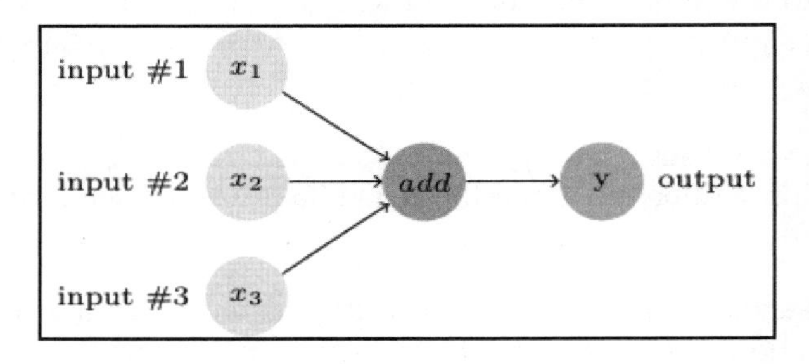

Computation graph

In TensorFlow, the add operation in the preceding image would correspond to the code `y = tf.add( x1 + x2 + x3 )`.

As we create the variables, constants, and placeholders, they get added to the graph. Then we create a session object to *execute* the operation objects and *evaluate* the tensor objects.

Let's build and execute a computation graph to calculate $y = w \times x + b$, as we already saw in the preceding example:

```
# Assume Linear Model y = w * x + b
# Define model parameters
w = tf.Variable([.3], tf.float32)
b = tf.Variable([-.3], tf.float32)
# Define model input and output
x = tf.placeholder(tf.float32)
y = w * x + b
output = 0
```

```
with tf.Session() as tfs:
    # initialize and print the variable y
    tf.global_variables_initializer().run()
    output = tfs.run(y,{x:[1,2,3,4]})
print('output : ',output)
```

Creating and using a session in the `with` block ensures that the session is automatically closed when the block is finished. Otherwise, the session has to be explicitly closed with the `tfs.close()` command, where `tfs` is the session name.

Order of execution and lazy loading

The nodes are executed in the order of dependency. If node *a* depends on node *b*, then *a* will be executed before *b* when the execution of *b* is requested. A node is not executed unless either the node itself or another node depending on it is not requested for execution. This is also known as lazy loading; namely, the node objects are not created and initialized until they are needed.

Sometimes, you may want to control the order in which the nodes are executed in a graph. This can be achieved with the `tf.Graph.control_dependencies()` function. For example, if the graph has nodes *a, b, c,* and *d* and you want to execute *c* and *d* before *a* and *b*, then use the following statement:

```
with graph_variable.control_dependencies([c,d]):
  # other statements here
```

This makes sure that any node in the preceding `with` block is executed only after nodes *c* and *d* have been executed.

Executing graphs across compute devices - CPU and GPGPU

A graph can be divided into multiple parts and each part can be placed and executed on separate devices, such as a CPU or GPU. You can list all the devices available for graph execution with the following command:

```
from tensorflow.python.client import device_lib
print(device_lib.list_local_devices())
```

We get the following output (your output would be different, depending on the compute devices in your system):

```
[name: "/device:CPU:0"
device_type: "CPU"
memory_limit: 268435456
locality {
}
incarnation: 12900903776306102093
, name: "/device:GPU:0"
device_type: "GPU"
memory_limit: 611319808
locality {
  bus_id: 1
}
incarnation: 2202031001192109390
physical_device_desc: "device: 0, name: Quadro P5000, pci bus id:
0000:01:00.0, compute capability: 6.1"
]
```

The devices in TensorFlow are identified with the string `/device:<device_type>:<device_idx>`. In the above output, the `CPU` and `GPU` denote the device type and `0` denotes the device index.

One thing to note about the above output is that it shows only one CPU, whereas our computer has 8 CPUs. The reason for that is TensorFlow implicitly distributes the code across the CPU units and thus by default `CPU:0` denotes all the CPU's available to TensorFlow. When TensorFlow starts executing graphs, it runs the independent paths within each graph in a separate thread, with each thread running on a separate CPU. We can restrict the number of threads used for this purpose by changing the number of `inter_op_parallelism_threads`. Similarly, if within an independent path, an operation is capable of running on multiple threads, TensorFlow will launch that specific operation on multiple threads. The number of threads in this pool can be changed by setting the number of `intra_op_parallelism_threads`.

Placing graph nodes on specific compute devices

Let us enable the logging of variable placement by defining a config object, set the `log_device_placement` property to `true`, and then pass this `config` object to the session as follows:

```
tf.reset_default_graph()

# Define model parameters
w = tf.Variable([.3], tf.float32)
b = tf.Variable([-.3], tf.float32)
# Define model input and output
x = tf.placeholder(tf.float32)
y = w * x + b

config = tf.ConfigProto()
config.log_device_placement=True

with tf.Session(config=config) as tfs:
   # initialize and print the variable y
   tfs.run(global_variables_initializer())
   print('output',tfs.run(y,{x:[1,2,3,4]}))
```

We get the following output in Jupyter Notebook console:

```
b: (VariableV2): /job:localhost/replica:0/task:0/device:GPU:0
b/read: (Identity): /job:localhost/replica:0/task:0/device:GPU:0
b/Assign: (Assign): /job:localhost/replica:0/task:0/device:GPU:0
w: (VariableV2): /job:localhost/replica:0/task:0/device:GPU:0
w/read: (Identity): /job:localhost/replica:0/task:0/device:GPU:0
mul: (Mul): /job:localhost/replica:0/task:0/device:GPU:0
add: (Add): /job:localhost/replica:0/task:0/device:GPU:0
w/Assign: (Assign): /job:localhost/replica:0/task:0/device:GPU:0
init: (NoOp): /job:localhost/replica:0/task:0/device:GPU:0
x: (Placeholder): /job:localhost/replica:0/task:0/device:GPU:0
b/initial_value: (Const): /job:localhost/replica:0/task:0/device:GPU:0
Const_1: (Const): /job:localhost/replica:0/task:0/device:GPU:0
w/initial_value: (Const): /job:localhost/replica:0/task:0/device:GPU:0
Const: (Const): /job:localhost/replica:0/task:0/device:GPU:0
```

Thus by default, the TensorFlow creates the variable and operations nodes on a device where it can get the highest performance. The variables and operations can be placed on specific devices by using `tf.device()` function. Let us place the graph on the CPU:

```
tf.reset_default_graph()

with tf.device('/device:CPU:0'):
    # Define model parameters
    w = tf.get_variable(name='w',initializer=[.3], dtype=tf.float32)
    b = tf.get_variable(name='b',initializer=[-.3], dtype=tf.float32)
    # Define model input and output
    x = tf.placeholder(name='x',dtype=tf.float32)
    y = w * x + b

config = tf.ConfigProto()
config.log_device_placement=True

with tf.Session(config=config) as tfs:
   # initialize and print the variable y
   tfs.run(tf.global_variables_initializer())
   print('output',tfs.run(y,{x:[1,2,3,4]}))
```

In the Jupyter console we see that now the variables have been placed on the CPU and the execution also takes place on the CPU:

```
b: (VariableV2): /job:localhost/replica:0/task:0/device:CPU:0
b/read: (Identity): /job:localhost/replica:0/task:0/device:CPU:0
b/Assign: (Assign): /job:localhost/replica:0/task:0/device:CPU:0
w: (VariableV2): /job:localhost/replica:0/task:0/device:CPU:0
w/read: (Identity): /job:localhost/replica:0/task:0/device:CPU:0
mul: (Mul): /job:localhost/replica:0/task:0/device:CPU:0
add: (Add): /job:localhost/replica:0/task:0/device:CPU:0
w/Assign: (Assign): /job:localhost/replica:0/task:0/device:CPU:0
init: (NoOp): /job:localhost/replica:0/task:0/device:CPU:0
x: (Placeholder): /job:localhost/replica:0/task:0/device:CPU:0
b/initial_value: (Const): /job:localhost/replica:0/task:0/device:CPU:0
Const_1: (Const): /job:localhost/replica:0/task:0/device:CPU:0
w/initial_value: (Const): /job:localhost/replica:0/task:0/device:CPU:0
Const: (Const): /job:localhost/replica:0/task:0/device:CPU:0
```

Simple placement

TensorFlow follows these simple rules, also known as the simple placement, for placing the variables on the devices:

```
If the graph was previously run,
    then the node is left on the device where it was placed earlier
Else If the tf.device() block is used,
    then the node is placed on the specified device
Else If the GPU is present
    then the node is placed on the first available GPU
Else If the GPU is not present
    then the node is placed on the CPU
```

Dynamic placement

The `tf.device()` can also be passed a function name instead of a device string. In such case, the function must return the device string. This feature allows complex algorithms for placing the variables on different devices. For example, TensorFlow provides a round robin device setter in `tf.train.replica_device_setter()` that we will discuss later in next section.

Soft placement

When you place a TensorFlow operation on the GPU, the TF must have the GPU implementation of that operation, known as the kernel. If the kernel is not present then the placement results in run-time error. Also if the GPU device you requested does not exist, you will get a run-time error. The best way to handle such errors is to allow the operation to be placed on the CPU if requesting the GPU device results in n error. This can be achieved by setting the following `config` value:

```
config.allow_soft_placement = True
```

GPU memory handling

When you start running the TensorFlow session, by default it grabs all of the GPU memory, even if you place the operations and variables only on one GPU in a multi-GPU system. If you try to run another session at the same time, you will get out of memory error. This can be solved in multiple ways:

- For multi-GPU systems, set the environment variable CUDA_VISIBLE_DEVICES=<list of device idx>

  ```
  os.environ['CUDA_VISIBLE_DEVICES']='0'
  ```

  The code executed after this setting will be able to grab all of the memory of only the visible GPU.

- When you do not want the session to grab all of the memory of the GPU, then you can use the config option per_process_gpu_memory_fraction to allocate a percentage of memory:

  ```
  config.gpu_options.per_process_gpu_memory_fraction = 0.5
  ```

  This will allocate 50% of the memory of all the GPU devices.

- You can also combine both of the above strategies, i.e. make only a percentage along with making only some of the GPU visible to the process.
- You can also limit the TensorFlow process to grab only the minimum required memory at the start of the process. As the process executes further, you can set a config option to allow the growth of this memory.

  ```
  config.gpu_options.allow_growth = True
  ```

  This option only allows for the allocated memory to grow, but the memory is never released back.

You will learn techniques for distributing computation across multiple compute devices and multiple nodes in later chapters.

Multiple graphs

You can create your own graphs separate from the default graph and execute them in a session. However, creating and executing multiple graphs is not recommended, as it has the following disadvantages:

- Creating and using multiple graphs in the same program would require multiple TensorFlow sessions and each session would consume heavy resources
- You cannot directly pass data in between graphs

Hence, the recommended approach is to have multiple subgraphs in a single graph. In case you wish to use your own graph instead of the default graph, you can do so with the `tf.graph()` command. Here is an example where we create our own graph, `g`, and execute it as the default graph:

```
g = tf.Graph()
output = 0

# Assume Linear Model y = w * x + b

with g.as_default():
 # Define model parameters
 w = tf.Variable([.3], tf.float32)
 b = tf.Variable([-.3], tf.float32)
 # Define model input and output
 x = tf.placeholder(tf.float32)
 y = w * x + b

with tf.Session(graph=g) as tfs:
 # initialize and print the variable y
 tf.global_variables_initializer().run()
 output = tfs.run(y,{x:[1,2,3,4]})

print('output : ',output)
```

TensorBoard

The complexity of a computation graph gets high even for moderately sized problems. Large computational graphs that represent complex machine learning models can become quite confusing and hard to understand. Visualization helps in easy understanding and interpretation of computation graphs, and thus accelerates the debugging and optimizations of TensorFlow programs. TensorFlow comes with a built-in tool that allows us to visualize computation graphs, namely, TensorBoard.

TensorBoard visualizes computation graph structure, provides statistical analysis and plots the values captured as summaries during the execution of computation graphs. Let's see how it works in practice.

A TensorBoard minimal example

1. Start by defining the variables and placeholders for our linear model:

```
# Assume Linear Model y = w * x + b
# Define model parameters
w = tf.Variable([.3], name='w',dtype=tf.float32)
b = tf.Variable([-.3], name='b', dtype=tf.float32)
# Define model input and output
x = tf.placeholder(name='x',dtype=tf.float32)
y = w * x + b
```

2. Initialize a session, and within the context of this session, do the following steps:
    - Initialize global variables
    - Create `tf.summary.FileWriter` that would create the output in the `tflogs` folder with the events from the default graph
    - Fetch the value of node `y`, effectively executing our linear model

```
with tf.Session() as tfs:
    tfs.run(tf.global_variables_initializer())
    writer=tf.summary.FileWriter('tflogs',tfs.graph)
    print('run(y,{x:3}) : ', tfs.run(y,feed_dict={x:3}))
```

3. We see the following output:

```
run(y,{x:3}) :  [ 0.60000002]
```

As the program executes, the logs are collected in the `tflogs` folder that would be used by TensorBoard for visualization. Open the command line interface, navigate to the folder from where you were running the `ch-01_TensorFlow_101` notebook, and execute the following command:

```
tensorboard --logdir='tflogs'
```

You would see an output similar to this:

```
Starting TensorBoard b'47' at http://0.0.0.0:6006
```

Open a browser and navigate to `http://0.0.0.0:6006`. Once you see the TensorBoard dashboard, don't worry about any errors or warnings shown and just click on the **GRAPHS** tab at the top. You will see the following screen:

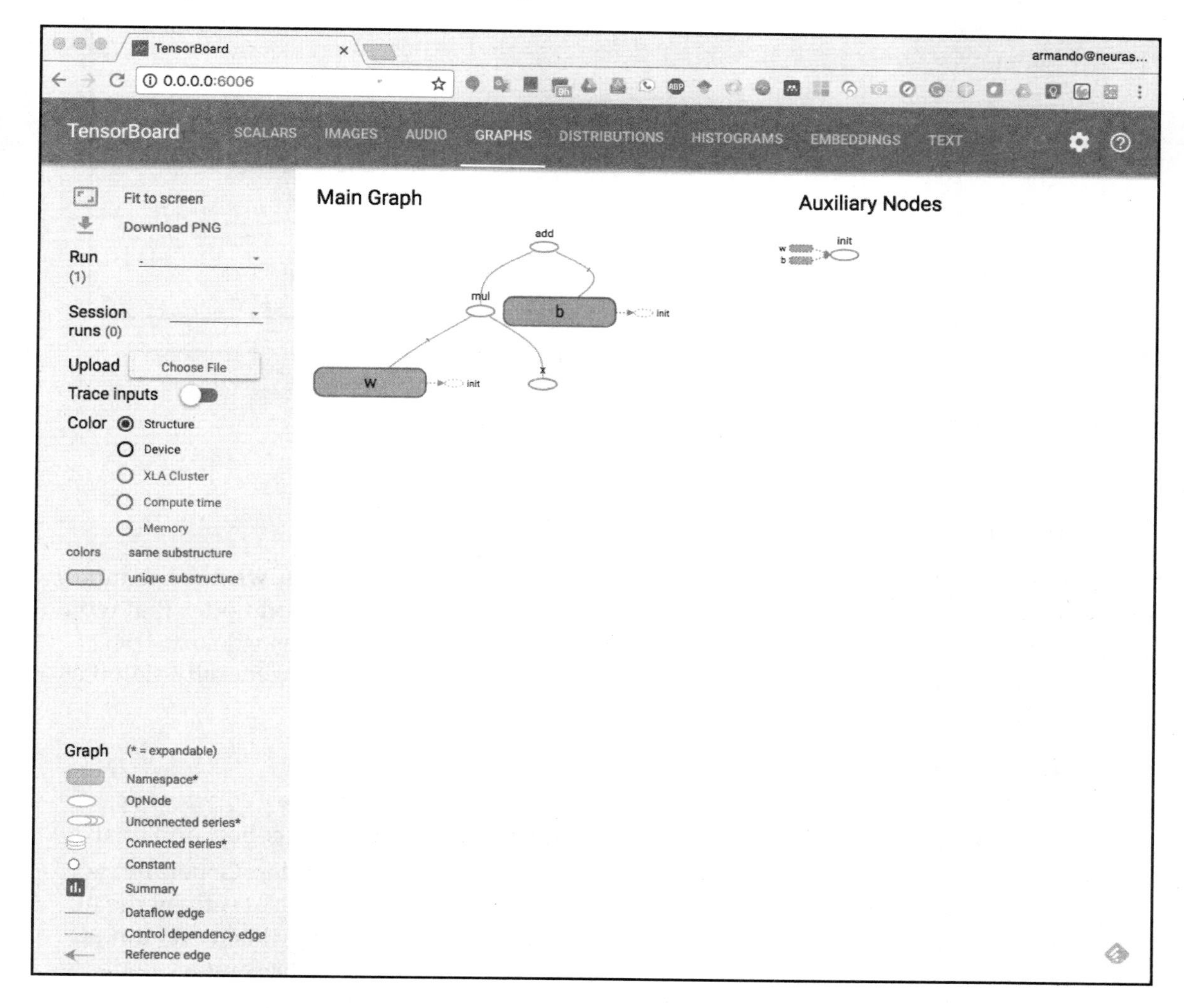

TensorBoard console

You can see that TensorBoard has visualized our first simple model as a computation graph:

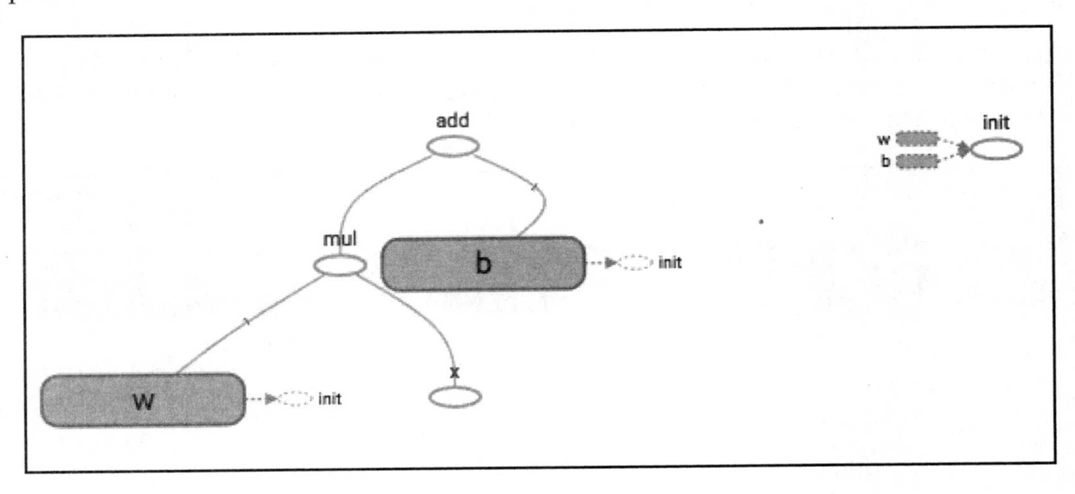

Computation graph in TensorBoard

Let's now try to understand how TensorBoard works in detail.

TensorBoard details

TensorBoard works by reading log files generated by TensorFlow. Thus, we need to modify the programming model defined here to incorporate additional operation nodes that would produce the information in the logs that we want to visualize using TensorBoard. The programming model or the flow of programs with TensorBoard can be generally stated as follows:

1. Create the computational graph as usual.
2. Create summary nodes. Attach summary operations from the `tf.summary` package to the nodes that output the values that you wish to collect and analyze.
3. Run the summary nodes along with running your model nodes. Generally, you would use the convenience function, `tf.summary.merge_all()`, to merge all the summary nodes into one summary node. Then executing this merged node would basically execute all the summary nodes. The merged summary node produces a serialized `Summary` ProtocolBuffers object containing the union of all the summaries.

4. Write the event logs to disk by passing the `Summary` ProtocolBuffers object to a `tf.summary.FileWriter` object.
5. Start TensorBoard and analyze the visualized data.

In this section, we did not create summary nodes but used TensorBoard in a very simple way. We will cover the advanced usage of TensorBoard later in this book.

Summary

In this chapter, we did a quick recap of the TensorFlow library. We learned about the TensorFlow data model elements, such as constants, variables, and placeholders, that can be used to build TensorFlow computation graphs. We learned how to create Tensors from Python objects. Tensor objects can also be generated as specific values, sequences, or random valued distributions from various library functions available in TensorFlow.

The TensorFlow programming model consists of building and executing computation graphs. The computation graphs have nodes and edges. The nodes represent operations and edges represent tensors that transfer data from one node to another. We covered how to create and execute graphs, the order of execution, and how to execute graphs on different compute devices, such as GPU and CPU. We also learned the tool to visualize the TensorFlow computation graphs, TensorBoard.

In the next chapter, we will explore some of the high-level libraries that are built on top of TensorFlow and allow us to build the models quickly.

2
High-Level Libraries for TensorFlow

There are several high-level libraries and interfaces (API) for TensorFlow that allow us to build and train models easily and with less amount of code such as TF Learn, TF Slim, Sonnet, PrettyTensor, Keras and recently released TensorFlow Estimators.

We will cover the following high-level libraries in this chapter while dedicating the next chapter to Keras:

- TF Estimator - previously TF Learn
- TF Slim
- TFLearn
- PrettyTensor
- Sonnet

We shall provide examples of building the models for MNIST dataset using all of the five libraries. Do not worry about understanding the details of the models yet as we cover the details of models from chapter 4 onwards.

You can follow the code examples in this chapter with the Jupyter Notebook `ch-02_TF_High_Level_Libraries` included in the code bundle. Try modifying the examples in the notebook to experiment and play around.

TF Estimator - previously TF Learn

TF Estimator is a high-level API that makes it simple to create and train models by encapsulating the functionalities for training, evaluating, predicting and exporting. TensorFlow recently re-branded and released the TF Learn package within TensorFlow under the new name TF Estimator, probably to avoid confusion with TFLearn package from tflearn.org. TF Estimator API has made significant enhancements to the original TF Learn package, that are described in the research paper presented in KDD 17 Conference, and can be found at the following link: `https://doi.org/10.1145/3097983.3098171`.

TF Estimator interface design is inspired from the popular machine learning library SciKit Learn, allowing to create the estimator object from different kinds of available models, and then providing four main functions on any kind of estimator:

- `estimator.fit()`
- `estimator.evaluate()`
- `estimator.predict()`
- `estimator.export()`

The names of the functions are self-explanatory. The estimator object represents the model, but the model itself is created from the model definition function provided to the estimator.

We can depict the estimator object and its interface in the following diagram:

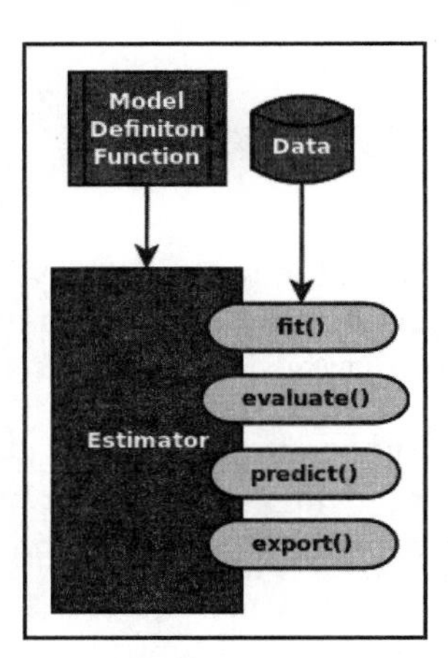

Using the Estimator API instead of building everything in core TensorFlow has the benefit of not worrying about graphs, sessions, initializing variables or other low-level details. at the time of writing this book, TensorFlow provides following pre-built estimators:

- `tf.contrib.learn.KMeansClustering`
- `tf.contrib.learn.DNNClassifier`
- `tf.contrib.learn.DNNRegressor`
- `tf.contrib.learn.DNNLinearCombinedRegressor`
- `tf.contrib.learn.DNNLinearCombinedClassifier`
- `tf.contrib.learn.LinearClassifier`
- `tf.contrib.learn.LinearRegressor`
- `tf.contrib.learn.LogisticRegressor`

The simple workflow in TF Estimator API is as follows:

1. Find the pre-built Estimator that is relevant to the problem you are trying to solve.
2. Write the function to import the dataset.
3. Define the columns in data that contain features.
4. Create the instance of the pre-built estimator that you selected in step 1.
5. Train the estimator.
6. Use the trained estimator to do evaluation or prediction.

Keras library discussed in the next chapter, provides a convenience function to convert Keras models to Estimators: `keras.estimator.model_to_estimator()`.

The complete code for the MNIST classification example is provided in the notebook `ch-02_TF_High_Level_Libraries`. The output from the TF Estimator MNIST example is as follows:

```
INFO:tensorflow:Using default config.
WARNING:tensorflow:Using temporary folder as model directory:
/tmp/tmprvcqgu07
INFO:tensorflow:Using config: {'_save_checkpoints_steps': None,
'_task_type': 'worker', '_save_checkpoints_secs': 600, '_service': None,
'_task_id': 0, '_master': '', '_session_config': None,
'_num_worker_replicas': 1, '_keep_checkpoint_max': 5, '_cluster_spec':
<tensorflow.python.training.server_lib.ClusterSpec object at
0x7ff9d15f5fd0>, '_keep_checkpoint_every_n_hours': 10000,
```

```
'_log_step_count_steps': 100, '_is_chief': True, '_save_summary_steps':
100, '_model_dir': '/tmp/tmprvcqgu07', '_num_ps_replicas': 0,
'_tf_random_seed': None}
INFO:tensorflow:Create CheckpointSaverHook.
INFO:tensorflow:Saving checkpoints for 1 into /tmp/tmprvcqgu07/model.ckpt.
INFO:tensorflow:loss = 2.4365, step = 1
INFO:tensorflow:global_step/sec: 597.996
INFO:tensorflow:loss = 1.47152, step = 101 (0.168 sec)
INFO:tensorflow:global_step/sec: 553.29
INFO:tensorflow:loss = 0.728581, step = 201 (0.182 sec)
INFO:tensorflow:global_step/sec: 519.498
INFO:tensorflow:loss = 0.89795, step = 301 (0.193 sec)
INFO:tensorflow:global_step/sec: 503.414
INFO:tensorflow:loss = 0.743328, step = 401 (0.202 sec)
INFO:tensorflow:global_step/sec: 539.251
INFO:tensorflow:loss = 0.413222, step = 501 (0.181 sec)
INFO:tensorflow:global_step/sec: 572.327
INFO:tensorflow:loss = 0.416304, step = 601 (0.174 sec)
INFO:tensorflow:global_step/sec: 543.99
INFO:tensorflow:loss = 0.459793, step = 701 (0.184 sec)
INFO:tensorflow:global_step/sec: 687.748
INFO:tensorflow:loss = 0.501756, step = 801 (0.146 sec)
INFO:tensorflow:global_step/sec: 654.217
INFO:tensorflow:loss = 0.666772, step = 901 (0.153 sec)
INFO:tensorflow:Saving checkpoints for 1000 into
/tmp/tmprvcqgu07/model.ckpt.
INFO:tensorflow:Loss for final step: 0.426257.
INFO:tensorflow:Starting evaluation at 2017-12-15-02:27:45
INFO:tensorflow:Restoring parameters from /tmp/tmprvcqgu07/model.ckpt-1000
INFO:tensorflow:Finished evaluation at 2017-12-15-02:27:45
INFO:tensorflow:Saving dict for global step 1000: accuracy = 0.8856,
global_step = 1000, loss = 0.40996

{'accuracy': 0.88559997, 'global_step': 1000, 'loss': 0.40995964}
```

You will see in chapter 5 how to create such models using core TensorFlow.

TF Slim

TF Slim is a lightweight library built on top of TensorFlow core for defining and training models. TF Slim can be used in conjunction with other TensorFlow low level and high-level libraries such as TF Learn. The TF Slim comes as part of the TensorFlow installation in the package: `tf.contrib.slim`. Run the following command to check if your TF Slim installation is working:

```
python3 -c 'import tensorflow.contrib.slim as slim; eval =
slim.evaluation.evaluate_once'
```

TF Slim provides several modules that can be picked and applied independently and mixed with other TensorFlow packages. For example, at the time of writing of this book TF Slim had following major modules:

TF Slim module	Module description
arg_scope	Provides a mechanism to apply elements to all graph nodes defined under a scope.
layers	Provides several different kinds of layers such as `fully_connected`, `conv2d`, and many more.
losses	Provides loss functions for training the optimizer
learning	Provides functions for training the models
evaluation	Provides evaluation functions
metrics	Provides metrics functions to be used for evaluating the models
regularizers	Provides functions for creating regularization methods
variables	Provides functions for variable creation
nets	Provides various pre-built and pre-trained models such as VGG16, InceptionV3, ResNet

The simple workflow in TF Slim is as follows:

1. Create the model using slim layers.
2. Provide the input to the layers to instantiate the model.
3. Use the logits and labels to define the loss.
4. Get the total loss using convenience function `get_total_loss()`.
5. Create an optimizer.

6. Create a training function using convenience function `slim.learning.create_train_op()`, `total_loss` and `optimizer`.
7. Run the training using the convenience function `slim.learning.train()` and training function defined in the previous step.

The complete code for the MNIST classification example is provided in the notebook `ch-02_TF_High_Level_Libraries`. The output from the TF Slim MNIST example is as follows:

```
INFO:tensorflow:Starting Session.
INFO:tensorflow:Saving checkpoint to path ./slim_logs/model.ckpt
INFO:tensorflow:global_step/sec: 0
INFO:tensorflow:Starting Queues.
INFO:tensorflow:global step 100: loss = 2.2669 (0.010 sec/step)
INFO:tensorflow:global step 200: loss = 2.2025 (0.010 sec/step)
INFO:tensorflow:global step 300: loss = 2.1257 (0.010 sec/step)
INFO:tensorflow:global step 400: loss = 2.0419 (0.009 sec/step)
INFO:tensorflow:global step 500: loss = 1.9532 (0.009 sec/step)
INFO:tensorflow:global step 600: loss = 1.8733 (0.010 sec/step)
INFO:tensorflow:global step 700: loss = 1.8002 (0.010 sec/step)
INFO:tensorflow:global step 800: loss = 1.7273 (0.010 sec/step)
INFO:tensorflow:global step 900: loss = 1.6688 (0.010 sec/step)
INFO:tensorflow:global step 1000: loss = 1.6132 (0.010 sec/step)
INFO:tensorflow:Stopping Training.
INFO:tensorflow:Finished training! Saving model to disk.
final loss=1.6131552457809448
```

As we see from the output, the convenience function `slim.learning.train()` saves the output of the training in checkpoint files in the specified log directory. If you restart the training, it will first check if the checkpoint exists and will resume the training from the checkpoint by default.

The documentation page for the TF Slim was found empty at the time of this writing at the following link: `https://www.tensorflow.org/api_docs/python/tf/contrib/slim`. However, some of the documentation can be found in the source code at the following link: `https://github.com/tensorflow/tensorflow/tree/r1.4/tensorflow/contrib/slim`.

We shall use TF Slim for learning how to use pre-trained models such as VGG16 and Inception V3 in later chapters.

TFLearn

TFLearn is a modular library in Python that is built on top of core TensorFlow.

TFLearn is different from the TensorFlow Learn package which is also known as TF Learn (with one space in between TF and Learn). TFLearn is available at the following link: `http://tflearn.org`, and the source code is available on GitHub at the following link: `https://github.com/tflearn/tflearn`.

TFLearn can be installed in Python 3 with the following command:

```
pip3 install tflearn
```

To install TFLearn in other environments or from source, please refer to the following link: `http://tflearn.org/installation/`.

The simple workflow in TFLearn is as follows:

1. Create an input layer first.
2. Pass the input object to create further layers.
3. Add the output layer.
4. Create the net using an estimator layer such as `regression`.
5. Create a model from the net created in the previous step.
6. Train the model with the `model.fit()` method.
7. Use the trained model to predict or evaluate.

Creating the TFLearn Layers

Let us learn how to create the layers of the neural network models in TFLearn:

1. Create an input layer first:

```
input_layer = tflearn.input_data(shape=[None,num_inputs]
```

2. Pass the input object to create further layers:

```
layer1 = tflearn.fully_connected(input_layer,10,
                                 activation='relu')
layer2 = tflearn.fully_connected(layer1,10,
                                 activation='relu')
```

3. Add the output layer:

```
output = tflearn.fully_connected(layer2,n_classes,
                                 activation='softmax')
```

4. Create the final net from the estimator layer such as `regression`:

```
net = tflearn.regression(output,
                         optimizer='adam',
                         metric=tflearn.metrics.Accuracy(),
                         loss='categorical_crossentropy'
                        )
```

The TFLearn provides several classes for layers that are described in following sub-sections.

TFLearn core layers

TFLearn offers the following layers in the `tflearn.layers.core` module:

Layer class	Description
`input_data`	**This layer is used to specify the input layer for the neural network.**
`fully_connected`	This layer is used to specify a layer where all the neurons are connected to all the neurons in the previous layer.
`dropout`	This layer is used to specify the dropout regularization. The input elements are scaled by `1/keep_prob` while keeping the expected sum unchanged.

Layer class	Description
custom_layer	This layer is used to specify a custom function to be applied to the input. This class wraps our custom function and presents the function as a layer.
reshape	This layer reshapes the input into the output of specified shape.
flatten	This layer converts the input tensor to a 2D tensor.
activation	This layer applies the specified activation function to the input tensor.
single_unit	This layer applies the linear function to the inputs.
highway	This layer implements the fully connected highway function.
one_hot_encoding	This layer converts the numeric labels to their binary vector one-hot encoded representations.
time_distributed	This layer applies the specified function to each time step of the input tensor.
multi_target_data	This layer creates and concatenates multiple placeholders, specifically used when the layers use targets from multiple sources.

TFLearn convolutional layers

TFLearn offers the following layers in the `tflearn.layers.conv` module:

Layer class	Description
conv_1d	This layer applies 1D convolutions to the input data
conv_2d	This layer applies 2D convolutions to the input data
conv_3d	This layer applies 3D convolutions to the input data
conv_2d_transpose	This layer applies transpose of conv2_d to the input data
conv_3d_transpose	This layer applies transpose of conv3_d to the input data
atrous_conv_2d	This layer computes a 2-D atrous convolution
grouped_conv_2d	This layer computes a depth-wise 2-D convolution
max_pool_1d	This layer computes 1-D max pooling
max_pool_2d	This layer computes 2D max pooling
avg_pool_1d	This layer computes 1D average pooling

avg_pool_2d	This layer computes 2D average pooling
upsample_2d	This layer applies the row and column wise 2-D repeat operation
upscore_layer	This layer implements the upscore as specified in http://arxiv.org/abs/1411.4038
global_max_pool	This layer implements the global max pooling operation
global_avg_pool	This layer implements the global average pooling operation
residual_block	This layer implements the residual block to create deep residual networks
residual_bottleneck	This layer implements the residual bottleneck block for deep residual networks
resnext_block	This layer implements the ResNeXt block

TFLearn recurrent layers

TFLearn offers the following layers in the `tflearn.layers.recurrent` module:

Layer class	Description
simple_rnn	This layer implements the simple recurrent neural network model
bidirectional_rnn	This layer implements the bi-directional RNN model
lstm	This layer implements the LSTM model
gru	This layer implements the GRU model

TFLearn normalization layers

TFLearn offers the following layers in the `tflearn.layers.normalization` module:

Layer class	Description
batch_normalization	This layer normalizes the output of activations of previous layers for each batch
local_response_normalization	This layer implements the LR normalization
l2_normalization	This layer applies the L2 normalization to the input tensors

TFLearn embedding layers

TFLearn offers only one layer in the `tflearn.layers.embedding_ops` module:

Layer class	Description
`embedding`	This layer implements the embedding function for a sequence of integer IDs or floats

TFLearn merge layers

TFLearn offers the following layers in the `tflearn.layers.merge_ops` module:

Layer class	Description
`merge_outputs`	This layer merges the list of tensors into a single tensor, generally used to merge the output tensors of the same shape
`merge`	This layer merges the list of tensors into a single tensor; you can specify the axis along which the merge needs to be done

TFLearn estimator layers

TFLearn offers only one layer in the `tflearn.layers.estimator` module:

Layer class	Description
`regression`	This layer implements the linear or logistic regression

While creating the regression layer, you can specify the optimizer and the loss and metric functions.

TFLearn offers the following optimizer functions as classes in the `tflearn.optimizers` module:

- `SGD`
- `RMSprop`
- `Adam`
- `Momentum`
- `AdaGrad`
- `Ftrl`

- `AdaDelta`
- `ProximalAdaGrad`
- `Nesterov`

You can create custom optimizers by extending the `tflearn.optimizers.Optimizer` base class.

TFLearn offers the following metric functions as classes or ops in the `tflearn.metrics` module:

- `Accuracy` or `accuracy_op`
- `Top_k` or `top_k_op`
- `R2` or `r2_op`
- `WeightedR2` or `weighted_r2_op`
- `binary_accuracy_op`

You can create custom metrics by extending the `tflearn.metrics.Metric` base class.

TFLearn provides the following loss functions, known as objectives, in the `tflearn.objectives` module:

- `softymax_categorical_crossentropy`
- `categorical_crossentropy`
- `binary_crossentropy`
- `weighted_crossentropy`
- `mean_square`
- `hinge_loss`
- `roc_auc_score`
- `weak_cross_entropy_2d`

While specifying the input, hidden, and output layers, you can specify the activation functions to be applied to the output. TFLearn provides the following activation functions in the `tflearn.activations` module:

- `linear`
- `tanh`
- `sigmoid`
- `softmax`
- `softplus`
- `softsign`
- `relu`
- `relu6`
- `leaky_relu`
- `prelu`
- `elu`
- `crelu`
- `selu`

Creating the TFLearn Model

Create the model from the net created in the previous step (step 4 in creating the TFLearn layers section):

```
model = tflearn.DNN(net)
```

Types of TFLearn models

The TFLearn offers two different classes of the models:

- `DNN` (Deep Neural Network) model: This class allows you to create a multilayer perceptron from the network that you have created from the layers
- `SequenceGenerator` model: This class allows you to create a deep neural network that can generate sequences

Training the TFLearn Model

After creating, train the model with the `model.fit()` method:

```
model.fit(X_train,
          Y_train,
          n_epoch=n_epochs,
          batch_size=batch_size,
          show_metric=True,
          run_id='dense_model')
```

Using the TFLearn Model

Use the trained model to predict or evaluate:

```
score = model.evaluate(X_test, Y_test)
print('Test accuracy:', score[0])
```

The complete code for the TFLearn MNIST classification example is provided in the notebook `ch-02_TF_High_Level_Libraries`. The output from the TFLearn MNIST example is as follows:

```
Training Step: 5499  | total loss: 0.42119 | time: 1.817s
| Adam | epoch: 010 | loss: 0.42119 - acc: 0.8860 -- iter: 54900/55000
Training Step: 5500  | total loss: 0.40881 | time: 1.820s
| Adam | epoch: 010 | loss: 0.40881 - acc: 0.8854 -- iter: 55000/55000
--
Test accuracy: 0.9029
```

You can get more information about TFLearn from the following link: `http://tflearn.org/`.

PrettyTensor

PrettyTensor provides a thin wrapper on top of TensorFlow. The objects provided by PrettyTensor support a chainable syntax to define neural networks. For example, a model could be created by chaining the layers as shown in the following code:

```
model = (X.
         flatten().
         fully_connected(10).
```

```
            softmax_classifier(n_classes, labels=Y))
```

PrettyTensor can be installed in Python 3 with the following command:

```
pip3 install prettytensor
```

PrettyTensor offers a very lightweight and extensible interface in the form of a method named `apply()`. Any additional function can be chained to PrettyTensor objects using the `.apply(function, arguments)` method. PrettyTensor will call the `function` and supply the current tensor as the first argument to the `function`.

User-created functions can be added using the `@prettytensor.register` decorator. Details can be found at `https://github.com/google/prettytensor`.

The workflow to define and train models in PrettyTensor is as follows:

1. Get the data.
2. Define hyperparameters and parameters.
3. Define the inputs and outputs.
4. Define the model.
5. Define the evaluator, optimizer, and trainer functions.
6. Create the runner object.
7. Within a TensorFlow session, train the model with the `runner.train_model()` method.
8. Within the same session, evaluate the model with the `runner.evaluate_model()` method.

The complete code for the PrettyTensor MNIST classification example is provided in the notebook `ch-02_TF_High_Level_Libraries`. The output from the PrettyTensor MNIST example is as follows:

```
[1] [2.5561881]
[600] [0.3553167]
Accuracy after 1 epochs 0.8799999952316284

[601] [0.47775066]
[1200] [0.34739292]
Accuracy after 2 epochs 0.8999999761581421

[1201] [0.19110668]
[1800] [0.17418651]
```

```
Accuracy after 3 epochs 0.8999999761581421

[1801] [0.27229539]
[2400] [0.34908807]
Accuracy after 4 epochs 0.8700000047683716

[2401] [0.40000191]
[3000] [0.30816519]
Accuracy after 5 epochs 0.8999999761581421

[3001] [0.29905257]
[3600] [0.41590339]
Accuracy after 6 epochs 0.8899999856948853

[3601] [0.32594997]
[4200] [0.36930788]
Accuracy after 7 epochs 0.8899999856948853

[4201] [0.26780865]
[4800] [0.2911002]
Accuracy after 8 epochs 0.8899999856948853

[4801] [0.36304188]
[5400] [0.39880857]
Accuracy after 9 epochs 0.8999999761581421

[5401] [0.1339224]
[6000] [0.14993289]
Accuracy after 10 epochs 0.8899999856948853
```

Sonnet

Sonnet is an object-oriented library written in Python. It was released by DeepMind in 2017. Sonnet intends to cleanly separate the following two aspects of building computation graphs from objects:

- The configuration of objects called modules
- The connection of objects to computation graphs

Sonnet can be installed in Python 3 with the following command:

```
pip3 install dm-sonnet
```

Sonnet can be installed from the source by following directions from the following link: https://github.com/deepmind/sonnet/blob/master/docs/INSTALL.md.

The modules are defined as sub-classes of the abstract class sonnet.AbstractModule. At the time of writing this book, the following modules are available in Sonnet:

Basic modules	AddBias, BatchApply, BatchFlatten, BatchReshape, FlattenTrailingDimensions, Linear, MergeDims, SelectInput, SliceByDim, TileByDim, **and** TrainableVariable
Recurrent modules	DeepRNN, ModelRNN, VanillaRNN, BatchNormLSTM, GRU, and LSTM
Recurrent + ConvNet modules	Conv1DLSTM and Conv2DLSTM
ConvNet modules	Conv1D, Conv2D, Conv3D, Conv1DTranspose, Conv2DTranspose, Conv3DTranspose, DepthWiseConv2D, InPlaneConv2D, and SeparableConv2D
ResidualNets	Residual, ResidualCore, and SkipConnectionCore
Others	BatchNorm, LayerNorm, clip_gradient, and scale_gradient

We can define our own new modules by creating a sub-class of sonnet.AbstractModule. An alternate non-recommended way of creating a module from a function is to create an object of the sonnet.Module class by passing the function to be wrapped as a module.

The workflow to build a model in the Sonnet library is as follows:

1. Create classes for the dataset and network architecture which inherit from sonnet.AbstractModule. In our example, we create an MNIST class and an MLP class.
2. Define the parameters and hyperparameters.
3. Define the test and train datasets from the dataset classes defined in the preceding step.
4. Define the model using the network class defined. As an example, model = MLP([20, n_classes]) in our case creates an MLP network with two layers of 20 and the n_classes number of neurons each.
5. Define the y_hat placeholders for the train and test sets using the model.

6. Define the loss placeholders for the train and test sets.
7. Define the optimizer using the train loss placeholder.
8. Execute the loss function in a TensorFlow session for the desired number of epochs to optimize the parameters.

The complete code for the Sonnet MNIST classification example is provided in the notebook `ch-02_TF_High_Level_Libraries`. The `__init__` method in each class initializes the class and the related superclass. The `_build` method creates and returns the dataset or the model objects when the class is called.The output from the Sonnet MNIST example is as follows:

```
Epoch : 0 Training Loss : 236.79913330078125
Epoch : 1 Training Loss : 227.3693084716797
Epoch : 2 Training Loss : 221.96337890625
Epoch : 3 Training Loss : 220.99142456054688
Epoch : 4 Training Loss : 215.5921173095703
Epoch : 5 Training Loss : 213.88958740234375
Epoch : 6 Training Loss : 203.7091064453125
Epoch : 7 Training Loss : 204.57427978515625
Epoch : 8 Training Loss : 196.17218017578125
Epoch : 9 Training Loss : 192.3954315185547
Test loss : 192.8847198486328
```

Your output could be a little varied due to the stochastic nature of the computations in the neural nets. This covers our overview of the Sonnet module.

For more details on Sonnet, you can explore the following link: `https://deepmind.github.io/sonnet/`.

Summary

In this chapter, we did a tour of some of the high-level libraries that are built on top of TensorFlow. We learned about TF Estimator, TF Slim, TFLearn, PrettyTensor, and Sonnet. We implemented the MNIST classification example for all five of them. If you could not understand the details of the models, do not worry, because the models built for MNIST example will be presented again in the following chapters.

We summarize the libraries and frameworks presented in this chapter in the following table:

High-Level Library	Documentation Link	Source Code Link	pip3 install package
TF Estimator	https://www.tensorflow.org/get_started/estimator	https://github.com/tensorflow/tensorflow/tree/master/tensorflow/python/estimator	preinstalled with TensorFlow
TF Slim	https://github.com/tensorflow/tensorflow/tree/r1.4/tensorflow/contrib/slim	https://github.com/tensorflow/tensorflow/tree/r1.4/tensorflow/contrib/slim	preinstalled with TensorFlow
TFLearn	http://tflearn.org/	https://github.com/tflearn/tflearn	tflearn
PrettyTensor	https://github.com/google/prettytensor/tree/master/docs	https://github.com/google/prettytensor	prettytensor
Sonnet	https://deepmind.github.io/sonnet/	https://github.com/deepmind/sonnet	dm-sonnet

In the next chapter, we shall learn about Keras, the most popular high-level library for creating and training TensorFlow models.

3
Keras 101

Keras is a high-level library that allows the use of TensorFlow as a backend deep learning library. TensorFlow team has included Keras in TensorFlow Core as module `tf.keras`. Apart from TensorFlow, Keras also supports Theano and CNTK at the time of writing this book.

The following guiding principles of Keras have made it very popular among the deep learning community:

- Minimalism to offer a consistent and simple API
- Modularity to allow the representation of various elements as pluggable modules
- Extensibility to add new modules as classes and functions
- Python-native for both code and model configuration
- Out-of-the-box common network architectures that support CNN, RNN, or a combination of both

Throughout the remainder of this book, we shall learn how to build different kinds of deep learning and machine learning models with both the low-level TensorFlow API and the high-level Keras API.

We shall cover the following topics in this chapter:

- Installing Keras
- Workflow for creating models in Keras
- Creating the Keras model with sequential and functional API
- Keras layers
- Creating and adding layers with sequential and functional API

- Compiling the Keras model
- Training the Keras model
- Predicting using the Keras model
- Additional modules in Keras
- Keras sequential model example for MNIST dataset

Installing Keras

Keras can be installed in Python 3 with the following command:

```
pip3 install keras
```

To install Keras in other environments or from source, please refer to the following link: `https://keras.io/#installation`.

Neural Network Models in Keras

Neural network models in Keras are defined as the graph of layers. The models in Keras can be created using the sequential or the functional APIs. Both the functional and sequential APIs can be used to build any kind of models. The functional API makes it easier to build the complex models that have multiple inputs, multiple outputs and shared layers.

Thus as a rule of thumb, we have seen engineers use the sequential API for simple models built from simple layers and the functional API for complex models involving branches and sharing of layers. We have also observed that building simple models with the functional API makes it easier to grow the models into complex models with branching and sharing. Hence for our work, we always use the functional API.

Workflow for building models in Keras

The simple workflow in Keras is as follows:

1. Create the model
2. Create and add layers to the model
3. Compile the model

4. Train the model
5. Use the model for prediction or evaluation

Let's check out each of these steps.

You can follow the code examples in this chapter with the Jupyter Notebook `ch-03_Keras_101` included in the code bundle. Try modifying the code in the notebook to explore various options.

Creating the Keras model

The Keras model can be created using the sequential API or functional API. The examples of creating models in both ways are given in the following subsections.

Sequential API for creating the Keras model

In the sequential API, create the empty model with the following code:

```
model = Sequential()
```

You can now add the layers to this model, which we will see in the next section.

Alternatively, you can also pass all the layers as a list to the constructor. As an example, we add four layers by passing them to the constructor using the following code:

```
model = Sequential([ Dense(10, input_shape=(256,)),
                     Activation('tanh'),
                     Dense(10),
                     Activation('softmax')
                    ])
```

Functional API for creating the Keras model

In the functional API, you create the model as an instance of the `Model` class that takes an input and output parameter. The input and output parameters represent one or more input and output tensors, respectively.

As an example, use the following code to instantiate a model from the functional API:

```
model = Model(inputs=tensor1, outputs=tensor2)
```

In the above code, `tensor1` and `tensor2` are either tensors or objects that can be treated like tensors, for example, Keras `layer` objects.

If there are more than one input and output tensors, they can be passed as a list, as shown in the following example:

```
model = Model(inputs=[i1,i2,i3], outputs=[o1,o2,o3])
```

Keras Layers

Keras provides several built-in layer classes for the easy construction of the network architecture. The following sections give a summary and description of the various types of layers provided by Keras 2 at the time of writing this book.

Keras core layers

The Keras core layers implement fundamental operations that are used in almost every kind of network architecture. The following tables give a summary and description of the layers provided by Keras 2:

Layer name	**Description**
`Dense`	This is a simple fully connected neural network layer. This layer produces the output of the following function: ***activation((inputs x weights)+bias)*** where *activation* refers to the activation function passed to the layer, which is `None` by default.
`Activation`	This layer applies the specified activation function to the output. This layer produces the output of the following function: ***activation(inputs)*** where *activation* refers to the activation function passed to the layer. The following activation functions are available to instantiate this layer: `softmax`, `elu`, `selu`, `softplus`, `softsign`, `relu`, `tanh`, `sigmoid`, `hard_sigmoid`, and `linear`
`Dropout`	This layer applies the dropout regularization to the inputs at a specified dropout rate.
`Flatten`	This layer flattens the input, that is, for a three-dimensional input, it flattens and produces a one-dimensional output.
`Reshape`	This layer converts the input to the specified shape.

Permute	This layer reorders the input dimensions as per the specified pattern.
RepeatVector	This layer repeats the input by the given number of times. Thus, if the input is a 2D tensor of shape (#samples, #features) and the layer is given *n* times to repeat, then the output will be a 3D tensor of shape (#samples, n, #features).
Lambda	This layer wraps the provided function as a layer. Thus, the inputs are passed through the custom function provided to produce the outputs. This layer provides ultimate extensibility to Keras users to add their own custom functions as layers.
ActivityRegularization	This layer applies L1 or L2, or a combination of both kinds of regularization to its inputs. This layer is applied to the output of an activation layer or to the output of a layer that has an activation function.
Masking	This layer masks or skips those time steps in the input tensor where all the values in the input tensor are equal to the mask value provided as an argument to the layer.

Keras convolutional layers

These layers implement the different type of convolution, sampling, and cropping operations for convolutional neural networks:

Layer Name	Description
Conv1D	This layer applies convolutions over a single spatial or temporal dimension to the inputs.
Conv2D	This layer applies two-dimensional convolutions to the inputs.
SeparableConv2D	This layer applies a depth-wise spatial convolution on each input channel, followed by a pointwise convolution that mixes together the resulting output channels.
Conv2DTranspose	This layer reverts the shape of convolutions to the shape of the inputs that produced those convolutions.
Conv3D	This layer applies three-dimensional convolutions to the inputs.
Cropping1D	This layer crops the input data along the temporal dimension.
Cropping2D	This layer crops the input data along the spatial dimensions, such as width and height in the case of an image.
Cropping3D	This layer crops the input data along the spatio-temporal, that is all three dimensions.
UpSampling1D	This layer repeats the input data by specified times along the time axis.

`UpSampling2D`	This layer repeats the row and column dimensions of the input data by specified times along the two dimensions.
`UpSampling3D`	This layer repeats the three dimensions of the input data by specified times along the three dimensions.
`ZeroPadding1D`	This layer adds zeros to the beginning and end of the time dimension.
`ZeroPadding2D`	This layer adds rows and columns of zeros to the top, bottom, left, or right of a 2D tensor.
`ZeroPadding3D`	This layer adds zeros to the three dimensions of a 3D tensor.

Keras pooling layers

These layers implement the different pooling operations for convolutional neural networks:

Layer Name	**Description**
`MaxPooling1D`	This layer implements the max pooling operation for one-dimensional input data.
`MaxPooling2D`	This layer implements the max pooling operation for two-dimensional input data.
`MaxPooling3D`	This layer implements the max pooling operation for three-dimensional input data.
`AveragePooling1D`	This layer implements the average pooling operation for two-dimensional input data.
`AveragePooling2D`	This layer implements the average pooling operation for two-dimensional input data.
`AveragePooling3D`	This layer implements the average pooling operation for three-dimensional input data.
`GlobalMaxPooling1D`	This layer implements the global max pooling operation for one-dimensional input data.
`GlobalAveragePooling1D`	This layer implements the global average pooling operation for one-dimensional input data.
`GlobalMaxPooling2D`	This layer implements the global max pooling operation for two-dimensional input data.
`GlobalAveragePooling2D`	This layer implements the global average pooling operation for two-dimensional input data.

Keras locally-connected layers

These layers are useful in convolutional neural networks:

Layer Name	Description
`LocallyConnected1D`	This layer applies convolutions over a single spatial or temporal dimension to the inputs, by applying a different set of filters at each different patch of the input, thus not sharing the weights.
`LocallyConnected2D`	This layer applies convolutions over two dimensions to the inputs, by applying a different set of filters at each different patch of the input, thus not sharing the weights.

Keras recurrent layers

These layers implement different variants of recurrent neural networks:

Layer Name	Description
`SimpleRNN`	This layer implements a fully connected recurrent neural network.
`GRU`	This layer implements a gated recurrent unit network.
`LSTM`	This layer implements a long short-term memory network.

Keras embedding layers

Presently, there is only one embedding layer option available:

Layer Name	Description
`Embedding`	This layer takes a 2D tensor of shape `(batch_size, sequence_length)` consisting of indexes, and produces a tensor consisting of dense vectors of shape `(batch_size, sequence_length, output_dim)`.

Keras merge layers

These layers merge two or more input tensors and produce a single output tensor by applying a specific operation that each layer represents:

Layer Name	Description
`Add`	This layer computes the element-wise addition of input tensors.
`Multiply`	This layer computes the element-wise multiplication of input tensors
`Average`	This layer computes the element-wise average of input tensors.
`Maximum`	This layer computes the element-wise maximum of input tensors.
`Concatenate`	This layer concatenates the input tensors along a specified axis.
`Dot`	This layer computes the dot product between samples in two input tensors.
`add`, `multiply`, `average`, `maximum`, `concatenate`, and `dot`	These functions represent the functional interface to the respective merge layers described in this table.

Keras advanced activation layers

These layers implement advanced activation functions that cannot be implemented as a simple underlying backend function. They operate similarly to the `Activation()` layer that we covered in the core layers section:

Layer Name	Description
`LeakyReLU`	This layer computes the leaky version of the `ReLU` activation function.
`PReLU`	This layer computes the parametric `ReLU` activation function.
`ELU`	This layer computes the exponential linear unit activation function.
`ThresholdedReLU`	This layer computes the thresholded version of the `ReLU` activation function.

Keras normalization layers

Presently, there is only one normalization layer available:

Layer name	Description
`BatchNormalization`	This layer normalizes the outputs of the previous layer at each batch, such that the output of this layer is approximated to have a mean close to zero and a standard deviation close to 1.

Keras noise layers

These layers can be added to the model to prevent overfitting by adding noise; they are also known as regularization layers. These layers operate the same way as the `Dropout()` and `ActivityRegularizer()` layers in the core layers section.

Layer name	Description
`GaussianNoise`	This layer applies additive zero-centered Gaussian noise to the inputs.
`GaussianDropout`	This layer applies multiplicative one-centered Gaussian noise to the inputs.
`AlphaDropout`	This layer drops a certain percentage of inputs, such that the mean and variance of the outputs after the dropout match closely with the mean and variance of the inputs.

Adding Layers to the Keras Model

All the layers mentioned in the previous section need to be added to the model we created earlier. In the following sections, we describe how to add the layers using the functional API and the sequential API.

Sequential API to add layers to the Keras model

In the sequential API, you can create layers by instantiating an object of one of the layer types given in the preceding sections. The created layers are then added to the model using the `model.add()` function. As an example, we will create a model and then add two layers to it:

```
model = Sequential()
model.add(Dense(10, input_shape=(256,))
model.add(Activation('tanh'))
model.add(Dense(10))
model.add(Activation('softmax'))
```

Functional API to add layers to the Keras Model

In the functional API, the layers are created first in a functional manner, and then while creating the model, the input and output layers are provided as tensor arguments, as we covered in the previous section.

Here is an example:

1. First, create the input layer:

   ```
   input = Input(shape=(64,))
   ```

2. Next, create the dense layer from the input layer in a functional way:

   ```
   hidden = Dense(10)(inputs)
   ```

3. In the same way, create further hidden layers building from the previous layers in a functional way:

   ```
   hidden = Activation('tanh')(hidden)
   hidden = Dense(10)(hidden)
   output = Activation('tanh')(hidden)
   ```

4. Finally, instantiate the model object with the input and output layers:

   ```
   model = Model(inputs=input, outputs=output)
   ```

For further in-depth details about creating sequential and functional Keras models, you can read the book titled *Deep Learning with Keras* by Antonio Gulli and Sujit Pal, Packt Publishing, 2017.

Compiling the Keras model

The model built in the previous sections needs to be compiled with the `model.compile()` method before it can be used for training and prediction. The full signature of the `compile()` method is as follows:

```
compile(self, optimizer, loss, metrics=None, sample_weight_mode=None)
```

The compile method takes three arguments:

- `optimizer`: You can specify your own function or one of the functions provided by Keras. This function is used to update the parameters in the optimization iterations. Keras offers the following built-in optimizer functions:
    - `SGD`
    - `RMSprop`
    - `Adagrad`
    - `Adadelta`
    - `Adam`
    - `Adamax`
    - `Nadam`
- `loss`: You can specify your own loss function or use one of the provided loss functions. The optimizer function optimizes the parameters so that the output of this loss function is minimized. Keras provides the following loss functions:
    - `mean_squared_error`
    - `mean_absolute_error`
    - `mean_absolute_pecentage_error`
    - `mean_squared_logarithmic_error`
    - `squared_hinge`
    - `hinge`
    - `categorical_hinge`
    - `sparse_categorical_crossentropy`
    - `binary_crossentropy`
    - `poisson`
    - `cosine proximity`
    - `binary_accuracy`

    - categorical_accuracy
    - sparse_categorical_accuracy
    - top_k_categorical_accuracy
    - sparse_top_k_categorical_accuracy

- metrics: The third argument is a list of metrics that need to be collected while training the model. If verbose output is on, then the metrics are printed for each iteration. The metrics are like loss functions; some are provided by Keras with the ability to write your own metrics functions. All the loss functions also work as the metric function.

Training the Keras model

Training a Keras model is as simple as calling the model.fit() method. The full signature of this method is as follows:

```
fit(self, x, y, batch_size=32, epochs=10, verbose=1, callbacks=None,
    validation_split=0.0, validation_data=None, shuffle=True,
    class_weight=None, sample_weight=None, initial_epoch=0)
```

We will not go into the details of the arguments of this method; you can read the details at the Keras website, https://keras.io/models/sequential/.

For the example model that we created earlier, train the model with the following code:

```
model.fit(x_data, y_labels)
```

Predicting with the Keras model

The trained model can be used either to predict the value with the model.predict() method or to evaluate the model with the model.evaluate() method.

The signatures of both the methods are as follows:

```
predict(self, x, batch_size=32, verbose=0)

evaluate(self, x, y, batch_size=32, verbose=1, sample_weight=None)
```

Additional modules in Keras

Keras provides several additional modules that supplement the basic workflow (described at the beginning of this chapter) with additional functionalities. Some of the modules are as follows:

- The `preprocessing` module provides several functions for the preprocessing of sequence, image, and text data.
- The `datasets` module provides several functions for quick access to several popular datasets, such as CIFAR10 images, CIFAR100 images, IMDB movie reviews, Reuters newswire topics, MNIST handwritten digits, and Boston housing prices.
- The `initializers` module provides several functions to set initial random weight parameters of layers, such as `Zeros`, `Ones`, `Constant`, `RandomNormal`, `RandomUniform`, `TruncatedNormal`, `VarianceScaling`, `Orthogonal`, `Identity`, `lecun_normal`, `lecun_uniform`, `glorot_normal`, `glorot_uniform`, `he_normal`, and `he_uniform`.
- The `models` module provides several functions to restore the model architectures and weights, such as `model_from_json`, `model_from_yaml`, and `load_model`. The model architectures can be saved using the `model.to_yaml()` and `model.to_json()` methods. The model weights can be saved by calling the `model.save()` method. The weights get saved in an HDF5 file.
- The `applications` module provides several pre-built and pre-trained models such as Xception, VGG16, VGG19, ResNet50, Inception V3, InceptionResNet V2, and MobileNet. We shall learn how to use the pre-built models to predict with our datasets. We shall also learn how to retrain the pre-trained models in the `applications` module with our datasets from a slightly different domain.

This concludes our brief tour of Keras, a high-level framework for TensorFlow. We will provide examples of building models with Keras throughout this book.

Keras sequential model example for MNIST dataset

The following is a small example of building a simple multilayer perceptron (covered in detail in Chapter 5) to classify handwritten digits from the MNIST set:

```
import keras
from keras.datasets import mnist
from keras.models import Sequential
from keras.layers import Dense, Dropout
from keras.optimizers import SGD
from keras import utils
import numpy as np

# define some hyper parameters
batch_size = 100
n_inputs = 784
n_classes = 10
n_epochs = 10

# get the data
(x_train, y_train), (x_test, y_test) = mnist.load_data()

# reshape the two dimensional 28 x 28 pixels
# sized images into a single vector of 784 pixels
x_train = x_train.reshape(60000, n_inputs)
x_test = x_test.reshape(10000, n_inputs)

# convert the input values to float32
x_train = x_train.astype(np.float32)
x_test = x_test.astype(np.float32)

# normalize the values of image vectors to fit under 1
x_train /= 255
x_test /= 255

# convert output data into one hot encoded format
y_train = utils.to_categorical(y_train, n_classes)
y_test = utils.to_categorical(y_test, n_classes)

# build a sequential model
model = Sequential()
# the first layer has to specify the dimensions of the input vector
model.add(Dense(units=128, activation='sigmoid', input_shape=(n_inputs,)))
# add dropout layer for preventing overfitting
model.add(Dropout(0.1))
```

```
model.add(Dense(units=128, activation='sigmoid'))
model.add(Dropout(0.1))
# output layer can only have the neurons equal to the number of outputs
model.add(Dense(units=n_classes, activation='softmax'))

# print the summary of our model
model.summary()

# compile the model
model.compile(loss='categorical_crossentropy',
              optimizer=SGD(),
              metrics=['accuracy'])

# train the model
model.fit(x_train, y_train,
          batch_size=batch_size,
          epochs=n_epochs)

# evaluate the model and print the accuracy score
scores = model.evaluate(x_test, y_test)

print('\n loss:', scores[0])
print('\n accuracy:', scores[1])
```

We get the following output from describing and training the Keras model:

```
_________________________________________________________________
Layer (type)                 Output Shape              Param #
=================================================================
dense_7 (Dense)              (None, 128)               100480
_________________________________________________________________
dropout_5 (Dropout)          (None, 128)               0
_________________________________________________________________
dense_8 (Dense)              (None, 128)               16512
_________________________________________________________________
dropout_6 (Dropout)          (None, 128)               0
_________________________________________________________________
dense_9 (Dense)              (None, 10)                1290
=================================================================
Total params: 118,282
Trainable params: 118,282
Non-trainable params: 0
_________________________________________________________________
Epoch 1/10
60000/60000 [========================] - 3s - loss: 2.3018 - acc: 0.1312
Epoch 2/10
60000/60000 [========================] - 2s - loss: 2.2395 - acc: 0.1920
Epoch 3/10
```

```
60000/60000 [========================] - 2s - loss: 2.1539 - acc: 0.2843
Epoch 4/10
60000/60000 [========================] - 2s - loss: 2.0214 - acc: 0.3856
Epoch 5/10
60000/60000 [========================] - 3s - loss: 1.8269 - acc: 0.4739
Epoch 6/10
60000/60000 [========================] - 2s - loss: 1.5973 - acc: 0.5426
Epoch 7/10
60000/60000 [========================] - 2s - loss: 1.3846 - acc: 0.6028
Epoch 8/10
60000/60000 [========================] - 3s - loss: 1.2133 - acc: 0.6502
Epoch 9/10
60000/60000 [========================] - 3s - loss: 1.0821 - acc: 0.6842
Epoch 10/10
60000/60000 [========================] - 3s - loss: 0.9799 - acc: 0.7157

loss: 0.859834249687
accuracy: 0.788
```

You can see how easy it is to build and train a model in Keras.

You can get more information about Keras from their very well-documented website, `https://keras.io`.

Summary

In this chapter, we learned about Keras. Keras is the most popular high-level library for TensorFlow. I personally prefer to use Keras for all the models that I develop for my commercial production work and also for academic research. We learned the workflow that we can follow to create and train the models in Keras, using both the functional and sequential APIs. We learned about various Keras layers and how to add the layers to the sequential and functional models. We also learned how to compile, train, and evaluate the Keras models. We also saw some of the additional modules provided by Keras.

Throughout the remaining chapters of the book, we shall be covering most of the examples in both core TensorFlow and Keras. In the next chapter, we will learn how to use TensorFlow for building traditional machine learning models for classification and regression.

4
Classical Machine Learning with TensorFlow

Machine learning is an area of computer science that involves research, development, and application of algorithms to make computing machines learn from data. The models learned by computing machines are used to make predictions and forecasts. Machine learning researchers and engineers achieve this goal by building models and then using these models for predictions. It's common knowledge now that machine learning has been used highly successfully in various areas such as natural language understanding, video processing, image recognition, speech, and vision.

Let's talk about models. All of the machine learning problems are abstracted to the following equation in one form or another:

$$y = f(x)$$

Here, y is the output or target and x is the input or features. If x is a collection of features, we also call it a feature vector and denote with X. When we say model, we mean to find the function f that maps features to targets. Thus once we find f, we can use the new value of x to predict values of y.

Machine learning is centered around finding the function f that can be used to predict y from values of x. As you might be able to recall from your high school mathematics days, the equation of the line is as follows:

$$y = mx + c$$

We can rewrite the preceding simple equation as follows:

$$y = Wx + b$$

Here, *W* is known as the weight and *b* is known as the bias. Don't worry about the terms weight and bias as of now, we will cover them later. For now, you can just think of *W* as the equivalent of *m* and *b* as the equivalent of *c* in the equation of the line. Thus, now the machine learning problem can be stated as a problem of finding *W* and *b* from current values of *X*, such that the equation can be used to predict the values of *y*.

Regression analysis or regression modeling refers to the methods and techniques used to estimate relationships among variables. The variables that are input to regression models are called independent variables or predictors or features and the output variable from regression models is called dependent variable or target. Regression models are defined as follows:

$$Y \approx f(X, \beta)$$

where Y is the target variable, X is a vector of features, and β is a vector of parameters

Often, we use a very simple form of regression known as simple linear regression to estimate the parameter β.

In machine learning problems, we have to learn the model parameters β_0 and β_1 from the given data so that we have an estimated model to predict the value of Y from future values of X. We use the term *weight* for β_1 and *bias* for β_0 and represent them with w and b in the code, respectively.

Thus the model becomes as follows:

$$y = X \times w + b$$

Classification is one of the classical problems in machine learning. Data under consideration could belong to one or other classes, for example, if the images provided are data, they could be pictures of cats or dogs. Thus the classes, in this case, are cats and dogs. Classification means to identify or recognize the label or class of the data or objects under consideration. Classification falls under the umbrella of supervised machine learning. In classification problems, the training dataset is provided that has features or inputs and their corresponding outputs or labels. Using this training dataset, a model is trained; in other words, parameters of the model are computed. The trained model is then used on new data to find its correct labels.

Classification problems can be of two types: **binary class** or **multiclass**. Binary class means the data is to be classified in two distinct and discrete labels, for example, the patient has cancer or the patient does not have cancer, the images are of cats or dogs. Multiclass means the data is to be classified among multiple classes, for example, an email classification problem will divide emails into social media emails, work-related emails, personal email, family-related emails, spam emails, shopping offer emails, and so on. Another example would be the example of pictures of digits; each picture could be labeled between 0 to 9, depending on what digit the picture represents. In this chapter, we will see an example of both kinds of classification.

In this chapter, we are going to further expand on the following topics:

- Regression
    - Simple linear regression
    - Multi regression
    - Regularized regression
    - Lasso regularization
    - Ridge regularization
    - ElasticNet regularization
- Classification
    - Classification using logistic regression
    - Binary classification
    - Multiclass classification

Simple linear regression

You might have used other machine learning libraries; now let's practice learning the simple linear regression model using TensorFlow. We will explain the concepts first using a generated dataset before moving on to domain-specific examples.

We will use generated datasets so that readers from all different domains can learn without getting overwhelmed with the details of the specific domain of the example.

You can follow along with the code in the Jupyter notebook `ch-04a_Regression`.

Data preparation

To generate the dataset, we use the `make_regression` function from the `datasets` module of the `sklearn` library:

```
from sklearn import datasets as skds
X, y = skds.make_regression(n_samples=200,
                            n_features=1,
                            n_informative=1,
                            n_targets=1,
                            noise = 20.0)
```

This generates a dataset for regression with 200 sample values for one feature and one target each, with some noise added. As we are generating only one target, the function generates y with a one-dimensional NumPy Array; thus, we reshape y to have two dimensions:

```
if (y.ndim == 1):
    y = y.reshape(len(y),1)
```

We plot the generated dataset to look at the data with the following code:

```
import matplotlib.pyplot as plt
plt.figure(figsize=(14,8))
plt.plot(X,y,'b.')
plt.title('Original Dataset')
plt.show()
```

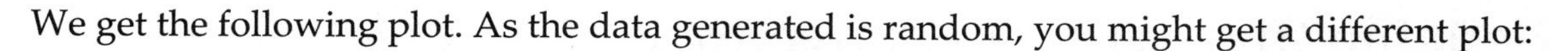

We get the following plot. As the data generated is random, you might get a different plot:

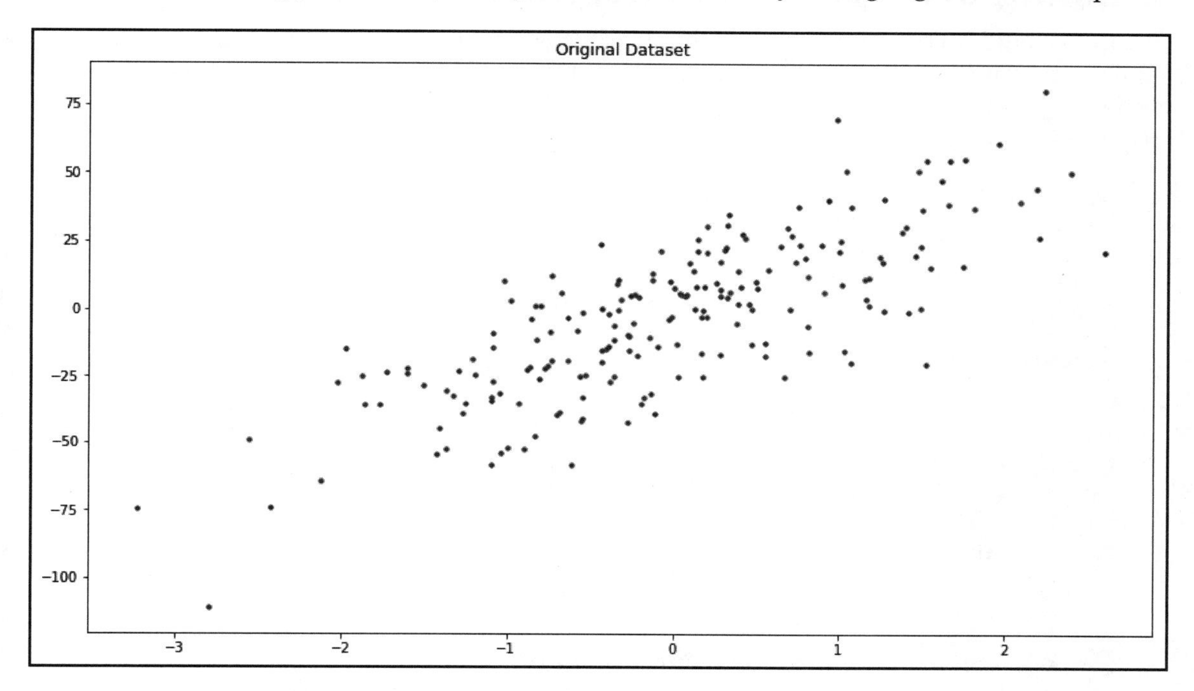

Now let's divide the data into train and test sets:

```
X_train, X_test, y_train, y_test = skms.train_test_split(X, y,
                                                          test_size=.4,
                                                          random_state=123)
```

Building a simple regression model

To build and train a regression model in TensorFlow, the following steps are taken in general:

1. Defining the inputs, parameters, and other variables.
2. Defining the model.
3. Defining the loss function.
4. Defining the optimizer function.
5. Training the model for a number of iterations known as epochs.

Defining the inputs, parameters, and other variables

Before we get into building and training the regression model using TensorFlow, let's define some important variables and operations. We find out the number of output and input variables from `X_train` and `y_train` and then use these numbers to define the *x*(`x_tensor`), *y* (`y_tensor`), *weights* (`w`), and *bias* (`b`):

```
num_outputs = y_train.shape[1]
num_inputs = X_train.shape[1]

x_tensor = tf.placeholder(dtype=tf.float32,
                          shape=[None, num_inputs],
                          name="x")
y_tensor = tf.placeholder(dtype=tf.float32,
                          shape=[None, num_outputs],
                          name="y")

w = tf.Variable(tf.zeros([num_inputs,num_outputs]),
                dtype=tf.float32,
                name="w")
b = tf.Variable(tf.zeros([num_outputs]),
                dtype=tf.float32,
                name="b")
```

- `x_tensor` is defined as having a shape of variable rows and `num_inputs` columns and the number of columns is only one in our example
- `y_tensor` is defined as having a shape of variable rows and `num_outputs` columns and the number of columns is only one in our example
- `w` is defined as a variable of dimensions `num_inputs` x `num_outputs`, which is **1 x 1** in our example
- `b` is defined as a variable of dimension `num_outputs`, which is one in our example

Defining the model

Next, we define the model as *(x_tensor × w) + b*:

```
model = tf.matmul(x_tensor, w) + b
```

Defining the loss function

Next, we define the loss function using the **mean squared error** (**MSE**). MSE is defined as follows:

$$\frac{1}{n}\sum(y_i - \hat{y}_i)^2$$

More details about MSE can be found from the following links:
`https://en.wikipedia.org/wiki/Mean_squared_error`
`http://www.statisticshowto.com/mean-squared-error/`

The difference in the actual and estimated value of y is known as **residual**. The loss function calculates the mean of squared residuals. We define it in TensorFlow in the following way:

```
loss = tf.reduce_mean(tf.square(model - y_tensor))
```

- `model - y_tensor` calculates the residuals
- `tf.square(model - y_tensor)` calculates the squares of each residual
- `tf.reduce_mean( ... )` finally calculates the mean of squares calculated in the preceding step

We also define the **mean squared error** (**mse**) and **r-squared** (**rs**) functions to evaluate the trained model. We use a separate `mse` function, because in the next chapters, the loss function will change but the `mse` function would remain the same.

```
# mse and R2 functions
mse = tf.reduce_mean(tf.square(model - y_tensor))
y_mean = tf.reduce_mean(y_tensor)
total_error = tf.reduce_sum(tf.square(y_tensor - y_mean))
unexplained_error = tf.reduce_sum(tf.square(y_tensor - model))
rs = 1 - tf.div(unexplained_error, total_error)
```

Defining the optimizer function

Next, we instantiate the `theGradientDescentOptimizer` function with a learning rate of 0.001 and set it to minimize the loss function:

```
learning_rate = 0.001
optimizer = tf.train.GradientDescentOptimizer(learning_rate).minimize(loss)
```

More details about gradient descent can be found at the following links:
`https://en.wikipedia.org/wiki/Gradient_descent`
`https://www.analyticsvidhya.com/blog/2017/03/introduction-to-gradient-descent-algorithm-along-its-variants/`

TensorFlow offers many other optimizer functions such as Adadelta, Adagrad, and Adam. We will cover some of them in the following chapters.

Training the model

Now that we have the model, loss function, and optimizer function defined, train the model to learn the parameters, *w*, and *b*. To train the model, define the following global variables:

- `num_epochs`: The number of iterations to run the training for. With every iteration, the model learns better parameters, as we will see in the plots later.
- `w_hat` and `b_hat`: To collect the estimated *w* and *b* parameters.
- `loss_epochs`, `mse_epochs`, `rs_epochs`: To collect the total error value on the training dataset, along with the mse and r-squared values of the model on the test dataset in every iteration.
- `mse_score` and `rs_score`: To collect mse and r-squared values of the final trained model.

```
num_epochs = 1500
w_hat = 0
b_hat = 0
loss_epochs = np.empty(shape=[num_epochs],dtype=float)
mse_epochs = np.empty(shape=[num_epochs],dtype=float)
rs_epochs = np.empty(shape=[num_epochs],dtype=float)

mse_score = 0
rs_score = 0
```

After initializing the session and the global variables, run the training loop for `num_epoch` times:

```
with tf.Session() as tfs:
    tf.global_variables_initializer().run()
    for epoch in range(num_epochs):
```

Within each iteration of the loop, run the optimizer on the training data:

```
tfs.run(optimizer, feed_dict={x_tensor: X_train, y_tensor: y_train})
```

Using the learned *w* and *b* values, calculate the error and save it in `loss_val` to plot it later:

```
loss_val = tfs.run(loss,feed_dict={x_tensor: X_train, y_tensor:
y_train})
loss_epochs[epoch] = loss_val
```

Calculate the mean squared error and r-squared value for the predicted values of the test data:

```
mse_score = tfs.run(mse,feed_dict={x_tensor: X_test, y_tensor: y_test})
mse_epochs[epoch] = mse_score

rs_score = tfs.run(rs,feed_dict={x_tensor: X_test, y_tensor: y_test})
rs_epochs[epoch] = rs_score
```

Finally, once the loop is finished, save the values of `w` and `b` to plot them later:

```
w_hat,b_hat = tfs.run([w,b])
w_hat = w_hat.reshape(1)
```

Let's print the model and final mean squared error on the test data after 2,000 iterations:

```
print('model : Y = {0:.8f} X + {1:.8f}'.format(w_hat[0],b_hat[0]))
print('For test data : MSE = {0:.8f}, R2 = {1:.8f} '.format(
    mse_score,rs_score))
```

This gives us the following output:

```
model : Y = 20.37448120 X + -2.75295663
For test data : MSE = 297.57995605, R2 = 0.66098368
```

Thus, the model that we trained is not a very good model, but we will see how to improve it using neural networks in later chapters.

The goal of this chapter is to introduce how to build and train regression models using TensorFlow without using neural networks.

Let's plot the estimated model along with the original data:

```
plt.figure(figsize=(14,8))
plt.title('Original Data and Trained Model')
x_plot = [np.min(X)-1,np.max(X)+1]
y_plot = w_hat*x_plot+b_hat
plt.axis([x_plot[0],x_plot[1],y_plot[0],y_plot[1]])
plt.plot(X,y,'b.',label='Original Data')
plt.plot(x_plot,y_plot,'r-',label='Trained Model')
plt.legend()
plt.show()
```

We get the following plot of the original data vs. the data from the trained model:

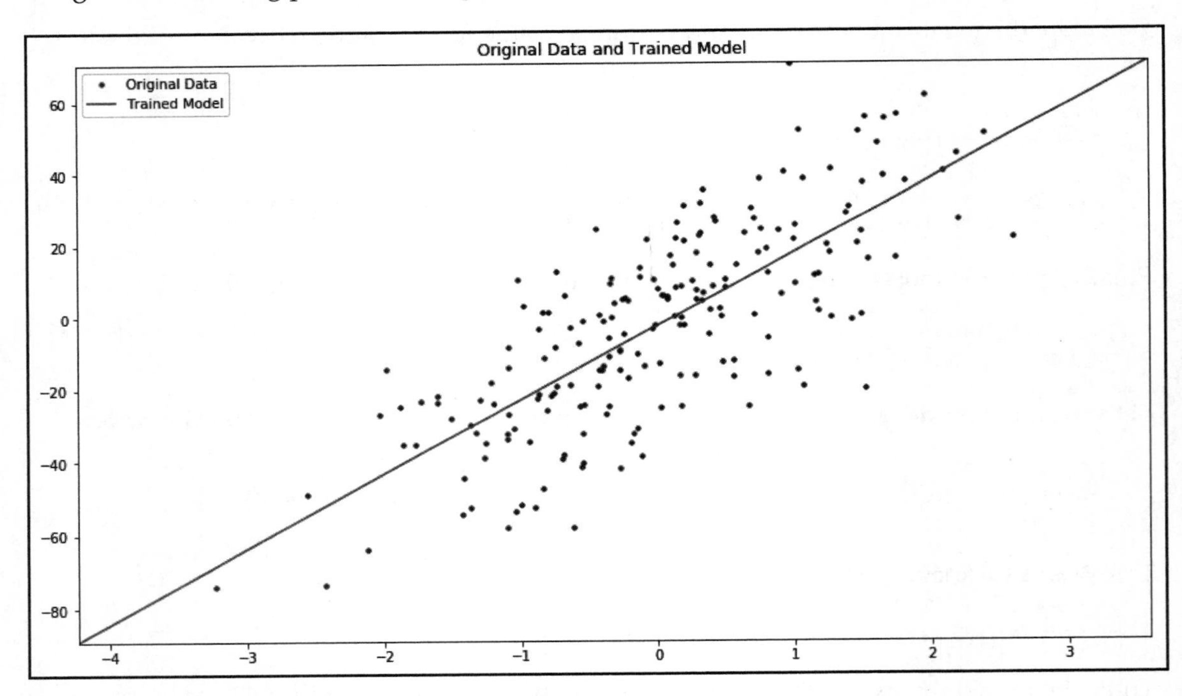

Let's plot the mean squared error for the training and test data in each iteration:

```
plt.figure(figsize=(14,8))

plt.axis([0,num_epochs,0,np.max(loss_epochs)])
plt.plot(loss_epochs, label='Loss on X_train')
plt.title('Loss in Iterations')
plt.xlabel('# Epoch')
plt.ylabel('MSE')

plt.axis([0,num_epochs,0,np.max(mse_epochs)])
plt.plot(mse_epochs, label='MSE on X_test')
plt.xlabel('# Epoch')
plt.ylabel('MSE')
plt.legend()

plt.show()
```

We get the following plot that shows that with each iteration, the mean squared error reduces and then remains at the same level near 500:

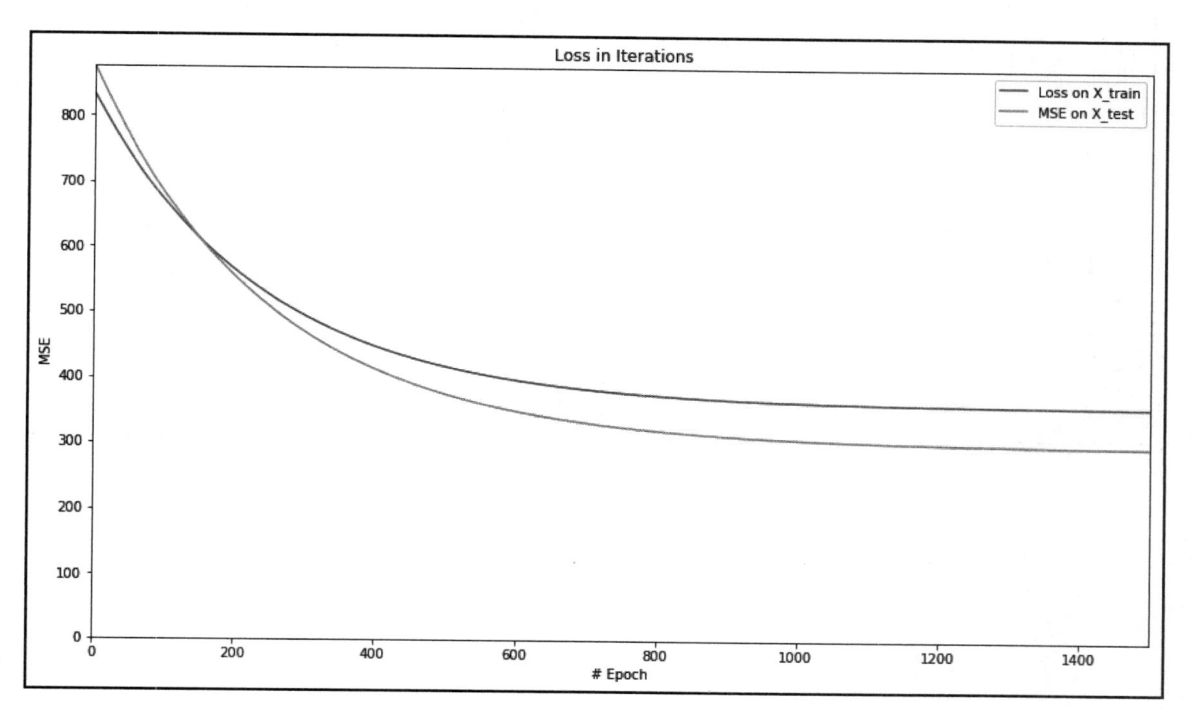

Let's plot the value of r-squared:

```
plt.figure(figsize=(14,8))
plt.axis([0,num_epochs,0,np.max(rs_epochs)])
plt.plot(rs_epochs, label='R2 on X_test')
plt.xlabel('# Epoch')
plt.ylabel('R2')
plt.legend()
plt.show()
```

We get the following plot when we plot the value of r-squared over epochs:

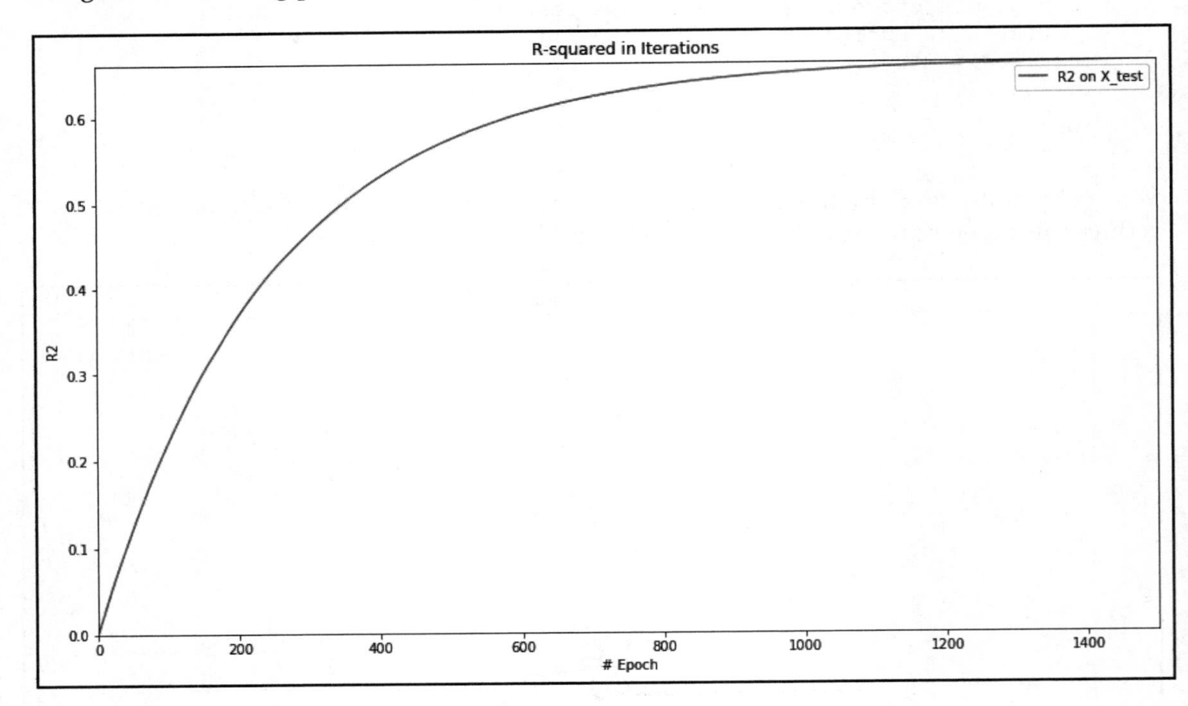

This basically shows that the model starts with a very low value of r-squared, but as the model gets trained and reduces the error, the value of r-squared starts getting higher and finally becomes stable at a point little higher than 0.6.

Plotting MSE and r-squared allows us to see how quickly our model is getting trained and where it starts becoming stable such that further training results in marginal or almost no benefits in reducing the error.

Using the trained model to predict

Now that you have the trained model, it can be used to make predictions about new data. The predictions from the linear model are made with the understanding of some minimum mean squared error that we saw in the previous plot because the straight line may not fit the data perfectly.

To get a better fitting model, we have to extend our model using different methods such as adding the linear combination of variables.

Multi-regression

Now that you have learned how to create a basic regression model with TensorFlow, let's try to run it on example datasets from different domains. The dataset that we generated as an example dataset is univariate, namely, the target was dependent only on one feature.

Most of the datasets, in reality, are multivariate. To emphasize a little more, the target depends on multiple variables or features, thus the regression model is called **multi-regression** or **multidimensional regression**.

We first start with the most popular Boston dataset. This dataset contains 13 attributes of 506 houses in Boston such as the average number of rooms per dwelling, nitric oxide concentration, weighted distances to five Boston employment centers, and so on. The target is the median value of owner-occupied homes. Let's dive into exploring a regression model for this dataset.

Load the dataset from the *sklearn* library and look at its description:

```
boston=skds.load_boston()
print(boston.DESCR)
X=boston.data.astype(np.float32)
y=boston.target.astype(np.float32)
if (y.ndim == 1):
    y = y.reshape(len(y),1)
X = skpp.StandardScaler().fit_transform(X)
```

We also extract `X`, a matrix of features, and `y`, a vector of targets in the preceding code. We reshape `y` to make it two-dimensional and scale the features in `x` to have a mean of zero and standard deviation of one. Now let's use this `X` and `y` to train the regression model, as we did in the previous example:

You may observe that the code for this example is similar to the code in the previous section on simple regression; however, we are using multiple features to train the model so it is called multi-regression.

```
X_train, X_test, y_train, y_test = skms.train_test_split(X, y,
    test_size=.4, random_state=123)
num_outputs = y_train.shape[1]
num_inputs = X_train.shape[1]

x_tensor = tf.placeholder(dtype=tf.float32,
    shape=[None, num_inputs], name="x")
y_tensor = tf.placeholder(dtype=tf.float32,
    shape=[None, num_outputs], name="y")

w = tf.Variable(tf.zeros([num_inputs,num_outputs]),
    dtype=tf.float32, name="w")
b = tf.Variable(tf.zeros([num_outputs]),
    dtype=tf.float32, name="b")

model = tf.matmul(x_tensor, w) + b
loss = tf.reduce_mean(tf.square(model - y_tensor))
# mse and R2 functions
mse = tf.reduce_mean(tf.square(model - y_tensor))
y_mean = tf.reduce_mean(y_tensor)
```

```
total_error = tf.reduce_sum(tf.square(y_tensor - y_mean))
unexplained_error = tf.reduce_sum(tf.square(y_tensor - model))
rs = 1 - tf.div(unexplained_error, total_error)

learning_rate = 0.001
optimizer = tf.train.GradientDescentOptimizer(learning_rate).minimize(loss)

num_epochs = 1500
loss_epochs = np.empty(shape=[num_epochs],dtype=np.float32)
mse_epochs = np.empty(shape=[num_epochs],dtype=np.float32)
rs_epochs = np.empty(shape=[num_epochs],dtype=np.float32)

mse_score = 0
rs_score = 0

with tf.Session() as tfs:
    tfs.run(tf.global_variables_initializer())
    for epoch in range(num_epochs):
        feed_dict = {x_tensor: X_train, y_tensor: y_train}
        loss_val, _ = tfs.run([loss, optimizer], feed_dict)
        loss_epochs[epoch] = loss_val

        feed_dict = {x_tensor: X_test, y_tensor: y_test}
        mse_score, rs_score = tfs.run([mse, rs], feed_dict)
        mse_epochs[epoch] = mse_score
        rs_epochs[epoch] = rs_score

print('For test data : MSE = {0:.8f}, R2 = {1:.8f} '.format(
    mse_score, rs_score))
```

We get the following output from the model:

```
For test data : MSE = 30.48501778, R2 = 0.64172244
```

Let's plot the MSE and R-squared values.

The following image shows the plotting of MSE:

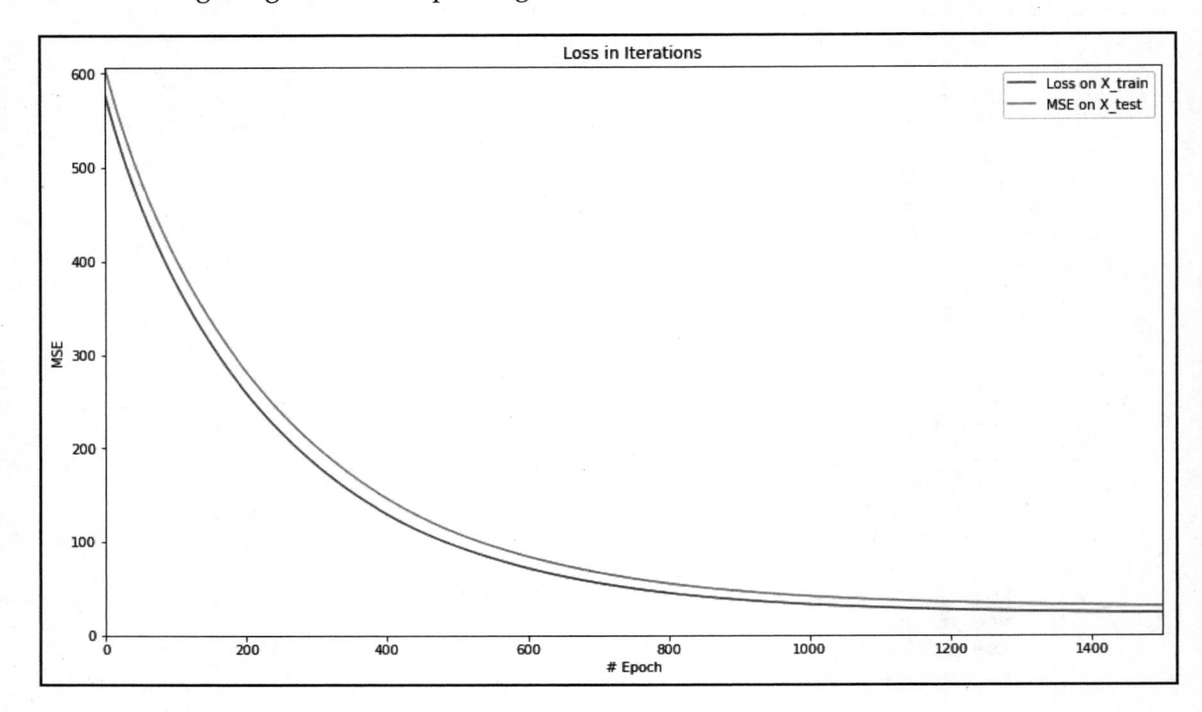

The following image shows the plotting of R-squared values:

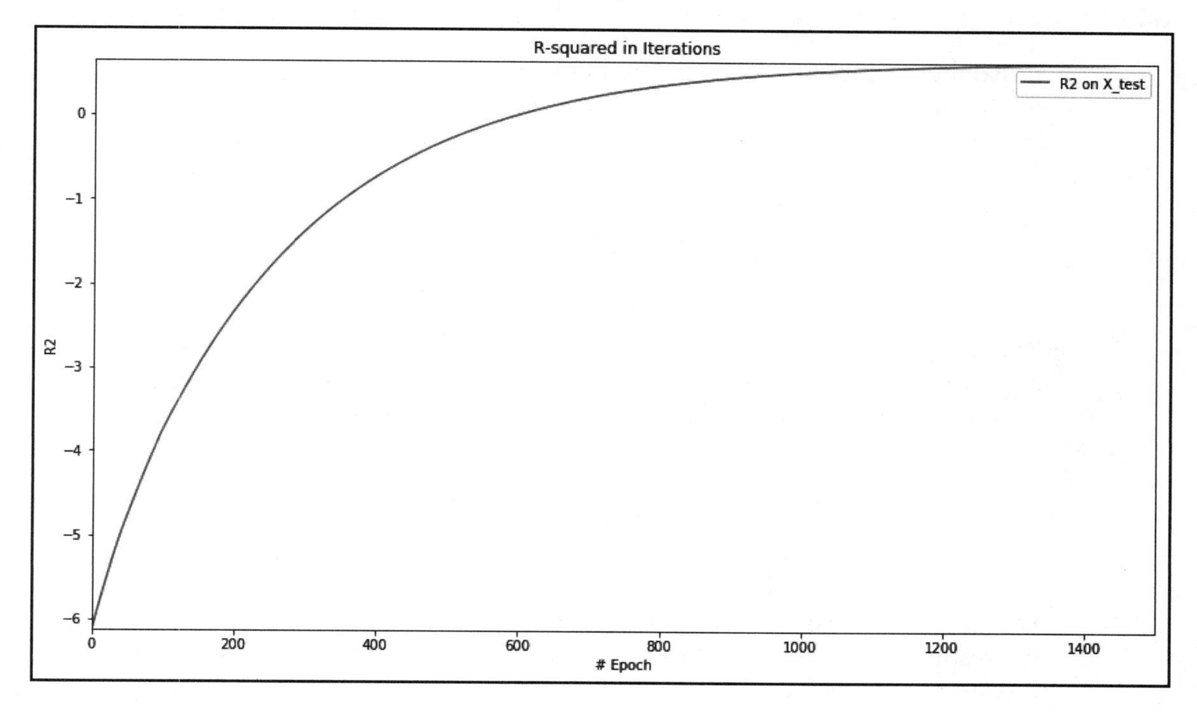

We see a similar pattern for MSE and r-squared, just as we saw for the univariate dataset.

Regularized regression

In linear regression, the model that we trained returns the best-fit parameters on the training data. However, finding the best-fit parameters on the training data may lead to overfitting.

Overfitting means that the model fits best to the training data but gives a greater error on the test data. Thus, we generally add a penalty term to the model to obtain a simpler model.

This penalty term is called a **regularization** term, and the regression model thus obtained is called a regularized regression model. There are three main types of regularization models:

- **Lasso regression**: In lasso regularization, also known as L1 regularization, the regularization term is the lasso parameter α multiplied with the sum of absolute values of the weights w. Thus, the loss function is as follows:

$$\frac{1}{n}\sum_{i=1}^{n}(y_i - \hat{y}_i)^2 + \alpha\frac{1}{n}\sum_{i=1}^{n}|w_i|$$

- **Ridge regression**: In ridge regularization, also known as L2 regularization, the regularization term is the ridge parameter α multiplied with the ith sum of the squares of the weights w. Thus, the loss function is as follows:

$$\frac{1}{n}\sum_{i=1}^{n}(y_i - \hat{y}_i)^2 + \alpha\frac{1}{n}\sum_{i=1}^{n}w_i^2$$

- **ElasticNet regression**: When we add both lasso and ridge regularization terms, the resulting regularization is known as the ElasticNet regularization. Thus, the loss function is as follows:

$$\frac{1}{n}\sum_{i=1}^{n}(y_i - \hat{y}_i)^2 + \alpha_1\frac{1}{n}\sum_{i=1}^{n}|w_i| + \alpha_2\frac{1}{n}\sum_{i=1}^{n}w_i^2$$

Refer to the following resources on the internet for further details on regularization:
`http://www.statisticshowto.com/regularization/`.

A simple rule of thumb is to use L1 or Lasso when we want to remove some features, thus reducing computation time, but at the cost of reduced accuracy.

Now let's see these regularization loss functions implemented in TensorFlow. We will continue with the Boston dataset that we used in the previous example.

Lasso regularization

We define the lasso parameter to have the value 0.8:

```
lasso_param = tf.Variable(0.8, dtype=tf.float32)
lasso_loss = tf.reduce_mean(tf.abs(w)) * lasso_param
```

Setting the lasso parameter as zero means no regularization as the term becomes zero. Higher the value of the regularization term, higher the penalty. The following is the complete code for lasso regularized regression to train the model in order to predict Boston house pricing:

The code below assumes that train and test datasets have been split as per the previous example.

```
num_outputs = y_train.shape[1]
num_inputs = X_train.shape[1]

x_tensor = tf.placeholder(dtype=tf.float32,
                          shape=[None, num_inputs], name='x')
y_tensor = tf.placeholder(dtype=tf.float32,
                          shape=[None, num_outputs], name='y')

w = tf.Variable(tf.zeros([num_inputs, num_outputs]),
                dtype=tf.float32, name='w')
b = tf.Variable(tf.zeros([num_outputs]),
                dtype=tf.float32, name='b')

model = tf.matmul(x_tensor, w) + b

lasso_param = tf.Variable(0.8, dtype=tf.float32)
lasso_loss = tf.reduce_mean(tf.abs(w)) * lasso_param

loss = tf.reduce_mean(tf.square(model - y_tensor)) + lasso_loss

learning_rate = 0.001
optimizer = tf.train.GradientDescentOptimizer(learning_rate).minimize(loss)

mse = tf.reduce_mean(tf.square(model - y_tensor))
y_mean = tf.reduce_mean(y_tensor)
total_error = tf.reduce_sum(tf.square(y_tensor - y_mean))
unexplained_error = tf.reduce_sum(tf.square(y_tensor - model))
rs = 1 - tf.div(unexplained_error, total_error)
```

```
num_epochs = 1500
loss_epochs = np.empty(shape=[num_epochs],dtype=np.float32)
mse_epochs = np.empty(shape=[num_epochs],dtype=np.float32)
rs_epochs = np.empty(shape=[num_epochs],dtype=np.float32)

mse_score = 0.0
rs_score = 0.0

num_epochs = 1500
loss_epochs = np.empty(shape=[num_epochs], dtype=np.float32)
mse_epochs = np.empty(shape=[num_epochs], dtype=np.float32)
rs_epochs = np.empty(shape=[num_epochs], dtype=np.float32)

mse_score = 0.0
rs_score = 0.0

with tf.Session() as tfs:
    tfs.run(tf.global_variables_initializer())
    for epoch in range(num_epochs):
        feed_dict = {x_tensor: X_train, y_tensor: y_train}
        loss_val,_ = tfs.run([loss,optimizer], feed_dict)
        loss_epochs[epoch] = loss_val

        feed_dict = {x_tensor: X_test, y_tensor: y_test}
        mse_score,rs_score = tfs.run([mse,rs], feed_dict)
        mse_epochs[epoch] = mse_score
        rs_epochs[epoch] = rs_score

print('For test data : MSE = {0:.8f}, R2 = {1:.8f} '.format(
    mse_score, rs_score))
```

We get the following output:

```
For test data : MSE = 30.48978233, R2 = 0.64166653
```

Let's plot the values of MSE and r-squared using the following code:

```
plt.figure(figsize=(14,8))

plt.axis([0,num_epochs,0,np.max([loss_epochs,mse_epochs])])
plt.plot(loss_epochs, label='Loss on X_train')
plt.plot(mse_epochs, label='MSE on X_test')
plt.title('Loss in Iterations')
plt.xlabel('# Epoch')
plt.ylabel('Loss or MSE')
plt.legend()

plt.show()
```

```
plt.figure(figsize=(14,8))

plt.axis([0,num_epochs,np.min(rs_epochs),np.max(rs_epochs)])
plt.title('R-squared in Iterations')
plt.plot(rs_epochs, label='R2 on X_test')
plt.xlabel('# Epoch')
plt.ylabel('R2')
plt.legend()

plt.show()
```

We get the following plot for loss:

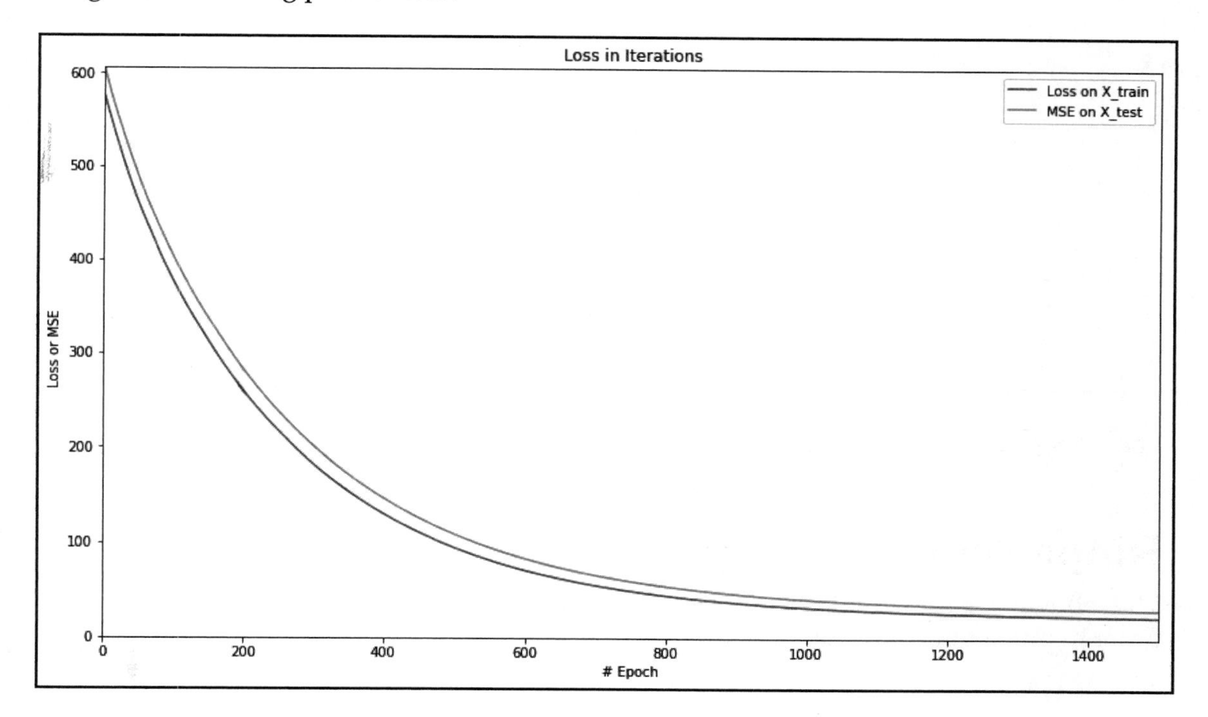

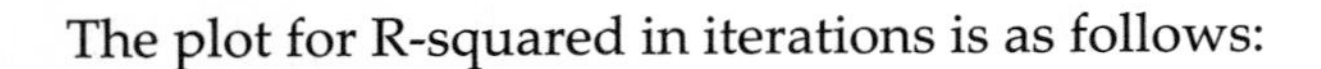

The plot for R-squared in iterations is as follows:

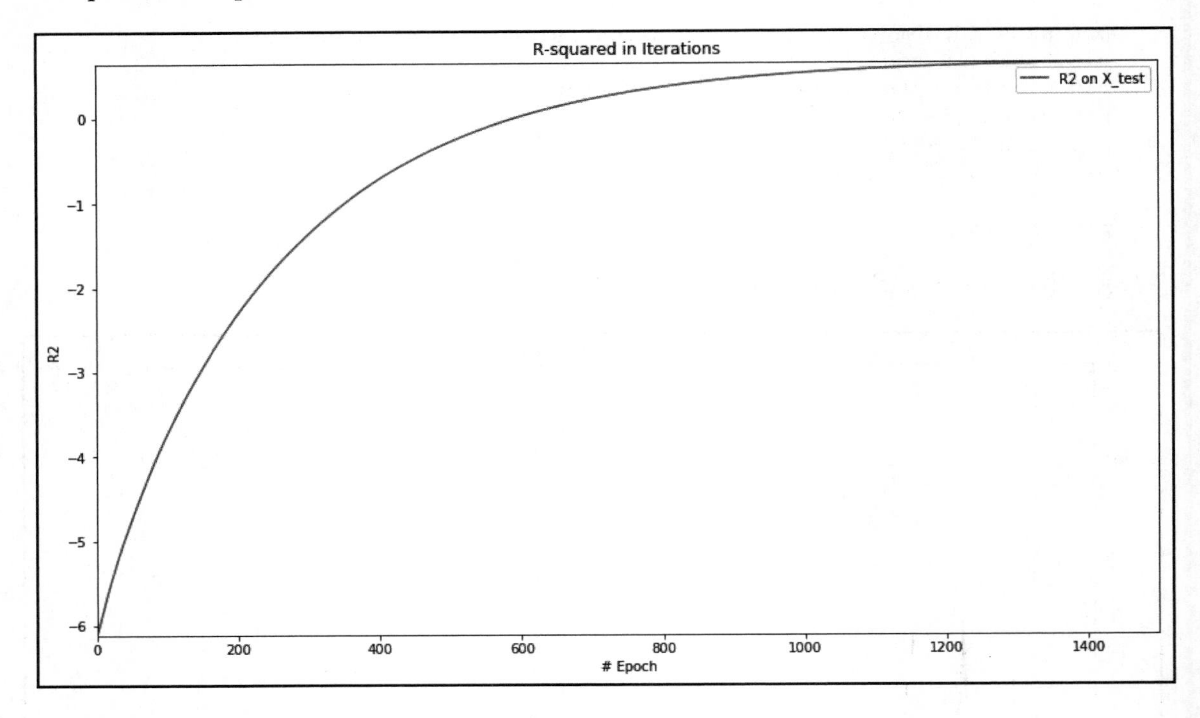

Let's repeat the same example with ridge regression.

Ridge regularization

The following is the complete code for ridge regularized regression to train the model in order to predict Boston house pricing:

```
num_outputs = y_train.shape[1]
num_inputs = X_train.shape[1]

x_tensor = tf.placeholder(dtype=tf.float32,
                          shape=[None, num_inputs], name='x')
y_tensor = tf.placeholder(dtype=tf.float32,
                          shape=[None, num_outputs], name='y')

w = tf.Variable(tf.zeros([num_inputs, num_outputs]),
                dtype=tf.float32, name='w')
b = tf.Variable(tf.zeros([num_outputs]),
```

```
                dtype=tf.float32, name='b')

model = tf.matmul(x_tensor, w) + b

ridge_param = tf.Variable(0.8, dtype=tf.float32)
ridge_loss = tf.reduce_mean(tf.square(w)) * ridge_param

loss = tf.reduce_mean(tf.square(model - y_tensor)) + ridge_loss

learning_rate = 0.001
optimizer = tf.train.GradientDescentOptimizer(learning_rate).minimize(loss)

mse = tf.reduce_mean(tf.square(model - y_tensor))
y_mean = tf.reduce_mean(y_tensor)
total_error = tf.reduce_sum(tf.square(y_tensor - y_mean))
unexplained_error = tf.reduce_sum(tf.square(y_tensor - model))
rs = 1 - tf.div(unexplained_error, total_error)

num_epochs = 1500
loss_epochs = np.empty(shape=[num_epochs],dtype=np.float32)
mse_epochs = np.empty(shape=[num_epochs],dtype=np.float32)
rs_epochs = np.empty(shape=[num_epochs],dtype=np.float32)

mse_score = 0.0
rs_score = 0.0

with tf.Session() as tfs:
    tfs.run(tf.global_variables_initializer())
    for epoch in range(num_epochs):
        feed_dict = {x_tensor: X_train, y_tensor: y_train}
        loss_val, _ = tfs.run([loss, optimizer], feed_dict=feed_dict)
        loss_epochs[epoch] = loss_val

        feed_dict = {x_tensor: X_test, y_tensor: y_test}
        mse_score, rs_score = tfs.run([mse, rs], feed_dict=feed_dict)
        mse_epochs[epoch] = mse_score
        rs_epochs[epoch] = rs_score

print('For test data : MSE = {0:.8f}, R2 = {1:.8f} '.format(
    mse_score, rs_score))
```

We get the following result:

```
For test data : MSE = 30.64177132, R2 = 0.63988018
```

Plotting the values of loss and MSE, we get the following plot for loss:

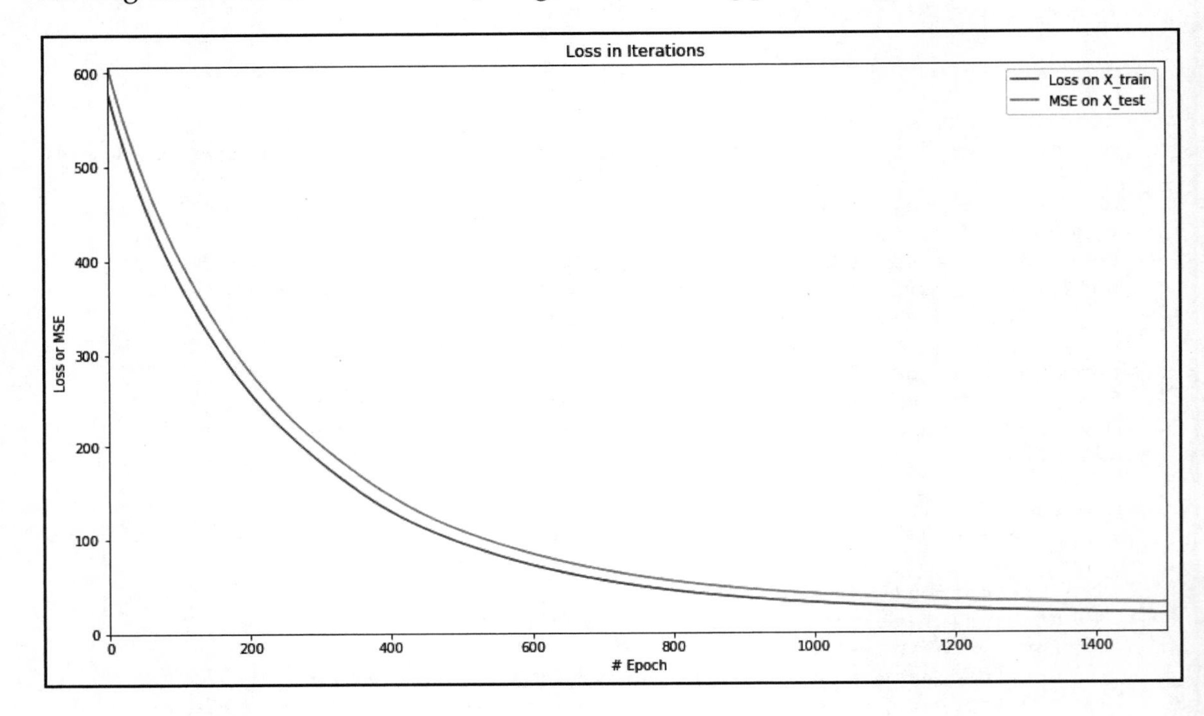

We get the following plot for R-squared:

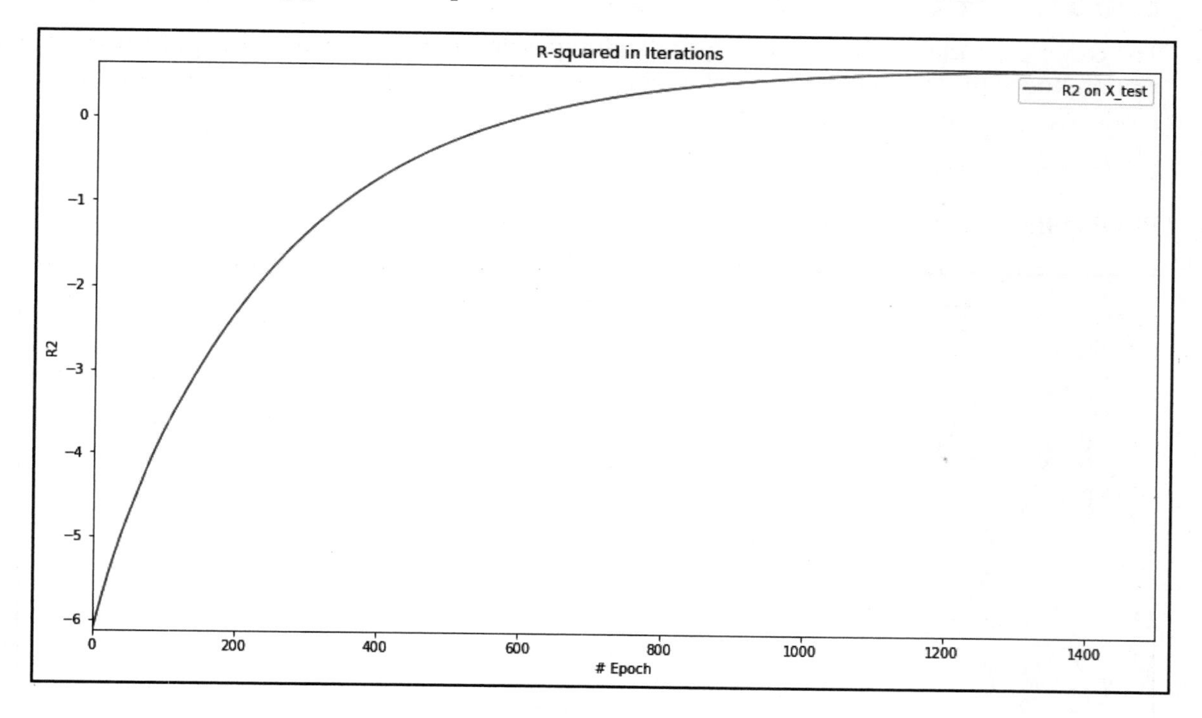

Let's look at the combination of lasso and ridge regularization methods.

ElasticNet regularization

The complete code for ElasticNet regularized regression to train the model to predict the Boston house pricing is provided in the notebook `ch-04a_Regression`. On running the model, we get the following result:

```
For test data : MSE = 30.64861488, R2 = 0.63979971
```

Plotting the values of loss and MSE, we get the following plots:

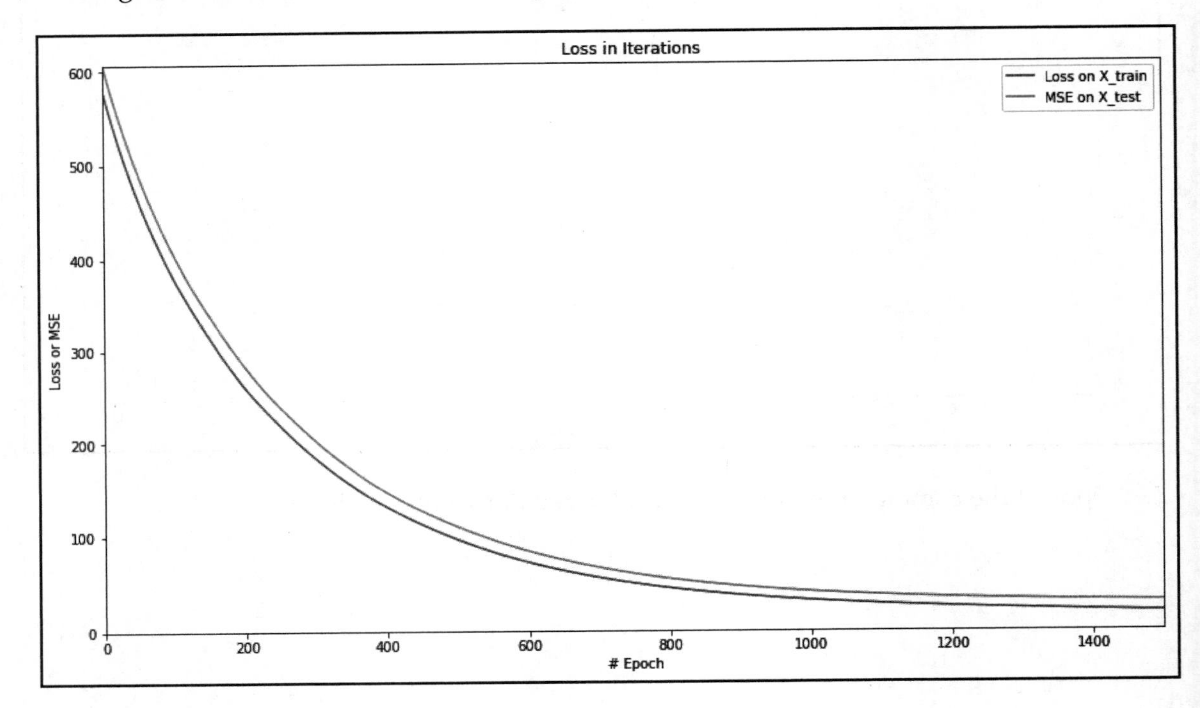

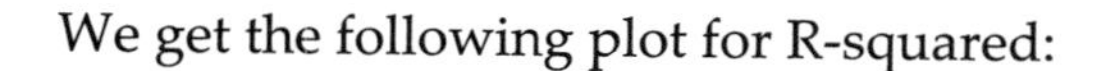

We get the following plot for R-squared:

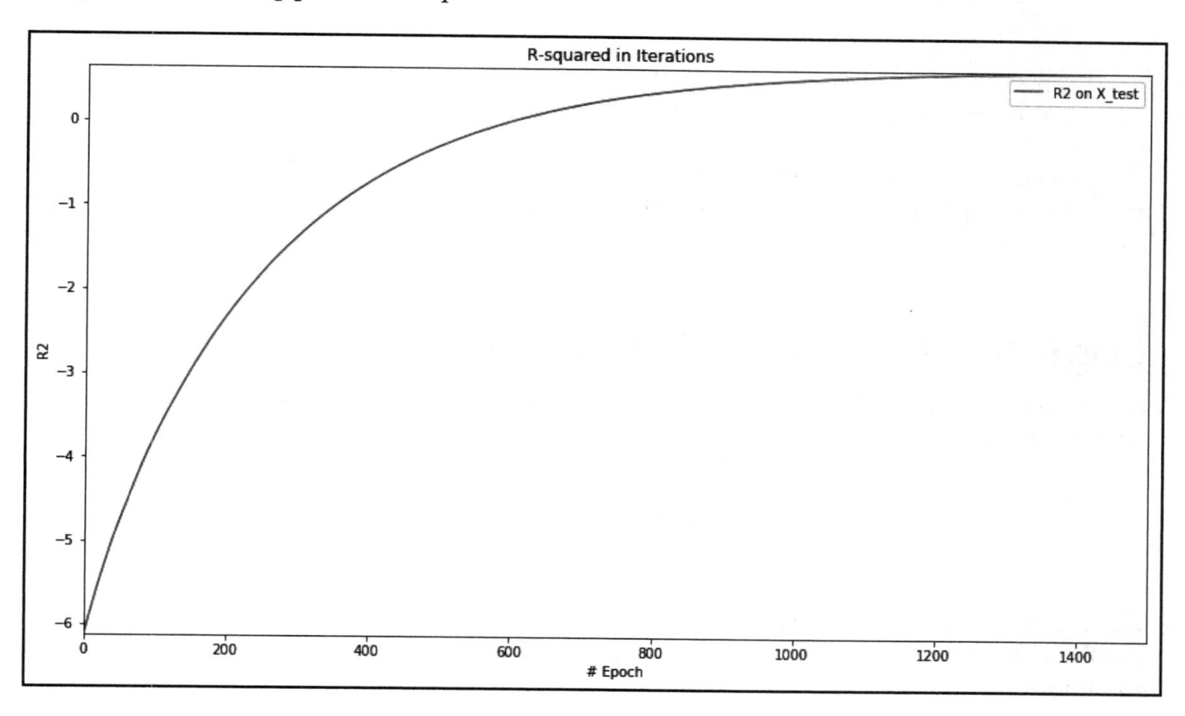

Classification using logistic regression

The most common method for classification is using logistic regression. Logistic regression is a probabilistic and linear classifier. The probability that vector of input features is a member of a specific class can be written formally as the following equation:

$$P(Y = i|x, w, b) = \phi(z)$$

In the above equation:

- Y represents the output,
- i represents one of the classes
- x represents the inputs

- w represents the weights
- b represents the biases
- z represents the regression equation $z = w \times x + b$
- ϕ represents the smoothing function or model in our case

The preceding equation represents that probability that x belongs to class i when w and b are given, is represented by function $\phi(z)$. Thus the model has to be trained to maximize the value of probability.

Logistic regression for binary classification

For binary classification, we define the model function $\phi(z)$ to be the sigmoid function, written as follows:

$$\phi(z) = \frac{1}{1+e^{-z}} = \frac{1}{1+e^{-(w \times x + b)}}$$

The sigmoid function produces the value of y to lie between the range [0,1]. Thus we can use the value of $y=\phi(z)$ to predict the class: if $y > 0.5$ then class is equal to 1, else class is equal to 0.

As we saw in the previous sections in this chapter that for linear regression, the model can be trained by finding parameters that minimize the loss function and loss function could be the sum of squared error or mean squared error. For logistic regression, we want to maximize the likelihood: $L(w) = P(y|x, w, b)$.

However, as it is easier to maximize the log-likelihood, thus we use the log-likelihood $l(w)$ as the cost function. The loss function ($J(w)$) is thus written as $-l(w)$ that can be minimized using the optimization algorithms such as gradient descent.

The loss function for binary logistic regression is written mathematically as follows:

$$J(w) = -\sum_{i=1}^{n}[(y_i \times log(\phi(z_i))) + ((1-y_i) \times (1 - log(\phi(z_i))))]$$

where $\phi(z)$ is the sigmoid function.

We will implement this loss function in the next section.

Logistic regression for multiclass classification

When there are more than two classes involved, the logistic regression is known multinomial logistic regression. In multinomial logistic regression, instead of sigmoid, we use softmax function that is one of the most popular functions. Softmax can be represented mathematically as follows:

$$softmax \;\; \phi_i(z) = \frac{e_i^z}{\sum_j e_j^z} = \frac{e_i^{(w \times x + b)}}{\sum_j e_j^{(w \times x + b)}} =$$

Softmax function produces the probabilities for each class, and the probabilities vector adds to 1. While predicting, the class with highest softmax value becomes the output or predicted class. The loss function, as we discussed earlier, is the negative log-likelihood function *-l(w)* that can be minimized by the optimizers such as gradient descent.

The loss function for multinomial logistic regression is written formally as follows:

$$J(w) = -\sum_{i=1}^{n} [y_i \times log(\phi(z_i))]$$

where $\phi(z)$ is the softmax function.

We will implement this loss function later in this chapter.

Let's dig into some examples in the next sections.

You can follow along with the code in the Jupyter notebook `ch-04b_Classification`.

Binary classification

Binary classification refers to problems with only two distinct classes. As we did in the previous chapter, we will generate a dataset using the convenience function, `make_classification()`, in the SciKit Learn library:

```
X, y = skds.make_classification(n_samples=200,
   n_features=2,
   n_informative=2,
   n_redundant=0,
   n_repeated=0,
   n_classes=2,
   n_clusters_per_class=1)
if (y.ndim == 1):
    y = y.reshape(-1,1)
```

The arguments to `make_classification()` are self-explanatory; `n_samples` is the number of data points to generate, `n_features` is the number of features to be generated, and `n_classes` is the number of classes, which is 2:

- `n_samples` is the number of data points to generate. We have kept it to 200 to keep the dataset small.
- `n_features` is the number of features to be generated; we are using only two features so that we can keep it a simple problem to understand the TensorFlow commands.
- `n_classes` is the number of classes, which is 2 as it is a binary classification problem.

Let's plot the data using the following code:

```
plt.scatter(X[:,0],X[:,1],marker='o',c=y)
plt.show()
```

We get the following plot; you might get a different plot as the data is generated randomly every time you run the data generation function:

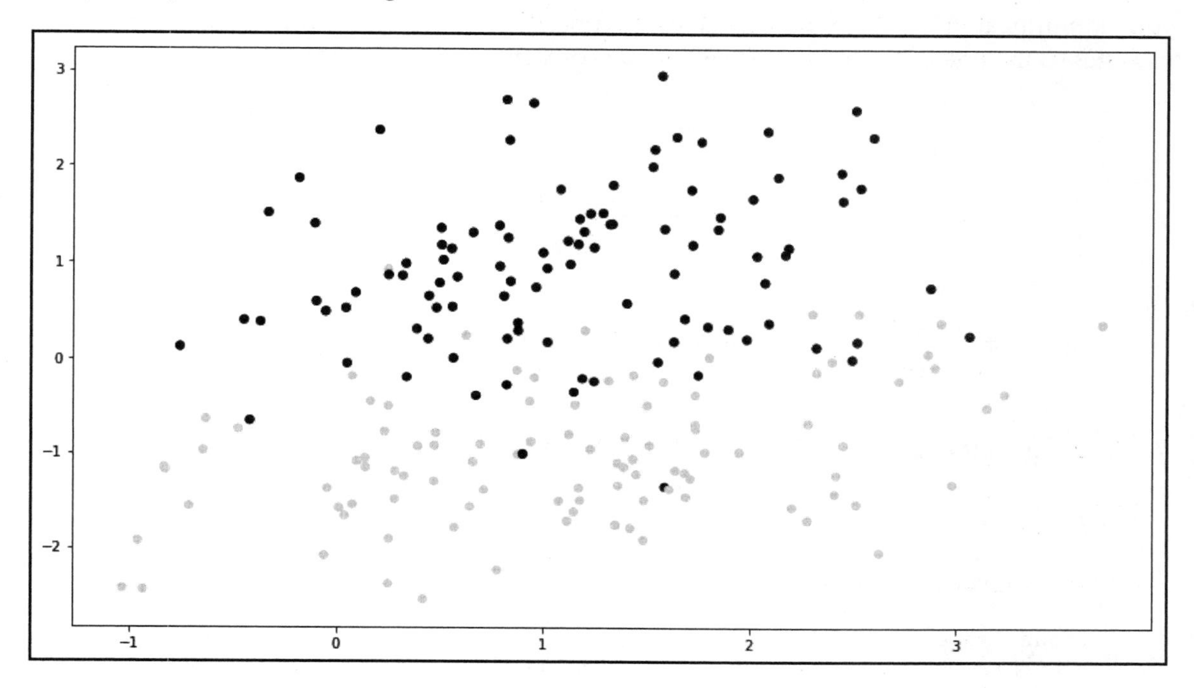

Then we use the NumPy eye function to convert y to one-hot encoded targets:

```
print(y[0:5])
y=np.eye(num_outputs)[y]
print(y[0:5])
```

The one-hot encoded targets appear as follows:

```
[1 0 0 1 0]
[[ 0.  1.]
 [ 1.  0.]
 [ 1.  0.]
 [ 0.  1.]
 [ 1.  0.]]
```

Divide the data into train and test categories:

```
X_train, X_test, y_train, y_test = skms.train_test_split(
    X, y, test_size=.4, random_state=42)
```

In classification, we use the sigmoid function to quantify the value of model such that the output value lies between the range [0,1]. The following equations denote the sigmoid function indicated by *$\phi(z)$*, where *z* is the equation $w \times x + b$. The loss function now changes to the one indicated by *$J(\theta)$*, where *θ* represents the parameters.

$$z_i = w_i \times x_i + b$$

$$\phi(z) = \frac{1}{1 + e^{-z}}$$

$$J(w) = -\sum_{i=1}^{n}[(y_i \times log(\phi(z_i))) + ((1 - y_i) \times (1 - log(\phi(z_i))))]$$

We implement the new model and loss function using the following code:

```
num_outputs = y_train.shape[1]
num_inputs = X_train.shape[1]

learning_rate = 0.001

# input images
x = tf.placeholder(dtype=tf.float32, shape=[None, num_inputs], name="x")
# output labels
y = tf.placeholder(dtype=tf.float32, shape=[None, num_outputs], name="y")

# model paramteres
w = tf.Variable(tf.zeros([num_inputs,num_outputs]), name="w")
b = tf.Variable(tf.zeros([num_outputs]), name="b")
model = tf.nn.sigmoid(tf.matmul(x, w) + b)

loss = tf.reduce_mean(-tf.reduce_sum(
    (y * tf.log(model)) + ((1 - y) * tf.log(1 - model)), axis=1))
optimizer = tf.train.GradientDescentOptimizer(
    learning_rate=learning_rate).minimize(loss)
```

Finally, we run our classification model:

```
num_epochs = 1
with tf.Session() as tfs:
    tf.global_variables_initializer().run()
    for epoch in range(num_epochs):
        tfs.run(optimizer, feed_dict={x: X_train, y: y_train})
        y_pred = tfs.run(tf.argmax(model, 1), feed_dict={x: X_test})
        y_orig = tfs.run(tf.argmax(y, 1), feed_dict={y: y_test})

        preds_check = tf.equal(y_pred, y_orig)
        accuracy_op = tf.reduce_mean(tf.cast(preds_check, tf.float32))
        accuracy_score = tfs.run(accuracy_op)
        print("epoch {0:04d} accuracy={1:.8f}".format(
            epoch, accuracy_score))

        plt.figure(figsize=(14, 4))
        plt.subplot(1, 2, 1)
        plt.scatter(X_test[:, 0], X_test[:, 1], marker='o', c=y_orig)
        plt.title('Original')
        plt.subplot(1, 2, 2)
        plt.scatter(X_test[:, 0], X_test[:, 1], marker='o', c=y_pred)
        plt.title('Predicted')
        plt.show()
```

We get a pretty good accuracy of about 96 percent and the original and predicted data graphs look like this:

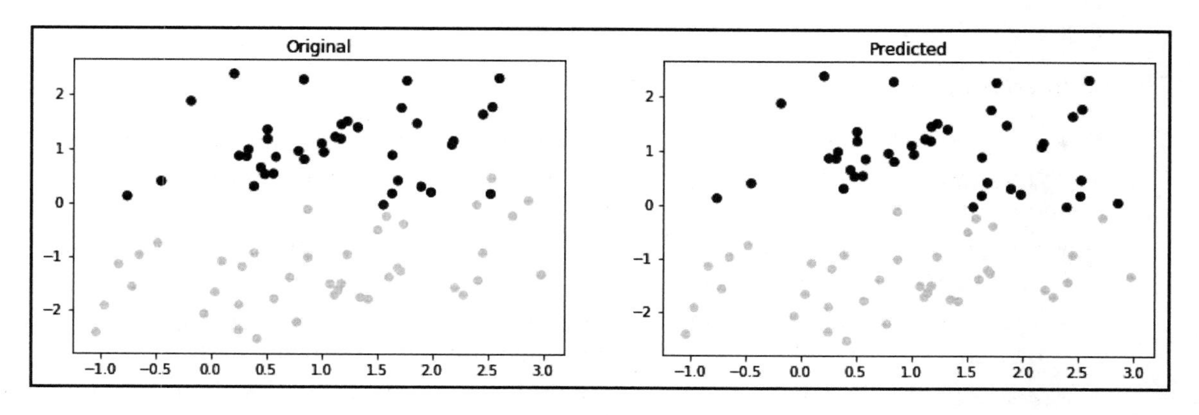

Pretty neat!! Now let's make our problem complicated and try to predict more than two classes.

Multiclass classification

One of the popular examples of multiclass classification is to label the images of handwritten digits. The classes or labels in this examples are {0,1,2,3,4,5,6,7,8,9}. In the following example, we will use MNIST. Let's load the MNIST images as we did in the earlier chapter with the following code:

```
from tensorflow.examples.tutorials.mnist import input_data
mnist = input_data.read_data_sets(os.path.join(
    datasetslib.datasets_root, 'mnist'), one_hot=True)
```

If the MNIST dataset is already downloaded as per instructions from an earlier chapter, then we would get the following output:

```
Extracting /Users/armando/datasets/mnist/train-images-idx3-ubyte.gz
Extracting /Users/armando/datasets/mnist/train-labels-idx1-ubyte.gz
Extracting /Users/armando/datasets/mnist/t10k-images-idx3-ubyte.gz
Extracting /Users/armando/datasets/mnist/t10k-labels-idx1-ubyte.gz
```

Now let's set some parameters, as shown in the following code:

```
num_outputs = 10 # 0-9 digits
num_inputs = 784 # total pixels

learning_rate = 0.001
num_epochs = 1
batch_size = 100
num_batches = int(mnist.train.num_examples/batch_size)
```

The parameters in the above code are as follows:

- `num_outputs`: As we have to predict that image represents which digit out of the ten digits, thus we set the number of outputs as 10. The digit is represented by the output that is turned on or set to one.
- `num_inputs`: We know that our input digits are 28 x 28 pixels, thus each pixel is an input to the model. Thus we have a total of 784 inputs.
- `learning_rate`: This parameter represents the learning rate for the gradient descent optimizer algorithm. We set the learning rate arbitrarily to 0.001.
- `num_epochs`: We will run our first example only for one iteration, hence we set the number of epochs to 1.

- `batch_size`: In the real world, we might have a huge dataset and loading the whole dataset in order to train the model may not be possible. Hence, we divide the input data into batches that are chosen randomly. We set the `batch_size` to 100 images that can be selected at a time using TensorFlow's inbuilt algorithm.
- `num_batches`: This parameter sets the number of times the batches should be selected from the total dataset; we set this to be equal to the number of items in the dataset divided by the number of items in a batch.

You are encouraged to experiment with different values of these parameters.

Now let's define the inputs, outputs, parameters, model, and loss function using the following code:

```
# input images
x = tf.placeholder(dtype=tf.float32, shape=[None, num_inputs], name="x")
# output labels
y = tf.placeholder(dtype=tf.float32, shape=[None, num_outputs], name="y")

# model paramteres
w = tf.Variable(tf.zeros([784, 10]), name="w")
b = tf.Variable(tf.zeros([10]), name="b")
model = tf.nn.softmax(tf.matmul(x, w) + b)

loss = tf.reduce_mean(-tf.reduce_sum(y * tf.log(model), axis=1))
optimizer = tf.train.GradientDescentOptimizer(
    learning_rate=learning_rate).minimize(loss)
```

The code is similar to the binary classification example with one significant difference: we use `softmax` instead of `sigmoid` function. Softmax is used for multiclass classification whereas sigmoid is used for binary class classification. Softmax function is a generalization of the sigmoid function that converts an n-dimensional vector z of arbitrary real values to an n-dimensional vector $\sigma(z)$ of real values in the range (0, 1] that add up to 1.

Now let's run the model and print the accuracy:

```
with tf.Session() as tfs:
    tf.global_variables_initializer().run()
    for epoch in range(num_epochs):
        for batch in range(num_batches):
            batch_x, batch_y = mnist.train.next_batch(batch_size)
            tfs.run(optimizer, feed_dict={x: batch_x, y: batch_y})
        predictions_check = tf.equal(tf.argmax(model, 1), tf.argmax(y, 1))
        accuracy_function = tf.reduce_mean(
            tf.cast(predictions_check, tf.float32))
        feed_dict = {x: mnist.test.images, y: mnist.test.labels}
        accuracy_score = tfs.run(accuracy_function, feed_dict)
        print("epoch {0:04d} accuracy={1:.8f}".format(
            epoch, accuracy_score))
```

We get the following accuracy:

```
epoch 0000  accuracy=0.76109999
```

Let's try training our model in multiple iterations, such that it learns with different batches in each iteration. We build two supporting functions to help us with this:

```
def mnist_batch_func(batch_size=100):
    batch_x, batch_y = mnist.train.next_batch(batch_size)
    return [batch_x, batch_y]
```

The preceding function takes the number of examples in a batch as input and uses the `mnist.train.next_batch()` function to return a batch of features (`batch_x`) and targets (`batch_y`):

```
def tensorflow_classification(num_epochs, num_batches, batch_size,
                              batch_func, optimizer, test_x, test_y):
    accuracy_epochs = np.empty(shape=[num_epochs], dtype=np.float32)
    with tf.Session() as tfs:
        tf.global_variables_initializer().run()
        for epoch in range(num_epochs):
            for batch in range(num_batches):
                batch_x, batch_y = batch_func(batch_size)
                feed_dict = {x: batch_x, y: batch_y}
                tfs.run(optimizer, feed_dict)
            predictions_check = tf.equal(
                tf.argmax(model, 1), tf.argmax(y, 1))
            accuracy_function = tf.reduce_mean(
                tf.cast(predictions_check, tf.float32))
            feed_dict = {x: test_x, y: test_y}
            accuracy_score = tfs.run(accuracy_function, feed_dict)
```

```
            accuracy_epochs[epoch] = accuracy_score
            print("epoch {0:04d} accuracy={1:.8f}".format(
                epoch, accuracy_score))

    plt.figure(figsize=(14, 8))
    plt.axis([0, num_epochs, np.min(
        accuracy_epochs), np.max(accuracy_epochs)])
    plt.plot(accuracy_epochs, label='Accuracy Score')
    plt.title('Accuracy over Iterations')
    plt.xlabel('# Epoch')
    plt.ylabel('Accuracy Score')
    plt.legend()
    plt.show()
```

The preceding function takes the parameters and performs the training iterations, printing the accuracy score for each iteration and prints the accuracy scores. It also saves the accuracy scores for each epoch in the `accuracy_epochs` array. Later, it plots the accuracy in each epoch. Let's run this function for 30 epochs using the parameters we set previously, using the following code:

```
num_epochs=30
tensorflow_classification(num_epochs=num_epochs,
    num_batches=num_batches,
    batch_size=batch_size,
    batch_func=mnist_batch_func,
    optimizer=optimizer,
    test_x=mnist.test.images,test_y=mnist.test.labels)
```

We get the following accuracy and graph:

```
epoch 0000  accuracy=0.76020002
epoch 0001  accuracy=0.79420000
epoch 0002  accuracy=0.81230003
epoch 0003  accuracy=0.82309997
epoch 0004  accuracy=0.83230001
epoch 0005  accuracy=0.83770001

--- epoch 6 to 24 removed for brevity ---

epoch 0025  accuracy=0.87930000
epoch 0026  accuracy=0.87970001
epoch 0027  accuracy=0.88059998
epoch 0028  accuracy=0.88120002
epoch 0029  accuracy=0.88180000
```

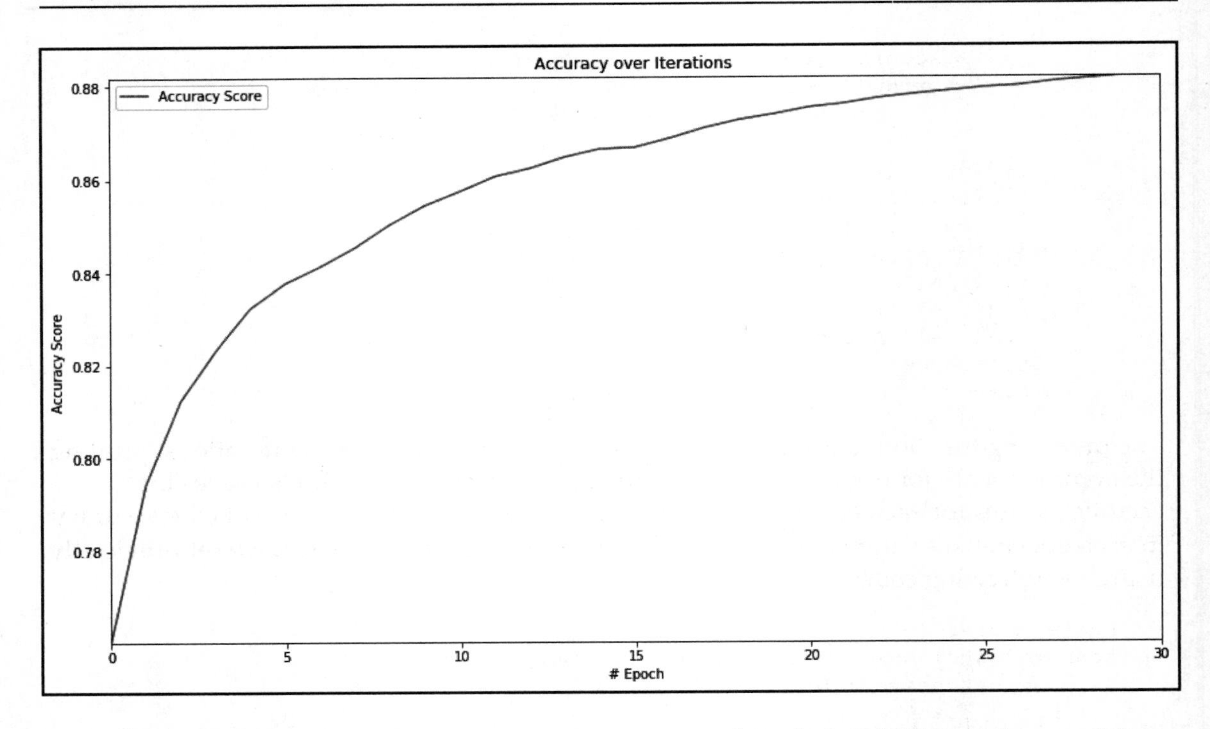

As we can see from the graph, accuracy improves very sharply in initial iterations and then the rate of improvement in accuracy slows down. Later, we will see how we can use the full power of neural networks in TensorFlow and bring this classification accuracy to a larger value.

Summary

In this chapter, we learned about applying classical machine learning algorithms in TensorFlow, without using neural networks. In the first section of the chapter, we learned about regression models. We explained how to train the models for linear regression with one or multiple features. We used TensorFlow to write the linear regression code. We also discussed that regularization is basically adding a penalty term so that the model does not overfit to the training data while learning the parameters in the training phase. We implemented Lasso, Ridge, and ElasticNet regularizations using TensorFlow. TensorFlow has some built-in regularization methods that we will study in the next chapters.

In the subsequent sections of this chapter, we learned about the classification problem in supervised machine learning. We discussed the model function, smoothing functions, and loss functions for binary class and multiclass classification. We used logistic regression in this chapter as that is the simplest method to implement classification. For binary classification, we used the sigmoid function and for multiclass classification, we used the softmax function to smooth the values of our linear model to produce the probabilities of output being in a specific class.

We implemented the logic for model and loss functions in TensorFlow and trained the model for binary classification and multiclass classification. Although we used the classical machine learning methods in this chapter and implemented them using TensorFlow, the full power of TensorFlow is unleashed when we implement neural networks and deep neural networks to solve machine learning problems. We will study such advanced methods in neural network-related chapters in this book.

You are encouraged to read the following books to learn more details on regression and classification:

Sebastian Raschka, *Python Machine Learning, 2nd Edition*. Packt Publishing, 2017

Trevor Hastie, Robert Tibshirani, Jerome Friedman, *The Elements of Statistical Learning*. Second Edition. Springer, 2013

5
Neural Networks and MLP with TensorFlow and Keras

The neural network is a modeling technique that was inspired by the structure and functioning of the brain. Just as the brain contains millions of tiny interconnected units known as neurons, the neural networks of today consist of millions of tiny interconnected computing units arranged in layers. Since the computing units of neural networks only exist in the digital world, as against the physical neurons of the brain, they are also called artificial neurons. Similarly, the **neural networks** (NN) are also known as the **artificial neural networks** (ANN).

In this chapter, we are going to further expand on the following topics:

- The perceptron (artificial neuron)
- Feed forward neural networks
- **MultiLayer Perceptron** (**MLP**) for image classification
    - TensorFlow-based MLP for MNIST image classification
    - Keras-based MLP for MNIST classification
    - TFLearn-based MLP for MNIST classification
- MLP for time series regression

The perceptron

Let's understand the most basic building block of a neural network, the **perceptron,** also known as the **artificial neuron**. The concept of the perceptron originated in the works of Frank Rosenblatt in 1962.

You may want to read the following work to explore the origins of neural networks:

Frank Rosenblatt, *Principles of Neurodynamics: Perceptrons and the Theory of Brain Mechanisms*. Spartan Books, 1962

In the most simplified view, a perceptron is modeled after the biological neurons such that it takes one or multiple inputs and combines them to generate output.

As shown in the following image, the perceptron takes three inputs and adds them to generate output *y*:

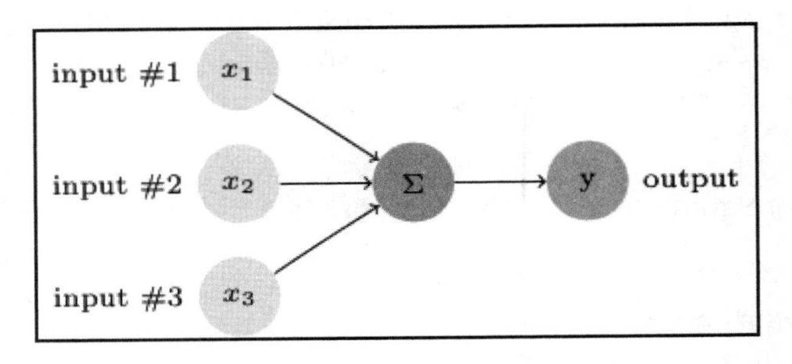

Simple perceptron

This perceptron is too simple to be of any practical use. Hence, it has been enhanced by adding the concept of weights, bias, and activation function. The weights are added to each input to get the weighted sum. If the weighted sum $\sum w_i x_i$ is less than the threshold value, then the output is 0, else output is 1:

$$y = \begin{cases} 0 & if \sum w_i x_i < threshold \\ 1 & if \sum w_i x_i \geq threshold \end{cases}$$

The threshold value is known as the **bias**. Let's move the bias to the left of the equation and denote it with b and represent $\Sigma w_i x_i$ with vector dot product of w and x. The equation for perceptron now becomes as follows:

$$y = \begin{cases} 0 & if \sum w \cdot x + b < 0 \\ 1 & if \sum w \cdot x + b \geq 0 \end{cases}$$

The perceptron now looks like the following image:

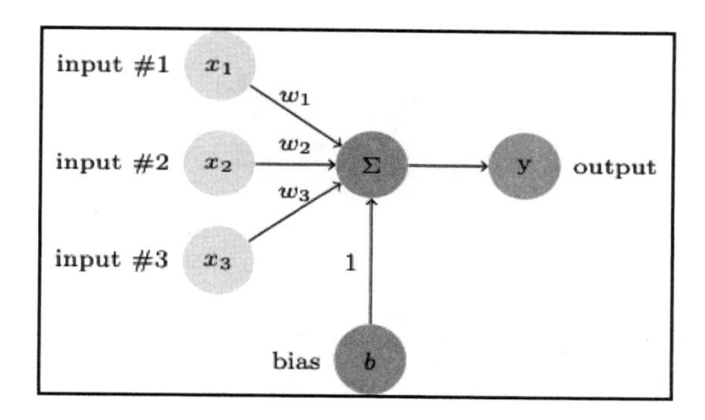

Simple perceptron with weights and bias

So far, the neuron is a linear function. In order to make this neuron produce a nonlinear decision boundary, run the output of summation through a nonlinear function known as the **activation** or transfer function. There are many popular activation functions available:

- `ReLU`: **Rectified Linear Unit**, smoothens the value to the range *(0,x)*, $ReLU(x) = max(0, x)$
- `sigmoid`: **Sigmoid** smoothens the value to the range *(0,1)*, $sigmoid(x) = \frac{1}{1 + e^{-x}} = \frac{e^x}{1 + e^x}$
- `tanh`: **Hyperbolic Tangent** smoothens the value to the range *(-1,1)*, $tanh(x) = \frac{e^x - e^{-x}}{e^x + e^{-x}}$

With the activation function, the equation for the perceptron becomes:

$$y = \varphi(w \cdot x + b)$$

where $\varphi(\cdot)$ is an activation function.

The neuron looks like the following image:

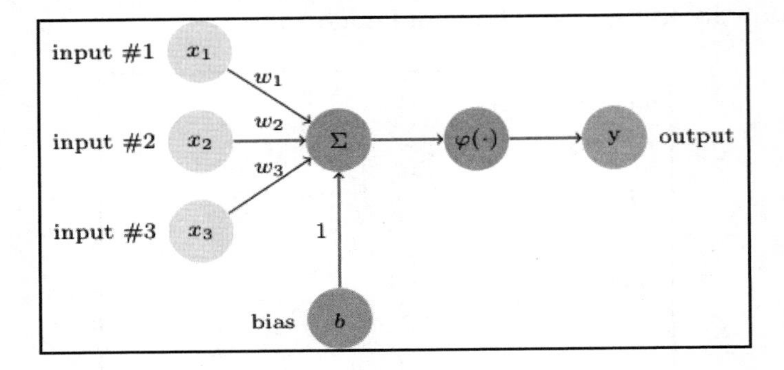

Simple perceptron with activation function, weights, and bias

MultiLayer Perceptron

When we connect the artificial neurons together, based on a well-defined structure, we call it a neural network. Here is the simplest neural network with one neuron:

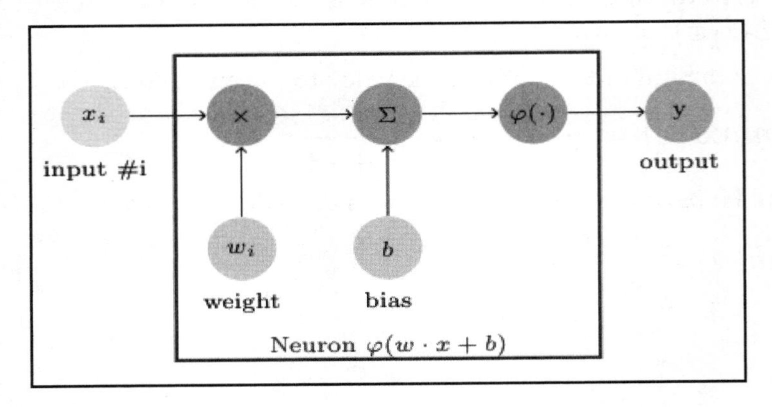

Neural network with one neuron

We connect the neurons such that the output of one layer becomes the input of the next layer, until the final layer's output becomes the final output. Such neural networks are called **feed forward neural networks (FFNN)**. As these FFNNs are made up of layers of neurons connected together, they are hence called **MultiLayer Perceptrons (MLP)** or **deep neural networks (DNN)**.

As an example, the MLP depicted in the following diagram has three features as inputs: two hidden layers of five neurons each and one output y. The neurons are fully connected to the neurons of the next layer. Such layers are also called dense layers or affine layers and such models are also known as sequential models.

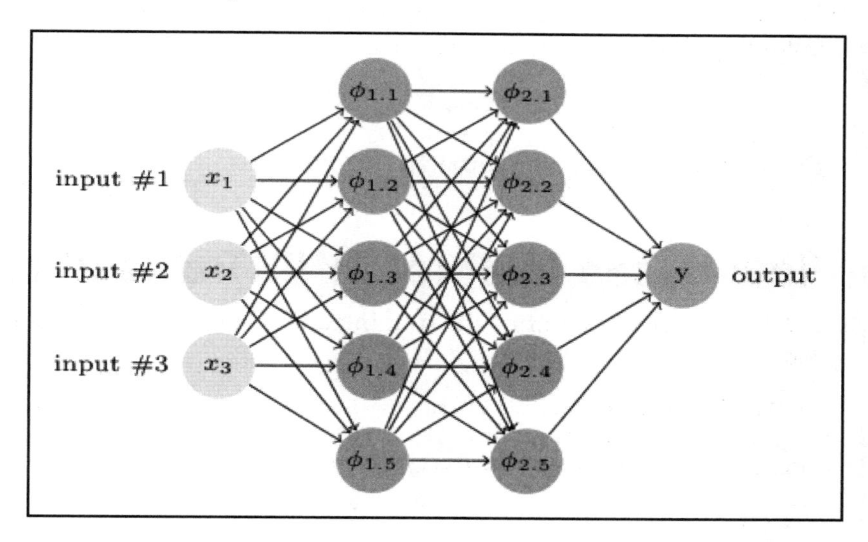

Let's revisit some of the example datasets that we explored earlier and build simple neural networks (MLP or DNN) in TensorFlow.

You can follow along with the code in the Jupyter notebook `ch-05_MLP`.

MLP for image classification

Let's build the MLP network for image classification using different libraries, such as TensorFlow, Keras, and TFLearn. We shall use the MNIST data set for the examples in this section.

The MNIST dataset contains the 28x28 pixel images of handwritten digits from 0 to 9, and their labels, 60K for the training set and 10K for the test set. The MNIST dataset is the most widely used data set, including in TensorFlow examples and tutorials.

The MNIST dataset and related documentation are available from the following link: `http://yann.lecun.com/exdb/mnist/`.

Let us start with the pure TensorFlow approach.

TensorFlow-based MLP for MNIST classification

First, load the MNIST dataset, and define the training and test features and the targets using the following code:

```
from tensorflow.examples.tutorials.mnist import input_data
mnist_home = os.path.join(datasetslib.datasets_root, 'mnist')
mnist = input_data.read_data_sets(mnist_home, one_hot=True)

X_train = mnist.train.images
X_test = mnist.test.images
Y_train = mnist.train.labels
Y_test = mnist.test.labels

num_outputs = 10 # 0-9 digits
num_inputs = 784 # total pixels
```

We create three helper functions that will help us create a simple MLP with only one hidden layer, followed by a larger MLP with multiple layers and multiple neurons in each layer.

The `mlp()` function builds the network layers with the following logic:

1. The `mlp()` function takes five inputs:
    - *x* is the input features tensor
    - `num_inputs` is the number of input features
    - `num_outputs` is the number of output targets
    - `num_layers` is the number of hidden layers required
    - `num_neurons` is the list containing the number of neurons for each layer
2. Set the weights and biases lists to empty:

```
w=[]
b=[]
```

3. Run a loop for the number of hidden layers to create weights and bias tensors and append them to their respective lists:
    - The tensors are given the names `w_<layer_num>` and `b_<layer_num>` respectively. Naming the tensors helps in the debugging and locating problems with the code.
    - The tensors are initialized with normal distribution using `tf.random_normal()`.
    - The first dimension of the weight tensor is the number of inputs from the previous layer. For the first hidden layer, the first dimension is `num_inputs`. The second dimension of the weights tensor is the number of neurons in the current layer.
    - The biases are all one-dimensional tensors, where the dimension equals the number of neurons in the current layer.

```
for i in range(num_layers):
   # weights
   w.append(tf.Variable(tf.random_normal(
        [num_inputs if i == 0 else num_neurons[i - 1],
         num_neurons[i]]),
        name="w_{0:04d}".format(i)
        ))
   # biases
   b.append(tf.Variable(tf.random_normal(
        [num_neurons[i]]),
        name="b_{0:04d}".format(i)
        ))
```

4. Create the weights and biases for the last hidden layer. In this case, the dimensions of the weights tensor are equal to the number of neurons in the last hidden layer and the number of output targets. The bias would be a tensor having a single dimension of the size of the number of output features:

```
w.append(tf.Variable(tf.random_normal(
     [num_neurons[num_layers - 1] if num_layers > 0 else num_inputs,
      num_outputs]), name="w_out"))
b.append(tf.Variable(tf.random_normal([num_outputs]),
    name="b_out"))
```

5. Now start defining the layers. First, treat `x` as the first most visible input layer:

```
# x is input layer
layer = x
```

6. Add the hidden layers in a loop. Each hidden layer represents the linear function `tf.matmul(layer, w[i]) + b[i]` being made nonlinear by the activation function `tf.nn.relu()`:

```
# add hidden layers
for i in range(num_layers):
    layer = tf.nn.relu(tf.matmul(layer, w[i]) + b[i])
```

7. Add the output layer. The one difference between the output layer and the hidden layer is the absence of activation function in the output layer:

```
layer = tf.matmul(layer, w[num_layers]) + b[num_layers]
```

8. Return the `layer` object that contains the MLP network:

```
return layer
```

The complete code of the entire MLP function is as follows:

```
def mlp(x, num_inputs, num_outputs, num_layers, num_neurons):
    w = []
    b = []
    for i in range(num_layers):
        # weights
        w.append(tf.Variable(tf.random_normal(
            [num_inputs if i == 0 else num_neurons[i - 1],
             num_neurons[i]]),
            name="w_{0:04d}".format(i)
        ))
        # biases
        b.append(tf.Variable(tf.random_normal(
            [num_neurons[i]]),
            name="b_{0:04d}".format(i)
        ))
    w.append(tf.Variable(tf.random_normal(
        [num_neurons[num_layers - 1] if num_layers > 0 else num_inputs,
         num_outputs]), name="w_out"))
    b.append(tf.Variable(tf.random_normal([num_outputs]), name="b_out"))

    # x is input layer
    layer = x
    # add hidden layers
    for i in range(num_layers):
        layer = tf.nn.relu(tf.matmul(layer, w[i]) + b[i])
    # add output layer
    layer = tf.matmul(layer, w[num_layers]) + b[num_layers]

    return layer
```

The helper function `mnist_batch_func()` wraps the TensorFlow's batch function for the MNIST dataset to provide the next batch of images:

```
def mnist_batch_func(batch_size=100):
    X_batch, Y_batch = mnist.train.next_batch(batch_size)
    return [X_batch, Y_batch]
```

This function is self-explanatory. TensorFlow provides this function for the MNIST dataset; however, for other datasets, we may have to write our own batch function.

The helper function, `tensorflow_classification()`, trains and evaluates the model.

1. The `tensorflow_classification()` function takes several inputs:

    - `n_epochs` is the number of training loops to run
    - `n_batches` is the number of randomly sampled batches for which the training in each cycle should be run
    - `batch_size` is the number of samples in each batch
    - `batch_func` is the function that takes the `batch_size` and returns the sample batch of `X` and `Y`
    - `model` is the actual neural network or layers with neurons
    - `optimizer` is the optimization function defined using TensorFlow
    - `loss` is the loss of cost function that the optimizer would optimize the parameters for
    - `accuracy_function` is the function that calculates the accuracy score
    - `X_test` and `Y_test` are the datasets for the testing

2. Start the TensorFlow session to run the training loop:

```
with tf.Session() as tfs:
    tf.global_variables_initializer().run()
```

3. Run the training for `n_epoch` cycles:

```
for epoch in range(n_epochs):
```

4. In each cycle, take the `n_batches` number of sample sets and train the model, calculate the loss for each batch, calculate the average loss for each epoch:

```
epoch_loss = 0.0
            for batch in range(n_batches):
                X_batch, Y_batch = batch_func(batch_size)
                feed_dict = {x: X_batch, y: Y_batch}
                _, batch_loss = tfs.run([optimizer, loss],
feed_dict)
                epoch_loss += batch_loss
            average_loss = epoch_loss / n_batches
            print("epoch: {0:04d} loss = {1:0.6f}".format(
                epoch, average_loss))
```

5. When all the epoch cycles are finished, calculate and print the accuracy score calculated with the `accuracy_function`:

```
feed_dict = {x: X_test, y: Y_test}
accuracy_score = tfs.run(accuracy_function,
                         feed_dict=feed_dict)
print("accuracy={0:.8f}".format(accuracy_score))
```

The complete code of `tensorflow_classification()` function is given below:

```
def tensorflow_classification(n_epochs, n_batches,
                              batch_size, batch_func,
                              model, optimizer, loss, accuracy_function,
                              X_test, Y_test):
    with tf.Session() as tfs:
        tfs.run(tf.global_variables_initializer())
        for epoch in range(n_epochs):
            epoch_loss = 0.0
            for batch in range(n_batches):
                X_batch, Y_batch = batch_func(batch_size)
                feed_dict = {x: X_batch, y: Y_batch}
                _, batch_loss = tfs.run([optimizer, loss], feed_dict)
                epoch_loss += batch_loss
            average_loss = epoch_loss / n_batches
            print("epoch: {0:04d} loss = {1:0.6f}".format(
                epoch, average_loss))
        feed_dict = {x: X_test, y: Y_test}
        accuracy_score = tfs.run(accuracy_function, feed_dict=feed_dict)
        print("accuracy={0:.8f}".format(accuracy_score))
```

Now let's define the input and output placeholders, *x* and *y*, and other hyper-parameters:

```
# input images
x = tf.placeholder(dtype=tf.float32, name="x",
                   shape=[None, num_inputs])
# target output
y = tf.placeholder(dtype=tf.float32, name="y",
                   shape=[None, num_outputs])
num_layers = 0
num_neurons = []
learning_rate = 0.01
n_epochs = 50
batch_size = 100
n_batches = int(mnist.train.num_examples/batch_size)
```

The parameters are described below:

- `num_layers` is the number of hidden layers. We first practice with no hidden layer, only the input, and output layers.
- `num_neurons` is the empty list because there are no hidden layers.
- `learning_rate` is 0.01, a randomly selected small number.
- `num_epochs` represents the 50 iterations to learn the parameters for the only neuron that connects the inputs to the output.
- `batch_size` is kept at 100, again a matter of choice. Larger batch size does not necessarily offer higher benefits. You might have to explore different batch sizes to find the optimum batch size for your neural networks.
- `n_batches`: Number of batches is calculated approximately to be the number of examples divided by the number of samples in a batch.

Now let's put everything together and define the network, `loss` function, `optimizer` function, and `accuracy` function using the variables defined so far.

```
model = mlp(x=x,
            num_inputs=num_inputs,
            num_outputs=num_outputs,
            num_layers=num_layers,
            num_neurons=num_neurons)

loss = tf.reduce_mean(
    tf.nn.softmax_cross_entropy_with_logits(logits=model, labels=y))
optimizer = tf.train.GradientDescentOptimizer(
    learning_rate=learning_rate).minimize(loss)

predictions_check = tf.equal(tf.argmax(model, 1), tf.argmax(y, 1))
accuracy_function = tf.reduce_mean(tf.cast(predictions_check, tf.float32))
```

In this code, we use a new tensorflow function to define the loss function:

```
tf.nn.softmax_cross_entropy_with_logits(logits=model, labels=y)
```

When the `softmax_cross_entropy_with_logits()` function is used, make sure that the output is unscaled and has not been passed through the `softmax` activation function. This function internally uses *softmax* to scale the output.

This function computes the softmax entropy between the model (the estimated value *y)* and the actual value of *y*. The entropy function is used when the output belongs to one class and not more than one class. As in our example, the image can only belong to one of the digits.

More information on this entropy function can be found at https://www.tensorflow.org/api_docs/python/tf/nn/softmax_cross_entropy_with_logits.

Once everything is defined, run the `tensorflow_classification` function to train and evaluate the model:

```
tensorflow_classification(n_epochs=n_epochs,
   n_batches=n_batches,
   batch_size=batch_size,
   batch_func=mnist_batch_func,
   model = model,
   optimizer = optimizer,
   loss = loss,
   accuracy_function = accuracy_function,
   X_test = mnist.test.images,
   Y_test = mnist.test.labels
   )
```

We get the following output from running the classification:

```
epoch: 0000    loss = 8.364567
epoch: 0001    loss = 4.347608
epoch: 0002    loss = 3.085622
epoch: 0003    loss = 2.468341
epoch: 0004    loss = 2.099220
epoch: 0005    loss = 1.853206

--- Epoch 06 to 45 output removed for brevity ---

epoch: 0046    loss = 0.684285
epoch: 0047    loss = 0.678972
epoch: 0048    loss = 0.673685
epoch: 0049    loss = 0.668717
accuracy=0.85720009
```

We see that the single neuron network slowly reduces the loss from 8.3 to 0.66 over 50 iterations, finally getting an accuracy of almost 85 percent. This is pretty bad accuracy for this specific example because this was only a demonstration of using TensorFlow for classification using MLP.

We ran the same code with more layers and neurons and got the following accuracy:

Number of Layers	Number of Neurons in Each Hidden Layer	Accuracy
0	0	0.857
1	8	0.616
2	256	0.936

Thus, by adding two rows and 256 neurons to each layer, we brought the accuracy up to 0.936. You are encouraged to try the code with different values of variables to observe how it affects the loss and accuracy.

Keras-based MLP for MNIST classification

Now let's build the same MLP network with Keras, a high-level library for TensorFlow. We keep all the parameters the same as we used for the TensorFlow example in this chapter, for example, the activation function for the hidden layers is kept as the ReLU function.

1. Import the required modules from the Keras:

```
import keras
from keras.models import Sequential
from keras.layers import Dense
from keras.optimizers import SGD
```

2. Define the hyper-parameters (we assume that the dataset has already been loaded into the `X_train`, `Y_train`, `X_test`, and `Y_test` variables):

```
num_layers = 2
num_neurons = []
for i in range(num_layers):
   num_neurons.append(256)
learning_rate = 0.01
n_epochs = 50
batch_size = 100
```

3. Create a sequential model:

```
model = Sequential()
```

4. Add the first hidden layer. Only in the first hidden layer, we have to specify the shape of the input tensor:

```
model.add(Dense(units=num_neurons[0], activation='relu',
    input_shape=(num_inputs,)))
```

5. Add the second layer:

```
model.add(Dense(units=num_neurons[1], activation='relu'))
```

6. Add the output layer with the activation function softmax:

```
model.add(Dense(units=num_outputs, activation='softmax'))
```

7. Print the model details:

```
model.summary()
```

We get the following output:

```
Layer (type)                 Output Shape              Param #
=================================================================
dense_1 (Dense)              (None, 256)               200960
_________________________________________________________________
dense_2 (Dense)              (None, 256)               65792
_________________________________________________________________
dense_3 (Dense)              (None, 10)                2570
=================================================================
Total params: 269,322
Trainable params: 269,322
Non-trainable params: 0
_________________________________________________________________
```

8. Compile the model with an SGD optimizer:

```
model.compile(loss='categorical_crossentropy',
   optimizer=SGD(lr=learning_rate),
   metrics=['accuracy'])
```

9. Train the model:

```
model.fit(X_train, Y_train,
   batch_size=batch_size,
   epochs=n_epochs)
```

As the model is being trained, we can observe the loss and accuracy of each training iteration:

```
Epoch 1/50
55000/55000 [========================] - 4s - loss: 1.1055 - acc: 0.7413
Epoch 2/50
55000/55000 [========================] - 3s - loss: 0.4396 - acc: 0.8833
Epoch 3/50
55000/55000 [========================] - 3s - loss: 0.3523 - acc: 0.9010
Epoch 4/50
55000/55000 [========================] - 3s - loss: 0.3129 - acc: 0.9112
Epoch 5/50
55000/55000 [========================] - 3s - loss: 0.2871 - acc: 0.9181

--- Epoch 6 to 45 output removed for brevity ---

Epoch 46/50
55000/55000 [========================] - 4s - loss: 0.0689 - acc: 0.9814
Epoch 47/50
55000/55000 [========================] - 4s - loss: 0.0672 - acc: 0.9819
Epoch 48/50
55000/55000 [========================] - 4s - loss: 0.0658 - acc: 0.9822
Epoch 49/50
55000/55000 [========================] - 4s - loss: 0.0643 - acc: 0.9829
Epoch 50/50
55000/55000 [========================] - 4s - loss: 0.0627 - acc: 0.9829
```

10. Evaluate the model and print the loss and accuracy:

```
score = model.evaluate(X_test, Y_test)
print('\n Test loss:', score[0])
print('Test accuracy:', score[1])
```

We get the following output:

```
Test loss: 0.089410082236
Test accuracy: 0.9727
```

The complete code for MLP for MNIST classification using Keras is provided in the notebook `ch-05_MLP`.

TFLearn-based MLP for MNIST classification

Now let's see how to implement the same MLP using TFLearn, another high-level library for TensorFlow:

1. Import the TFLearn library:

```
import tflearn
```

2. Define the hyper-parameters (we assume that the dataset has already been loaded into the `X_train`, `Y_train`, `X_test`, and `Y_test` variables):

```
num_layers = 2
num_neurons = []
for i in range(num_layers):
num_neurons.append(256)

learning_rate = 0.01
n_epochs = 50
batch_size = 100
```

3. Build the input layer, two hidden layers, and the output layer (the same architecture as examples in TensorFlow and Keras sections):

```
# Build deep neural network
input_layer = tflearn.input_data(shape=[None, num_inputs])
dense1 = tflearn.fully_connected(input_layer, num_neurons[0],
    activation='relu')
dense2 = tflearn.fully_connected(dense1, num_neurons[1],
    activation='relu')
softmax = tflearn.fully_connected(dense2, num_outputs,
    activation='softmax')
```

4. Define the optimizer function, neural network, and MLP model (known as DNN in TFLearn) using the DNN built in the last step (in the variable `softmax`):

```
optimizer = tflearn.SGD(learning_rate=learning_rate)
net = tflearn.regression(softmax, optimizer=optimizer,
                         metric=tflearn.metrics.Accuracy(),
                         loss='categorical_crossentropy')
model = tflearn.DNN(net)
```

5. Train the model:

```
model.fit(X_train, Y_train, n_epoch=n_epochs,
          batch_size=batch_size,
          show_metric=True, run_id="dense_model")
```

We get the following output once the training is finished:

```
Training Step: 27499  | total loss: 0.11236 | time: 5.853s
| SGD | epoch: 050 | loss: 0.11236 - acc: 0.9687 -- iter: 54900/55000
Training Step: 27500  | total loss: 0.11836 | time: 5.863s
| SGD | epoch: 050 | loss: 0.11836 - acc: 0.9658 -- iter: 55000/55000
--
```

6. Evaluate the model and print the accuracy score:

```
score = model.evaluate(X_test, Y_test)
print('Test accuracy:', score[0])
```

We get the following output:

```
Test accuracy: 0.9637
```

We get a pretty comparable accuracy from using TFLearn as well.

The complete code for MLP for MNIST classification using TFLearn is provided in the notebook `ch-05_MLP`.

Summary of MLP with TensorFlow, Keras, and TFLearn

In the previous sections, we learned how to build a simple MLP architecture using TensorFLow and its high-level libraries. We got an accuracy of about 0.93-0.94 with pure TensorFlow, 0.96-0.98 with Keras, and 0.96-0.97 with TFLearn. Even though all the examples of our code use TensorFlow underneath, the difference in accuracy for the same architecture and parameters can be attributed to the fact that although we initialized some important hyper-parameters, the high-level libraries and TensorFlow abstract away many other hyper-parameters that we did not modify from their default values.

We observe that the code in TensorFlow is very detailed and lengthy as compared to Keras and TFLearn. The high-level libraries make it easier for us to build and train neural network models.

MLP for time series regression

We have seen examples of classification for image data; now let's look at regression for time series data. We shall build and use MLP for a smaller univariate time series dataset known as the international airline passengers dataset. This dataset contains the total number of passengers over the years. The dataset is available at the following links:

- https://www.kaggle.com/andreazzini/international-airline-passengers/data
- https://datamarket.com/data/set/22u3/international-airline-passengers-monthly-totals-in-thousands-jan-49-dec-60

Let us start by preparing our dataset.

1. First, load the dataset using the following code:

```
filename = os.path.join(datasetslib.datasets_root,
                        'ts-data',
                        'international-airline-passengers-
cleaned.csv')
dataframe = pd.read_csv(filename,usecols=[1],header=0)
dataset = dataframe.values
dataset = dataset.astype('float32')
```

2. With a utility function from the `datasetslib`, we split the dataset into test and train sets. For time series datasets, we have a separate function that does not shuffle the observations because for time series regression we need to maintain the order of the observations. We use 67 percent data for training and 33 percent for testing. You may want to try the example with a different ratio.

```
train,test=dsu.train_test_split(dataset,train_size=0.67)
```

3. For time series regression, we convert the dataset to build a supervised data set. We use a lag of two time steps in this example. We set `n_x` to 2 and the `mvts_to_xy()` function returns the input and output (`X` and `Y`) train and test sets such that X has values for time {t-1,t} in two columns and Y has values for time {t+1} in one column. Our learning algorithm assumes that values at time t+1 can be learned by finding the relationship between values for time {t-1, t, t+1}.

```
# reshape into X=t-1,t and Y=t+1
n_x=2
n_y=1
X_train, Y_train, X_test, Y_test = tsd.mvts_to_xy(train,
                                        test,n_x=n_x,n_y=n_y)
```

More information on converting time series datasets as supervised learning problems can be found at the following link: `http://machinelearningmastery.com/convert-time-series-supervised-learning-problem-python/`.

Now we build and train the model on our train dataset:

1. Import the required Keras modules:

```
from keras.models import Sequential
from keras.layers import Dense
from keras.optimizers import SGD
```

2. Set the hyper-parameters required to build the model:

```
num_layers = 2
num_neurons = [8,8]
n_epochs = 50
batch_size = 2
```

Note that we use a batch size of two as the dataset is very small. We use a two-layer MLP with only eight neurons in each layer because of the small size of our example problem.

3. Build, compile, and train the model:

```
model = Sequential()
model.add(Dense(num_neurons[0], activation='relu',
    input_shape=(n_x,)))
model.add(Dense(num_neurons[1], activation='relu'))
model.add(Dense(units=1))
model.summary()

model.compile(loss='mse', optimizer='adam')

model.fit(X_train, Y_train,
   batch_size=batch_size,
   epochs=n_epochs)
```

Note that instead of SGD, we use the Adam optimizer. You may want to try out the different optimizers available in TensorFlow and Keras.

4. Evaluate the model and print the Mean Square Error (MSE) and the Root Mean Square Error (RMSE):

```
score = model.evaluate(X_test, Y_test)
print('\nTest mse:', score)
print('Test rmse:', math.sqrt(score))
```

We get the following output:

```
Test mse: 5619.24934188
Test rmse: 74.96165247566114
```

5. Predict the values using our model and plot them, both for test and train datasets:

```
# make predictions
Y_train_pred = model.predict(X_train)
Y_test_pred = model.predict(X_test)

# shift train predictions for plotting
Y_train_pred_plot = np.empty_like(dataset)
Y_train_pred_plot[:, :] = np.nan
Y_train_pred_plot[n_x-1:len(Y_train_pred)+n_x-1, :] = Y_train_pred

# shift test predictions for plotting
Y_test_pred_plot = np.empty_like(dataset)
Y_test_pred_plot[:, :] = np.nan
Y_test_pred_plot[len(Y_train_pred)+(n_x*2)-1:len(dataset)-1, :] = \
    Y_test_pred

# plot baseline and predictions
plt.plot(dataset,label='Original Data')
plt.plot(Y_train_pred_plot,label='Y_train_pred')
plt.plot(Y_test_pred_plot,label='Y_test_pred')
plt.legend()
plt.show()
```

We get the following plot for our original and predicted time series values:

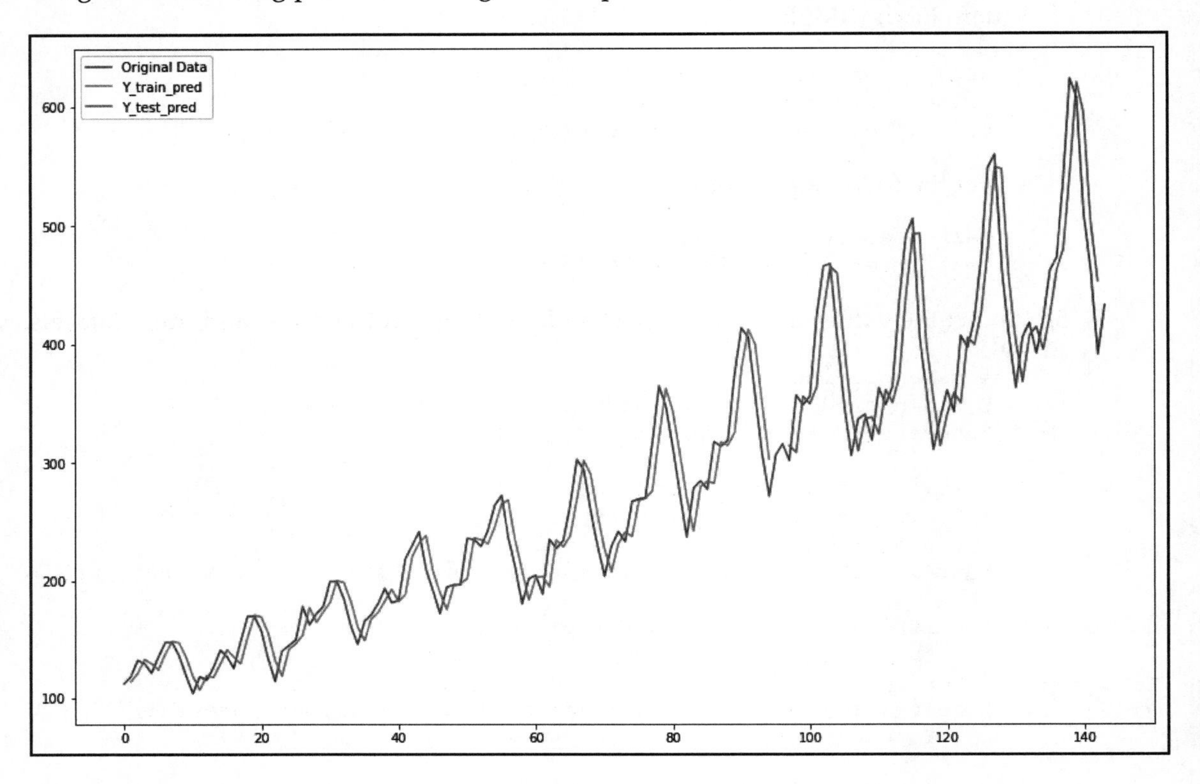

As you can see, it's a pretty good estimation. However, in real life, the data is multivariate and complex in nature. Hence, we shall see recurrent neural network architectures for timeseries data, in the following chapters.

Summary

In this chapter, we learned about multilayer perceptrons. We explained how to build and train MLP models for classification and regression problems. We built MLP models with pure TensorFlow, Keras, and TFLearn. For classification, we used image data, and for regression, we used the time series data.

The techniques to build and train MLP network models are the same for any other kind of data, such as numbers or text. However, for image datasets, the CNN architectures have proven to be the best architectures, and for sequence datasets, such as time series and text, the RNN models have proven to be the best architectures.

While we only used simple dataset examples to demonstrate the MLP architecture in this chapter, in the further chapters, we shall cover CNN and RNN architectures with some large and advanced datasets.

6
RNN with TensorFlow and Keras

In problems involving ordered sequences of data, such as **time series Forecasting** and **natural language processing**, the context is very valuable to predict the output. The context for such problems can be determined by ingesting the whole sequence, not just one last data point. Thus, the previous output becomes part of the current input, and when repeated, the last output turns out to be the results of all the previous inputs along with the last input. **Recurrent Neural Network** (**RNN**) architecture is a solution for handling machine learning problems that involve sequences.

Recurrent Neural Network (**RNN**) is a specialized neural network architecture for handling sequential data. The sequential data could be the sequence of observations over a period of time, as in time series data, or sequence of characters, words, and sentences, as in textual data.

One of the assumptions for the standard neural network is that the input data is arranged in a way that one input has no dependency on another. However, for time series data and textual data, this assumption does not hold true, since the values appearing later in the sequence are often influenced by the values that appeared before.

In order to achieve that, RNN extends the standard neural networks in the following ways:

- RNN adds the ability to use the output of one layer as an input to the same or previous layer, by adding loops or cycles in the computation graph.
- RNN adds the memory unit to store previous inputs and outputs that can be used in the current computation.

In this chapter, we cover the following topics to learn about RNN:

- Simple Recurrent Neural Networks
- RNN variants
- Long Short-Term Memory networks
- Gated Recurrent Unit networks
- TensorFlow for RNN
- Keras for RNN
- RNN in Keras for MNIST data

The next two chapters will cover practical examples of building RNN models in TensorFlow and Keras for time series and text (NLP) data.

Simple Recurrent Neural Network

Here is what a simple neural network with loops looks like:

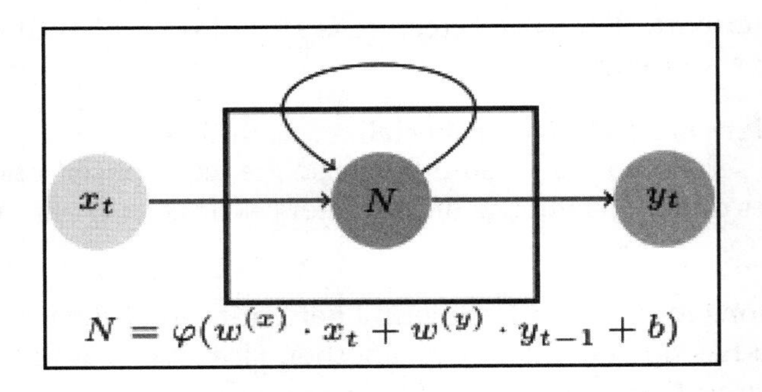

RNN Network

In this diagram, a Neural Network *N* takes input x_t to produce output y_t. Due to the loop, at the next time step $t+1$, it takes the input y_t along with input x_{t+1} to produce output y_{t+1}. Mathematically, we represent this as the following equation:

$$y_t = \varphi(w^{(x)} \cdot x_t + w^{(y)} \cdot y_{t-1} + b)$$

When we unroll the loop, the RNN architecture looks as follows at time step t_1:

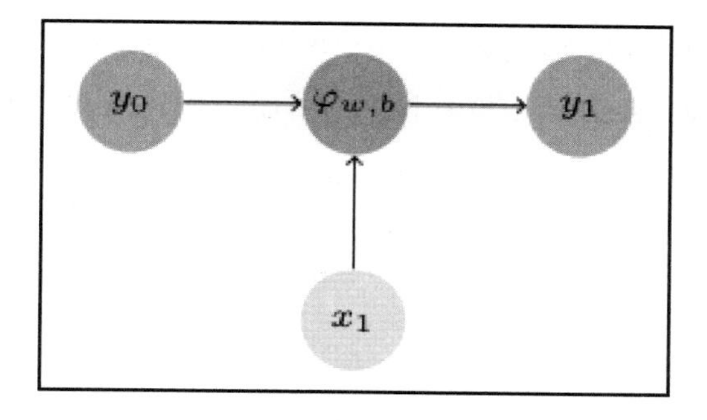

Unrolled RNN at timestep t_1

As the time steps evolve, this loop unrolls as follows at time step 5:

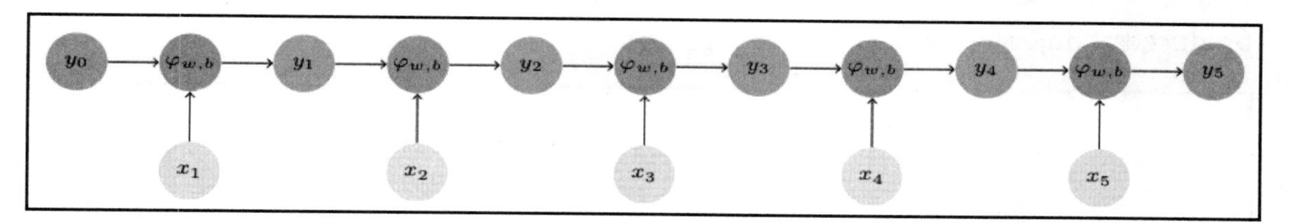

Unrolled RNN at timestep t_5

At every time step, the same learning function, $\varphi(\cdot)$, and the same parameters, w and b, are used.

The output y is not always produced at every time step. Instead, an output h is produced at every time step, and another activation function is applied to this output h to produce the output y. The equations for the RNN look like this now:

$$h_t = \varphi(w^{(hx)} \cdot x_t + w^{(hh)} \cdot h_{t-1} + b^{(h)})$$

$$y_t = \varphi(w^{(yh)} \cdot h_t + b^{(y)})$$

where,

- $w^{(hx)}$ is the weight vector for x inputs that are connected to the hidden layer
- $w^{(hh)}$ is the weight vector for the value of h from the previous time step
- $w^{(yh)}$ is the weight vector for layer connecting the hidden layer to the output layer
- The function used for h_t is usually a nonlinear function, such as tanh or ReLU

In RNN, same parameters ($w^{(hx)}, w^{(hh)}, w^{(yh)}, b^{(h)}, b^{(y)}$) are used at every time step. This fact greatly reduces the number of parameters we need to learn for sequence-based models.

With this, the RNN unrolls as follows at time step t_5, assuming that the output y is only produced at time step t_5:

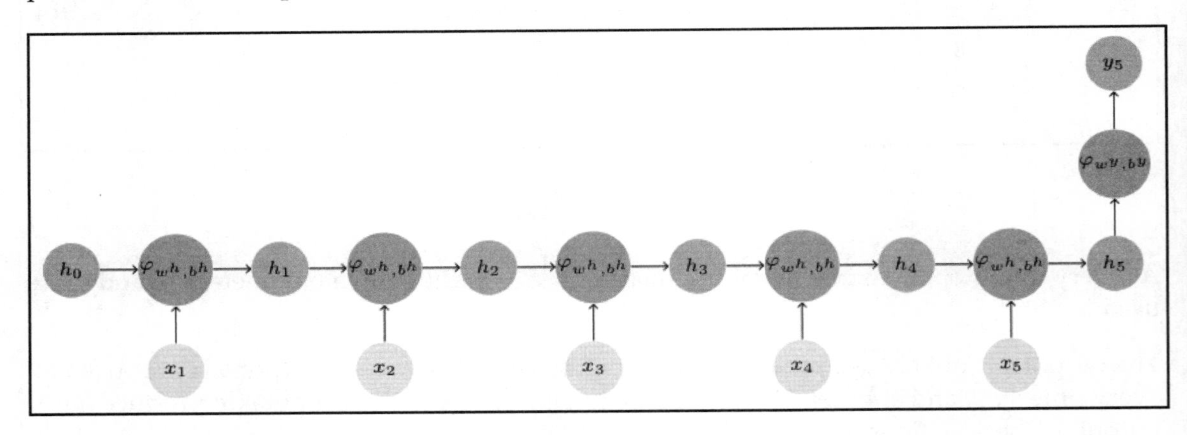

Unrolled RNN with one output at timestep t_5

Simple RNN was introduced by Elman in 1990, thus it is also known as Elman network. However, simple RNN falls short of our processing needs today, hence we will learn about the variants of the RNN in the next section.

Read the Elman's original research paper to learn about origins of RNN architecture:

J. L. Elman, Finding Structure in Time, Cogn. Sci., vol. 14, no. 2, pp. 179–211, 1990.

RNN variants

The RNN architecture has been extended in many ways to accommodate the extra needs in certain problems and to overcome the shortcomings of simple RNN models. We list some of the major extensions to the RNN architecture below.

- **Bidirectional RNN** (**BRNN**) is used when the output depends on both the previous and future elements of a sequence. BRNN is implemented by stacking two RNNs, known as forward and backward Layer, and the output is the result of the hidden state of both the RNNs. In the forward layer, the memory state h flows from time step *t* to time step *t+1* and in the backward layer the memory state flows from time step *t* to time step *t-1*. Both the layers take same input x_t at time step *t*, but they jointly produce the output at time step *t*.

- **Deep Bidirectional RNN** (**DBRNN**) extends the BRNN further by adding multiple layers. The BRNN has hidden layers or cells across the time dimensions. However, by stacking BRNN, we get the hierarchical presentation in DBRNN. One of the significant difference is that in BRNN we use the same parameters for each cell in the same layer, but in DBRNN we use different parameters for each stacked layer.
- **Long Short-Term Memory** (**LSTM**) network extends the RNN by using an architecture that involves multiple nonlinear functions instead of one simple nonlinear function to compute the hidden state. The LSTM is composed of black boxes called **cells** that take the three inputs: the working memory at time $t-1$ (h_{t-1}), current input (x_t) and long-term memory at time $t-1$ (c_{t-1}), and produce the two outputs: updated working memory (h_t) and long-term memory (c_t). The cells use the functions known as gates, to make decisions about saving and erasing the content selectively from the memory. We describe the LSTM in detail in the sections below.

Read the following research paper on LSTM to get more information about origins of LSTM:

S. Hochreiter and J. Schmidhuber, Long Short-Term Memory, Neural Comput., vol. 9, no. 8, pp. 1735–1780, 1997. http://www.bioinf.jku.at/publications/older/2604.pdf

- **Gated Recurrent Unit** (**GRU**) network is a simplified variation of LSTM. It combines the function of the *forget* and the *input* gates in a simpler *update* gate. It also combines the *hidden state* and *cell state* into one single state. Hence, GRU is computationally less expensive as compared to LSTM. We describe the GRU in detail in the sections below.

Read the following research papers to explore more details on GRU:

K. Cho, B. van Merrienboer, C. Gulcehre, D. Bahdanau, F. Bougares, H. Schwenk, and Y. Bengio, Learning Phrase Representations using RNN Encoder-Decoder for Statistical Machine Translation, 2014. `https://arxiv.org/abs/1406.1078`

J. Chung, C. Gulcehre, K. Cho, and Y. Bengio, Empirical Evaluation of Gated Recurrent Neural Networks on Sequence Modeling, pp. 1–9, 2014. `https://arxiv.org/abs/1412.3555`

- The **seq2seq** model combines the encoder-decoder architecture with RNN architectures. In seq2seq architecture, the model is trained on sequences of data, such as text data or time series data, and then the model is used to generate the output sequences. For example, train the model on English text and then generate Spanish text from the model. The seq2seq model consists of an encoder and a decoder model, both of them built with the RNN architecture. The seq2seq models can be stacked to build hierarchical multi-layer models.

LSTM network

When RNNs are trained over very long sequences of data, the gradients tend to become either very large or very small that they vanish to almost zero. **Long Short-Term Memory** (**LSTM**) networks address the vanishing/exploding gradient problem by adding gates for controlling the access to past information. LSTM concept was first introduced by Hochreiter and Schmidhuber in 1997.

Read the following research paper on LSTM to get more information about origins of LSTM:

S. Hochreiter and J. Schmidhuber, Long Short-Term Memory, Neural Comput., vol. 9, no. 8, pp. 1735–1780, 1997. `http://www.bioinf.jku.at/publications/older/2604.pdf`

In RNN, a single neural network layer of repeatedly used learning function φ is used, whereas, in LSTM, a repeating module consisting of four main functions is used. The module that builds the LSTM network is called the **cell**. The LSTM cell helps train the model more effectively when long sequences are passed, by selectively learning or erasing information. The functions composing the cell are also known as gates as they act as gatekeeper for the information that is passed in and out of the cell.

The LSTM model has two kinds of memory:

- working memory denoted with $\boldsymbol{h}$ (hidden state) and
- long-term memory denoted with with $\boldsymbol{c}$ (cell state).

The cell state or long-term memory flows from cell to cell with only two linear interactions. The LSTM adds information to the long term memory, or removes information from the long-term memory, through gates.

Following diagram depicts the LSTM cell:

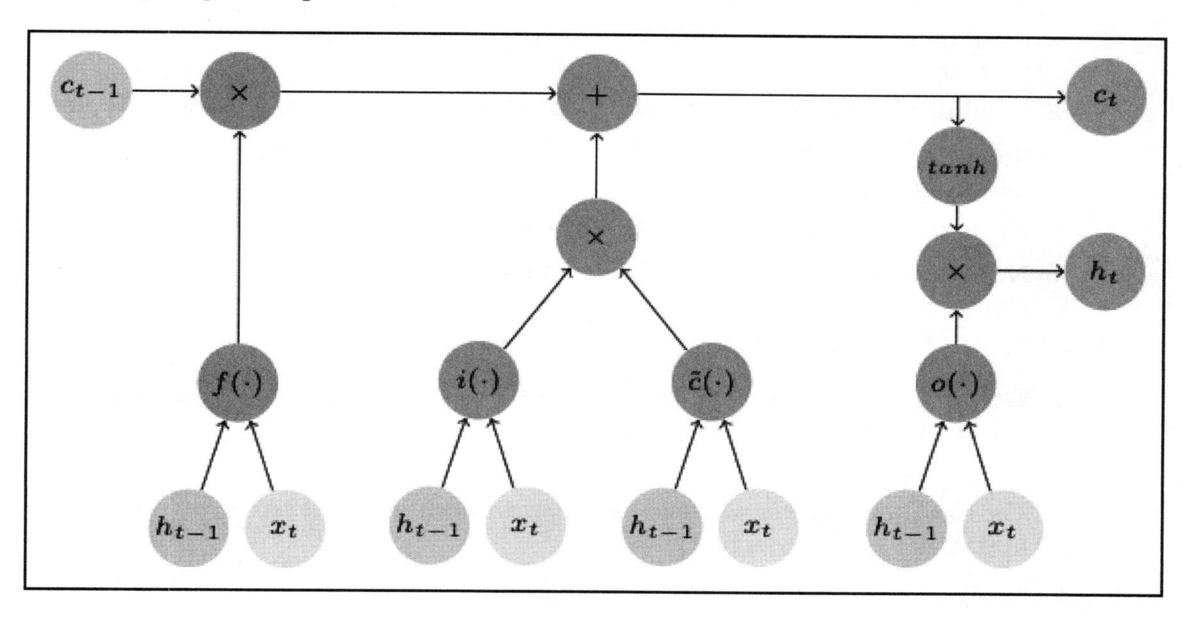

The LSTM Cell

The internal flow through the gates in the LSTM cell is as follows:

1. **Forget Gate f( ) (or remember gate)**: The h_{t-1} and x_t flows as input to *f( )* gate as per the following equation: $f(\cdot) = \sigma(w^{(fx)} \cdot x_t + w^{(fh)} \cdot h_{t-1} + b^{(f)})$
   The function of *forget gate* is to decide which information to forget and which information to remember. The *sigmoid* activation function is used here, so that an output of 1 represents that the information is carried over to the next step within the cell, and an output of 0 represents that the information is selectively discarded.

2. **Input Gate i( ) (or save gate)**: The h_{t-1} and x_t flows as input to *i( )* gate as per the following equation:
   $i(\cdot) = \sigma(w^{(ix)} \cdot x_t + w^{(ih)} \cdot h_{t-1} + b^{(i)})$
   The function of *input gate* is to decide whether to save or discard the input. The input function also allows the cell to learn which part of candidate memory to keep or discard.
3. **Candidate Long-Term Memory**: The candidate long-term memory is computed from h_{t-1} and x_t using an activation function, which is mostly *tanh*, as per the following equation:
   $\tilde{c}(\cdot) = tanh(w^{(\tilde{c}x)} \cdot x_t + w^{(\tilde{c}h)} \cdot h_{t-1} + b^{(\tilde{c})})$
4. Next, the preceding three calculations are combined to get the update long-term memory, denoted by c_t as per the following equation:
   $c_t = c_{t-1} \times f(\cdot) + i(\cdot) \times \tilde{c}(\cdot)$
5. **Output o( ) (or focus/attention gate)**: The h_{t-1} and x_t flows as input to the *o( )* gate as per the following equation:
   $o(\cdot) = \sigma(w^{(ox)} \cdot x_t + w^{(oh)} \cdot h_{t-1} + b^{(o)})$
   The function of *output gate* is to decide how much information can be used to update the working memory.
6. Next, working memory h_t is updated from the long-term memory c_t and the focus/attention vector as per the following equation:
   $h_t = \varphi(c_t) \times o(\cdot)$
   where $\varphi(\cdot)$ is an activation function, that is usually *tanh*.

GRU network

LSTM Network is computationally expensive, hence, researchers found an almost equally effective configuration of RNNs, known as **Gated Recurrent Unit** (**GRU**) architecture.

In GRU, instead of a working and a long-term memory, only one kind of memory is used, indicated with ***h*** (hidden state). The GRU cell adds information to this state memory or removes information from this state memory through **reset** and **update** gates.

Following diagram depicts the GRU cell (explanation follows the diagram):

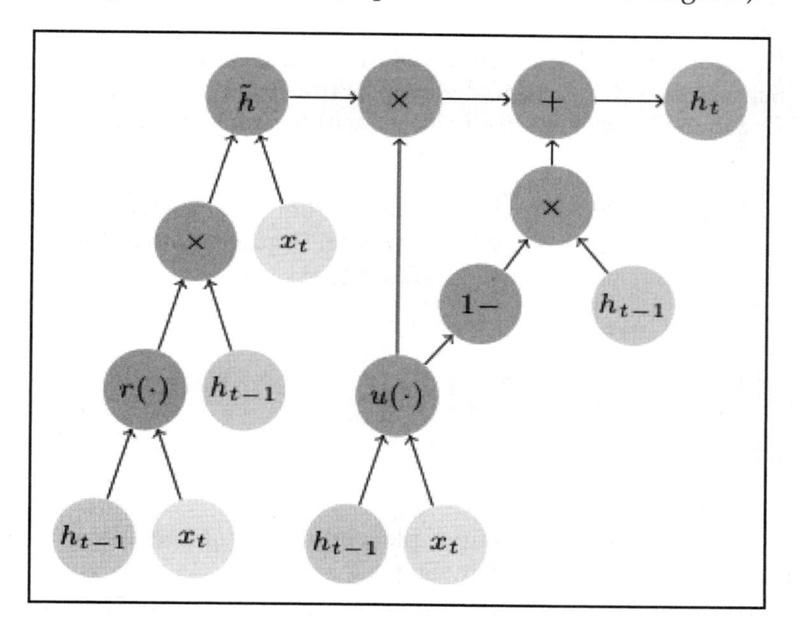

The GRU Cell

The internal flow through the gates in the GRU cell is as follows:

1. **Update gate u( )**: The input h_{t-1} and x_t flows to the *u( )* gate as per the following equation:
   $$u(\cdot) = \sigma(w^{(ux)} \cdot x_t + w^{(uh)} \cdot h_{t-1} + b^{(u)})$$
2. **Reset Gate r( )**: The input h_{t-1} and x_t flows to the *r( )* gate as per the following equation:
   $$r(\cdot) = \sigma(w^{(rx)} \cdot x_t + w^{(rh)} \cdot h_{t-1} + b^{(r)})$$

3. **Candidate State Memory**: The candidate long-term memory is computed from the output of the *r( )* gate, h_{t-1}, and x_t, as per the following equation:
$\tilde{h}(\cdot) = tanh(w^{(\tilde{h}x)} \cdot x_t + w^{(\tilde{h}h)} \cdot (r_t \cdot h_{t-1}) + b^{(\tilde{h})})$
4. Next, the preceding three calculations are combined to get the updated state memory, denoted by h_t, as per following equation:
$h_t = (u_t \cdot \tilde{h}_t) + ((1 - u_t) \cdot h_{t-1})$

Read the following research papers to explore more details on GRU:

K. Cho, B. van Merrienboer, C. Gulcehre, D. Bahdanau, F. Bougares, H. Schwenk, and Y. Bengio, Learning Phrase Representations using RNN Encoder-Decoder for Statistical Machine Translation, 2014. `https://arxiv.org/abs/1406.1078`

J. Chung, C. Gulcehre, K. Cho, and Y. Bengio, Empirical Evaluation of Gated Recurrent Neural Networks on Sequence Modeling, pp. 1–9, 2014. `https://arxiv.org/abs/1412.3555`

TensorFlow for RNN

The basic workflow for creating RNN models in low-level TensorFlow library is almost the same as MLP:

- First create the input and output placeholders of shape (None, # TimeSteps, # Features) or (Batch Size, # TimeSteps, # Features)
- From the input placeholder, create a list of length # TimeSteps, containing Tensors of Shape (None, #Features) or (Batch Size, # Features)
- Create a cell of the desired RNN type from the `tf.rnn.rnn_cell` module
- Use the cell and the input tensor list created previously to create a static or dynamic RNN
- Create the output weights and bias variables, and define the loss and optimizer functions
- For the required number of epochs, train the model using the loss and optimizer functions

This basic workflow would be demonstrated with the example code in the next chapter. Let us look at the various classes available to support the previous workflow.

TensorFlow RNN Cell Classes

The `tf.nn.rnn_cell` module contains the following classes for creating different kinds of cells in TensorFlow:

Class	Description
BasicRNNCell	Provides simple RNN cell
BasicLSTMCell	Provides simple LSTM RNN cell, based on `http://arxiv.org/abs/1409.2329`
LSTMCell	Provides LSTM RNN cell, based on `http://deeplearning.cs.cmu.edu/pdfs/Hochreiter97_lstm.pdf` and `https://research.google.com/pubs/archive/43905.pdf`
GRUCell	Provides GRU RNN cell, based on `http://arxiv.org/abs/1406.1078`
MultiRNNCell	Provides RNN cell made of multiple simple cells joined sequentially

The `tf.contrib.rnn` module provides the following additional classes for creating different kinds of cells in TensorFlow:

Class	Description
LSTMBlockCell	Provides the block LSTM RNN cell, based on `http://arxiv.org/abs/1409.2329`
LSTMBlockFusedCell	Provides the block fused LSTM RNN cell, based on `http://arxiv.org/abs/1409.2329`
GLSTMCell	Provides the group LSTM cell, based on `https://arxiv.org/abs/1703.10722`
GridLSTMCell	Provides the grid LSTM RNN cell, based on `http://arxiv.org/abs/1507.01526`
GRUBlockCell	Provides the block GRU RNN cell, based on `http://arxiv.org/abs/1406.1078`
BidirectionalGridLSTMCell	Provides bidirectional grid LSTM with bi-direction only in frequency and not in time

NASCell	Provides neural architecture search RNN cell, based on `https://arxiv.org/abs/1611.01578`
UGRNNCell	Provides update gate RNN cell, based on `https://arxiv.org/abs/1611.09913`

TensorFlow RNN Model Construction Classes

TensorFlow provides classes to create RNN models from the RNN cell objects. The static RNN classes add unrolled cells for time steps at the compile time, while dynamic RNN classes add unrolled cells for time steps at the run time.

- `tf.nn.static_rnn`
- `tf.nn.static_state_saving_rnn`
- `tf.nn.static_bidirectional_rnn`
- `tf.nn.dynamic_rnn`
- `tf.nn.bidirectional_dynamic_rnn`
- `tf.nn.raw_rnn`
- `tf.contrib.rnn.stack_bidirectional_dynamic_rnn`

TensorFlow RNN Cell Wrapper Classes

TensorFlow also provides classes that wrap other cell classes:

- `tf.contrib.rnn.LSTMBlockWrapper`
- `tf.contrib.rnn.DropoutWrapper`
- `tf.contrib.rnn.EmbeddingWrapper`
- `tf.contrib.rnn.InputProjectionWrapper`
- `tf.contrib.rnn.OutputProjectionWrapper`
- `tf.contrib.rnn.DeviceWrapper`
- `tf.contrib.rnn.ResidualWrapper`

Latest documentation on RNN in TensorFlow at the following link: `https://www.tensorflow.org/api_guides/python/contrib.rnn.`

Keras for RNN

Creating RNN in Keras is much easier as compared to the TensorFlow. As you learned in chapter 3, Keras offers both functional and sequential API for creating the recurrent networks. To build the RNN model, you have to add layers from the `kera.layers.recurrent` module. Keras provides the following kinds of recurrent layers in the `keras.layers.recurrent` module:

- SimpleRNN
- LSTM
- GRU

Stateful Models

Keras recurrent layers also support RNN models that save state between the batches. You can create a stateful RNN, LSTM, or GRU model by passing `stateful` parameters as `True`. For stateful models, the batch size specified for the inputs has to be a fixed value. In stateful models, the hidden state learnt from training a batch is reused for the next batch. If you want to reset the memory at some point during training, it can be done with extra code by calling the `model.reset_states()` or `layer.reset_states()` functions.

We shall see examples of building RNNs using Keras in the next chapter.

The latest documentation on Recurrent Layers in Keras can be found at the following link: `https://keras.io/layers/recurrent/`.

Application areas of RNNs

Some of the application areas where RNNs are used more often are as follows:

- **Natural Language Modeling**: The RNN models have been used in natural language processing (NLP) for natural language understanding and natural language generation tasks. In NLP, an RNN model is given a sequence of words and it predicts another sequence of words. Thus, the trained models can be used for generating the sequence of words, a field known as Text Generation. For example, generating stories, and screenplays. Another area of NLP is language translation, where given a sequence of words in one language, the model predicts a sequence of words in another language.

- **Voice and Speech Recognition**: The RNN models have great use in building models for learning from the audio data. In speech recognition, an RNN model is given audio data and it predicts a sequence of phonetic segments. It can be used to train the models to recognize the voice commands, or even for conversation with speech based chatbots.
- **Image/Video Description or Caption Generation**: The RNN models can be used in combination with CNN to generate the descriptions of elements found in images and videos. Such descriptions can also be used to generate captions for images and videos.
- **TimeSeries Data**: Most importantly, RNNs are very useful for TimeSeries data. Most of the sensors and systems generate data where the temporal order is important. The RNN models fit very well to finding patterns and forecasting such data.

Explore more information about RNN at the following links:

`http://karpathy.github.io/2015/05/21/rnn-effectiveness/`
`http://colah.github.io/posts/2015-08-Understanding-LSTMs/`
`http://www.wildml.com/2015/09/recurrent-neural-networks-tutorial-part-1-introduction-to-rnns/`
`https://r2rt.com/written-memories-understanding-deriving-and-extending-the-lstm.html`

RNN in Keras for MNIST data

Although RNN is mostly used for sequence data, it can also be used for image data. We know that images have minimum two dimensions - height and width. Now think of one of the dimensions as time steps, and other as features. For MNIST, the image size is 28 x 28 pixels, thus we can think of an MNIST image as having 28 time steps with 28 features in each timestep.

We would give examples from time series and text data in next chapters, but let us build and train an RNN for MNIST in Keras to quickly glance over the process of building and training the RNN models.

You can follow along with the code in the Jupyter notebook `ch-06_RNN_MNIST_Keras`.

Import the required modules:

```
import keras
from keras.models import Sequential
from keras.layers import Dense, Activation
from keras.layers.recurrent import SimpleRNN
from keras.optimizers import RMSprop
from keras.optimizers import SGD
```

Get the MNIST data and transform the data from 784 pixels in 1-D to 28 x 28 pixels in 2-D:

```
from tensorflow.examples.tutorials.mnist import input_data
mnist = input_data.read_data_sets(os.path.join(datasetslib.datasets_root,
                                               'mnist'),
                                  one_hot=True)
X_train = mnist.train.images
X_test = mnist.test.images
Y_train = mnist.train.labels
Y_test = mnist.test.labels
n_classes = 10
n_classes = 10
X_train = X_train.reshape(-1,28,28)
X_test = X_test.reshape(-1,28,28)
```

Build the SimpleRNN model in Keras:

```
# create and fit the SimpleRNN model
model = Sequential()
model.add(SimpleRNN(units=16, activation='relu', input_shape=(28,28)))
model.add(Dense(n_classes))
model.add(Activation('softmax'))

model.compile(loss='categorical_crossentropy',
              optimizer=RMSprop(lr=0.01),
              metrics=['accuracy'])
model.summary()
```

The model appears as follows:

```
_________________________________________________________________
Layer (type)                 Output Shape              Param #
=================================================================
simple_rnn_1 (SimpleRNN)     (None, 16)                720
_________________________________________________________________
dense_1 (Dense)              (None, 10)                170
_________________________________________________________________
activation_1 (Activation)    (None, 10)                0
=================================================================
Total params: 890
Trainable params: 890
Non-trainable params: 0
_________________________________________________________________
```

Train the model and print the accuracy of test data set:

```
model.fit(X_train, Y_train,
          batch_size=100, epochs=20)

score = model.evaluate(X_test, Y_test)
print('\nTest loss:', score[0])
print('Test accuracy:', score[1])
```

We get the following results:

```
Test loss: 0.520945608187
Test accuracy: 0.8379
```

Summary

In this chapter, we learned about Recurrent Neural Networks (RNNs). We learned about the various variants of RNN and described two of them in detail: Long Short-Term Memory (LSTM) networks and Gated Recurrent Unit (GRU) networks. We also described the classes available for constructing RNN cells, models, and layers in TensorFlow and Keras. We built a simple RNN network for classifying the digits of the MNIST dataset.

In next chapter, we shall learn how to build and train the RNN models for time series data.

7
RNN for Time Series Data with TensorFlow and Keras

Time series data is a sequence of values, recorded or measured at different time intervals. Being a sequence, the RNN architecture is the best method to train models from such data. In this chapter, we will use a sample time series data set to showcase how to use TensorFlow and Keras to build RNN models.

We will cover the following topics in this chapter:

- Airline passengers time series dataset:
    - Description and downloading of the dataset
    - Visualizing the dataset
- Preprocessing the dataset for RNN in TensorFlow
- RNN in TensorFlow for time series data:
    - SimpleRNN in TensorFlow
    - LSTM in TensorFlow
    - GRU in TensorFlow
- Preprocessing the dataset for RNN in Keras
- RNN in Keras for time series data:
    - SimpleRNN in Keras
    - LSTM in Keras
    - GRU in Keras

Let's start by learning about the sample dataset.

You can follow along with the code in the Jupyter notebook `ch-07a_RNN_TimeSeries_TensorFlow`.

Airline Passengers dataset

For the sake of brevity and simplicity, we selected a very small dataset called International Airline Passengers (airpass). The data contains the number of total passengers every month, from January 1949 to December 1960. The numbers in the dataset refer to the amount in thousands. This dataset was originally used by Box and Jenkins in their work in 1976. It was collected as part of **TimeSeries Dataset Library** (**TSDL**) along with various other time series datasets by Professor Rob Hyndman at Monash University, Australia. Later, the TSDL was moved to DataMarket (`http://datamarket.com`).

You can download the dataset from the following link: `https://datamarket.com/data/set/22u3/international-airline-passengers-monthly-totals-in-thousands-jan-49-dec-60`.

Loading the airpass dataset

We keep the dataset as a CSV file in the `ts-data` folder in the datasets root (~/datasets) and use the following commands to load the data in a pandas data frame:

```
filepath = os.path.join(datasetslib.datasets_root,
                        'ts-data',
                        'international-airline-passengers-cleaned.csv'
                       )
dataframe = pd.read_csv(filepath,usecols=[1],header=0)
dataset = dataframe.values
dataset = dataset.astype(np.float32)
```

Extract the values from the data frame in a NumPy array and convert to `np.float32`:

```
dataset = dataframe.values
dataset = dataset.astype(np.float32)
```

Visualizing the airpass dataset

Let's visualize how the dataset looks:

```
plt.plot(dataset,label='Original Data')
plt.legend()
plt.show()
```

The plot of the `airpass` dataset looks as follows:

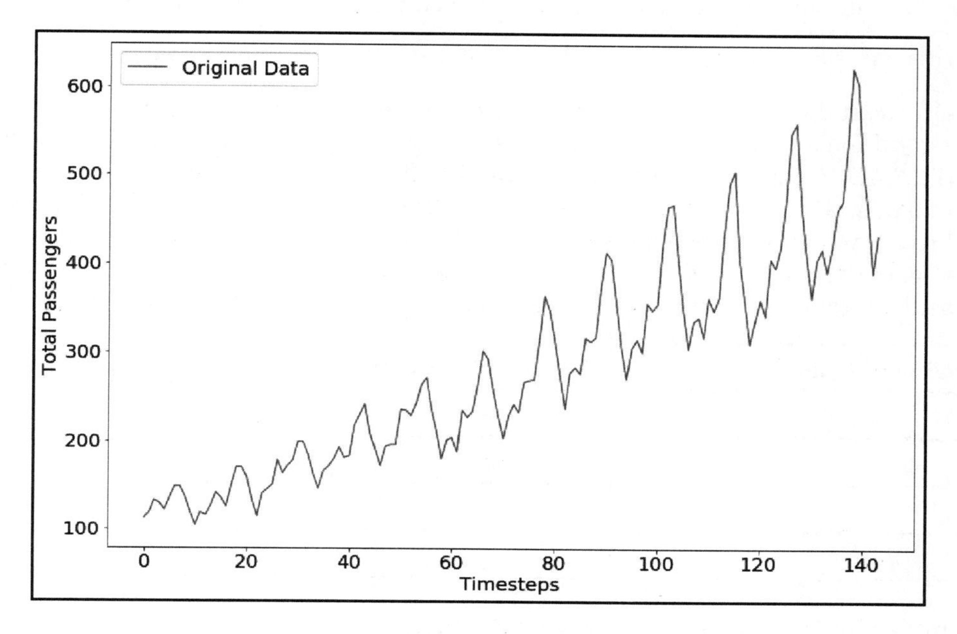

Airline Passengers Dataset

Preprocessing the dataset for RNN models with TensorFlow

In order to make it ready for the learning models, normalize the dataset by applying MinMax scaling that brings the dataset values between 0 and 1. You can try applying different scaling methods to the data depending on the nature of your data.

```
# normalize the dataset
scaler = skpp.MinMaxScaler(feature_range=(0, 1))
normalized_dataset = scaler.fit_transform(dataset)
```

We use our homegrown utility function to split the dataset into train and test datasets. The data has to be split without shuffling the dataset as shuffling the dataset breaks the sequence. Maintaining the sequence of the data is important to train the time series models.

```
train,test=tsu.train_test_split(normalized_dataset,train_size=0.67)
```

Then we convert the train and test datasets into supervised machine learning sets. Let's try to understand the meaning of a supervised learning set. Let's say that we have a sequence of data: 1,2,3,4,5. We want to learn the probability distribution that generated the dataset. In order to do this, we can assume that the value at time step *t* is a result of values from time steps *t-1* to *t-k*, where *k* is the window size. To simplify things, let's say the window size is 1. Thus, the value at time step *t*, known as input features, is the result of value at time step *t-1*, known as the target. Let's repeat this for all time steps, and we get the following table:

Input value or feature	Output value or target
1	2
2	3
3	4
4	5

The example we showed has only one variable value, which gets converted to features and targets. When a target value depends on one variable, it is called univariate time series. This same logic could be applied to multivariate time series where the target depends on multiple variables. We use x to represent the input features and y to represent the output targets.

With this background in mind, for the purpose of converting `airpass` data to supervised machine learning data, we set the following hyper-parameters:

1. Set the number of past timesteps that are used to learn or predict the next timestep:

```
n_x=1
```

2. Set the number of future time steps to learn or predict:

```
n_y=1
```

3. Set the number of *x* variables that are used to learn; as the current example is univariate, this is set to 1:

```
n_x_vars = 1
```

4. Set the number of *y* variables that are to be predicted; as the current example is univariate, this is set to 1:

```
n_y_vars = 1
```

5. Finally, we convert the train and test datasets into *X* and *Y* sets by applying the logic we described in the beginning of this section:

```
X_train, Y_train, X_test, Y_test = tsu.mvts_to_xy(train,
                                   test,n_x=n_x,n_y=n_y)
```

Now that the data is preprocessed and available to be fed into our models, let's prepare a SimpleRNN model using TensorFlow.

Simple RNN in TensorFlow

The workflow to define and train a simple RNN in TensorFlow is as follows:

1. Define the hyper-parameters for the model:

```
state_size = 4
n_epochs = 100
n_timesteps = n_x
learning_rate = 0.1
```

The new hyper-parameter here is the `state_size`. The `state_size` represents the number of weight vectors of an RNN cell.

2. Define the placeholders for *X* and *Y* parameters for the model. The shape of *X* placeholder is `(batch_size, number_of_input_timesteps, number_of_inputs)` and the shape of *Y* placeholder is `(batch_size, number_of_output_timesteps, number_of_outputs)`. For `batch_size`, we use `None` so that we can input the batch of any size later.

```
X_p = tf.placeholder(tf.float32, [None, n_timesteps, n_x_vars],
    name='X_p')
Y_p = tf.placeholder(tf.float32, [None, n_timesteps, n_y_vars],
    name='Y_p')
```

3. Transform the input placeholder `X_p` into a list of tensors of length equal to the number of time steps, which is `n_x` or 1 in this example:

```
# make a list of tensors of length n_timesteps
rnn_inputs = tf.unstack(X_p,axis=1)
```

4. Create a simple RNN cell using `tf.nn.rnn_cell.BasicRNNCell`:

```
cell = tf.nn.rnn_cell.BasicRNNCell(state_size)
```

5. TensorFlow provides the `static_rnn` and `dynamic_rnn` convenience methods (among others) to create a static and dynamic RNN respectively. Create a static RNN:

```
rnn_outputs, final_state = tf.nn.static_rnn(cell,
                                             rnn_inputs,
                                             dtype=tf.float32
                                            )
```

Static RNN creates the cells, i.e. unrolls the loop, at the compile time. Dynamic RNN creates the cells, namely unrolls the loop, at the runtime. In this chapter, we showcase example with `static_rnn` only, but you should explore `dynamic_rnn` once you have gained expertise in static RNN.

The `static_rnn` method takes the following parameters:

- `cell`: The basic RNN cell object that we defined earlier. It could be another kind of cell, as we will see further in the chapter.
- `rnn_inputs`: The list of Tensors of shape `(batch_size, number_of_inputs)`.
- `dtype`: The data type of initial state and expected outputs.

6. Define the weight and bias parameters for the predictions layer:

```
W = tf.get_variable('W', [state_size, n_y_vars])
b = tf.get_variable('b', [n_y_vars],
    initializer=tf.constant_initializer(0.0))
```

7. Define the predictions layer as a dense linear layer:

```
predictions = [tf.matmul(rnn_output, W) + b \
               for rnn_output in rnn_outputs]
```

8. The output Y is in the shape of Tensors; convert it to a list of tensors:

```
y_as_list = tf.unstack(Y_p, num=n_timesteps, axis=1)
```

9. Define the loss function as mean squared error between the predicted and actual labels:

```
mse = tf.losses.mean_squared_error
losses = [mse(labels=label, predictions=prediction)
          for prediction, label in zip(predictions, y_as_list)
         ]
```

10. Define the total loss as the average of losses of all the predicted time steps:

```
total_loss = tf.reduce_mean(losses)
```

11. Define the optimizer to minimize the `total_loss`:

```
optimizer =
tf.train.AdagradOptimizer(learning_rate).minimize(total_loss)
```

12. Now that we have the model, loss, and optimizer function defined, let's train the model and calculate the train loss:

```
with tf.Session() as tfs:
    tfs.run(tf.global_variables_initializer())
    epoch_loss = 0.0
    for epoch in range(n_epochs):
        feed_dict={X_p: X_train.reshape(-1, n_timesteps,
                                        n_x_vars),
                   Y_p: Y_train.reshape(-1, n_timesteps,
                                        n_x_vars)
                  }
        epoch_loss,y_train_pred,_=tfs.run([total_loss,predictions,
                              optimizer], feed_dict=feed_dict)
    print("train mse = {}".format(epoch_loss))
```

We get the following value:

```
train mse = 0.0019413739209994674
```

13. Let's test the model on test data:

```
feed_dict={X_p: X_test.reshape(-1, n_timesteps,n_x_vars),
           Y_p: Y_test.reshape(-1, n_timesteps,n_y_vars)
          }
test_loss, y_test_pred = tfs.run([total_loss,predictions],
                                  feed_dict=feed_dict
                                 )
print('test mse = {}'.format(test_loss))
print('test rmse = {}'.format(math.sqrt(test_loss)))
```

We get the following mse and rmse (root mean square error) on the test data:

```
test mse = 0.008790395222604275
test rmse = 0.09375710758446143
```

This is pretty impressive.

This was a very simple case of predicting one time step ahead with just one variable value. In real life, the output is influenced by more than one feature and it is required to predict more than one time step ahead. The latter kind of problems are called multivariate multi-time-step-ahead forecasting problems. Such problems are an area of active research to get better predictions using Recurrent Neural Networks.

Now let's rescale the predictions and original value and plot against the original values (please look for the code in the notebook).

We get the following plot:

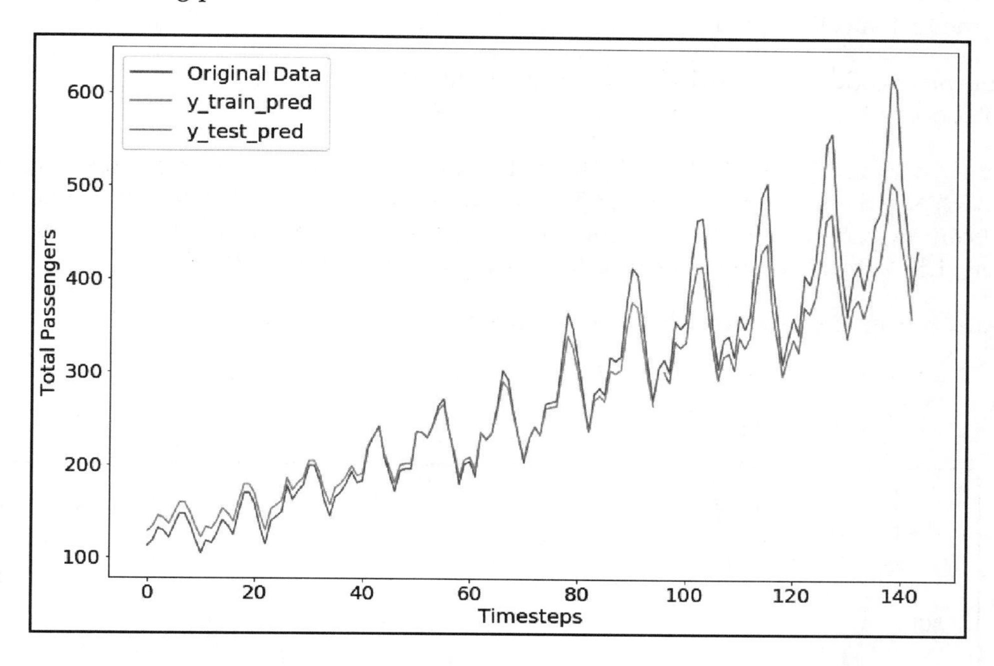

It is pretty impressive to see that the predicted data almost matches the original data in our simple example. One possible explanation for such an accurate forecast is that the predictions for a single time step are based on the predictions of a single variable from the last time step, hence they are always in the neighborhood of the previous value.

Nonetheless, the aim of the preceding example was to showcase the method of creating an RNN in TensorFlow. Now let's recreate the same example using the RNN variants.

LSTM in TensorFlow

Simple RNN architecture does not always work due to the problem of exploding and vanishing gradients, hence improved RNN architectures are used such as LSTM networks. TensorFlow provides API to create LSTM RNN architectures.

In the example showcased in the previous section, to change the Simple RNN to the LSTM network, all we have to do is change the cell type as follows:

```
cell = tf.nn.rnn_cell.LSTMCell(state_size)
```

The rest of the code remains the same as TensorFlow does the work of creating the gates inside the LSTM cell for you.

The complete code for the LSTM model is provided in the notebook `ch-07a_RNN_TimeSeries_TensorFlow`.

However, with LSTM, we had to run the code for 600 epochs in order to get results closer to a basic RNN. The reason being that LSTM has more parameters to learn, hence it needs more training iterations. For our simple example, it seems like overkill, but for larger datasets, LSTM has shown better results as compared to a simple RNN.

The output from the model with LSTM architecture is as follows:

```
train mse = 0.0020806745160371065
test mse = 0.014992355369031433
test rmse = 0.12244327408653947
```

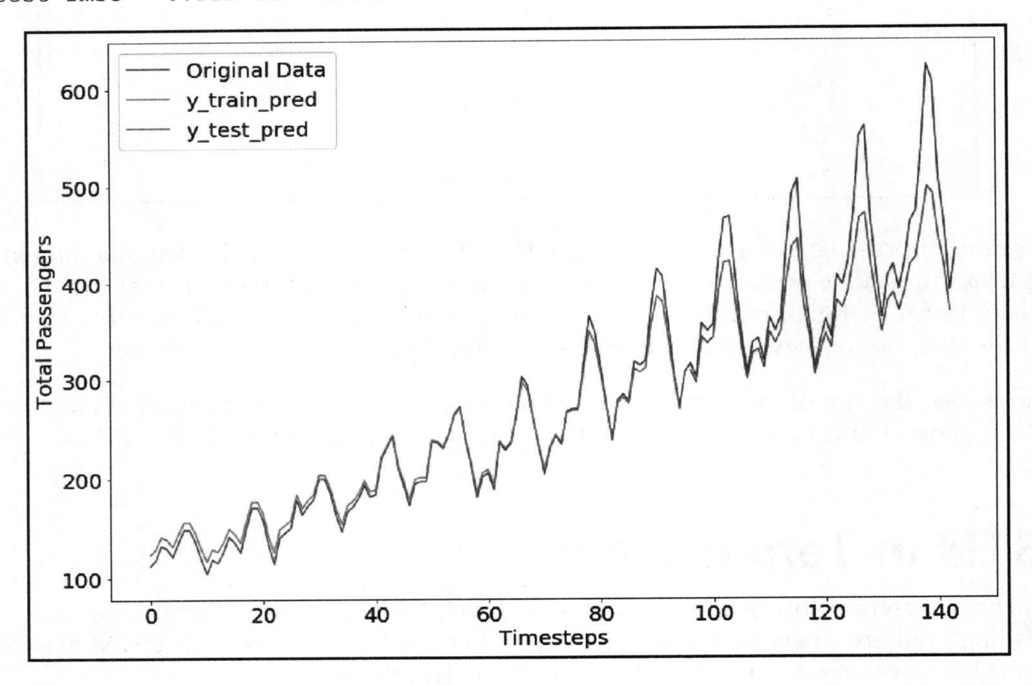

GRU in TensorFlow

To change the LSTM example in the last section to the GRU network, change the cell type as follows and the TensorFlow takes care of the rest for you:

```
cell = tf.nn.rnn_cell.GRUCell(state_size)
```

The complete code for the GRU model is provided in the notebook `ch-07a_RNN_TimeSeries_TensorFlow`.

For the small `airpass` dataset, the GRU has shown better performance for the same number of epochs. In practice, GRU and LSTM have shown comparable performance. In terms of execution speed, the GRU model trains and predicts faster as compared to the LSTM.

The complete code for the GRU model is provided in the Jupyter notebook. The results from the GRU model are as follows:

```
train mse = 0.0019633215852081776
test mse = 0.014307591132819653
test rmse = 0.11961434334066987
```

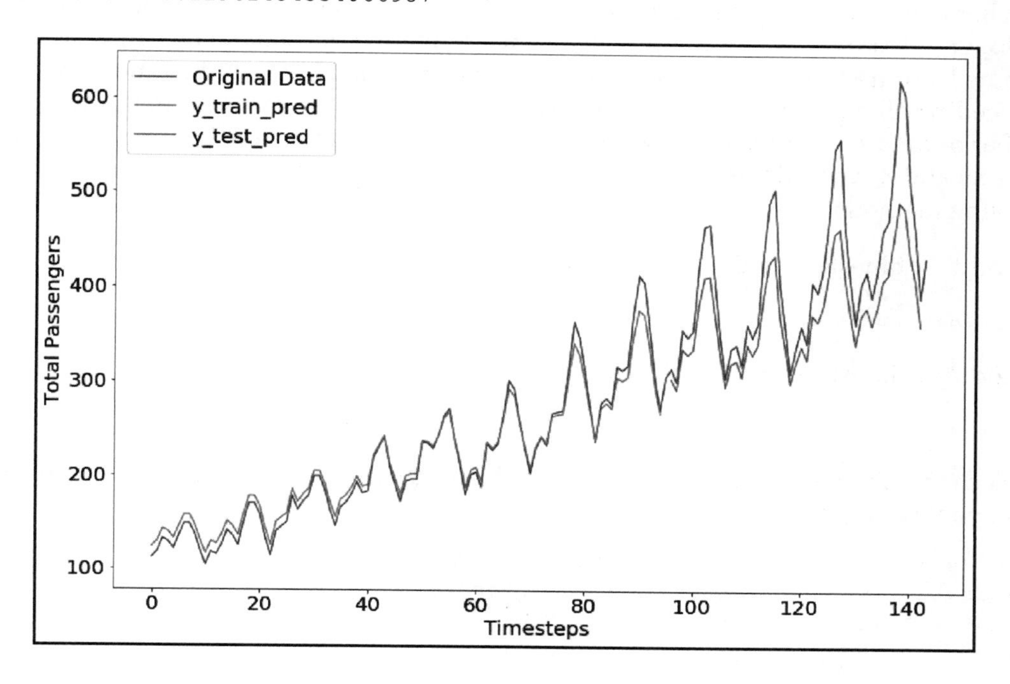

We encourage you to explore other options available in TensorFlow to create recurrent neural networks. Now let's try the same example in one of the high-level libraries for TensorFlow.

For the next sections, you can follow along with the code in the Jupyter notebook `ch-07b_RNN_TimeSeries_Keras`.

Preprocessing the dataset for RNN models with Keras

Building an RNN network in Keras is much simpler as compared to building using lower=level TensorFlow classes and methods. For Keras, we preprocess the data, as described in the previous sections, to get the supervised machine learning time series datasets: `X_train, Y_train, X_test, Y_test`.

From here onwards, the preprocessing differs. For Keras, the input has to be in the shape `(samples, time steps, features)`. As we converted our data to the supervised machine learning format, while reshaping the data, we can either set the time steps to 1, thus feeding all input time steps as features, or we can set the time steps to the actual number of time steps, thus feeding the feature set for each time step. In other words, the `X_train` and `X_test` datasets that we obtained earlier could be reshaped as one of the following methods:

Method 1 : *n* timesteps with *1* feature:

```
X_train.reshape(X_train.shape[0], X_train.shape[1],1)
```

Method 2 : *1* timestep with *n* features:

```
X_train.reshape(X_train.shape[0], 1, X_train.shape[1])
```

In this chapter, we will shape the datasets with a feature size of 1 because we are using only one variable as input:

```
# reshape input to be [samples, time steps, features]
X_train = X_train.reshape(X_train.shape[0], X_train.shape[1],1)
X_test = X_test.reshape(X_test.shape[0], X_train.shape[1], 1)
```

Simple RNN with Keras

An RNN model can be easily built in Keras by adding the SimpleRNN layer with the number of internal neurons and the shape of input tensor, excluding the number of samples dimension. The following code creates, compiles, and fits the SimpleRNN:

```
# create and fit the SimpleRNN model
model = Sequential()
model.add(SimpleRNN(units=4, input_shape=(X_train.shape[1],
    X_train.shape[2])))
model.add(Dense(1))
model.compile(loss='mean_squared_error', optimizer='adam')
model.fit(X_train, Y_train, epochs=20, batch_size=1)
```

As our dataset is small, we use a `batch_size` of 1 and train for 20 iterations, but for larger datasets, you would need to tune the value of these and other hyper-parameters.

The model is structured as follows:

```
_________________________________________________________________
Layer (type)                 Output Shape              Param #
=================================================================
simple_rnn_1 (SimpleRNN)     (None, 4)                 24
_________________________________________________________________
dense_1 (Dense)              (None, 1)                 5
=================================================================
Total params: 29
Trainable params: 29
Non-trainable params: 0
```

The output from the training is as follows:

```
Epoch 1/20
95/95 [==============================] - 0s - loss: 0.0161
Epoch 2/20
95/95 [==============================] - 0s - loss: 0.0074
Epoch 3/20
95/95 [==============================] - 0s - loss: 0.0063
Epoch 4/20
95/95 [==============================] - 0s - loss: 0.0051

-- epoch 5 to 14 removed for the sake of brevity --

Epoch 14/20
95/95 [==============================] - 0s - loss: 0.0021
Epoch 15/20
95/95 [==============================] - 0s - loss: 0.0020
```

```
Epoch 16/20
95/95 [==============================] - 0s - loss: 0.0020
Epoch 17/20
95/95 [==============================] - 0s - loss: 0.0020
Epoch 18/20
95/95 [==============================] - 0s - loss: 0.0020
Epoch 19/20
95/95 [==============================] - 0s - loss: 0.0020
Epoch 20/20
95/95 [==============================] - 0s - loss: 0.0020
```

The loss starts at 0.0161 and plateaus at 0.0020. Let's make predictions and rescale the predictions and originals. We use the functions provided by Keras to calculate the root mean square error:

```
from keras.losses import mean_squared_error as k_mse
from keras.backend import sqrt as k_sqrt
import keras.backend as K

# make predictions
y_train_pred = model.predict(X_train)
y_test_pred = model.predict(X_test)

# invert predictions
y_train_pred = scaler.inverse_transform(y_train_pred)
y_test_pred = scaler.inverse_transform(y_test_pred)

#invert originals
y_train_orig = scaler.inverse_transform(Y_train)
y_test_orig = scaler.inverse_transform(Y_test)

# calculate root mean squared error
trainScore = k_sqrt(k_mse(y_train_orig[:,0],
                          y_train_pred[:,0])
                   ).eval(session=K.get_session())
print('Train Score: {0:.2f} RMSE'.format(trainScore))

testScore = k_sqrt(k_mse(y_test_orig[:,0],
                         y_test_pred[:,0])
                  ).eval(session=K.get_session())
print('Test Score: {0:.2f} RMSE'.format(testScore))
```

We get the following results:

```
Train Score: 23.27 RMSE
Test Score: 54.13 RMSE
```

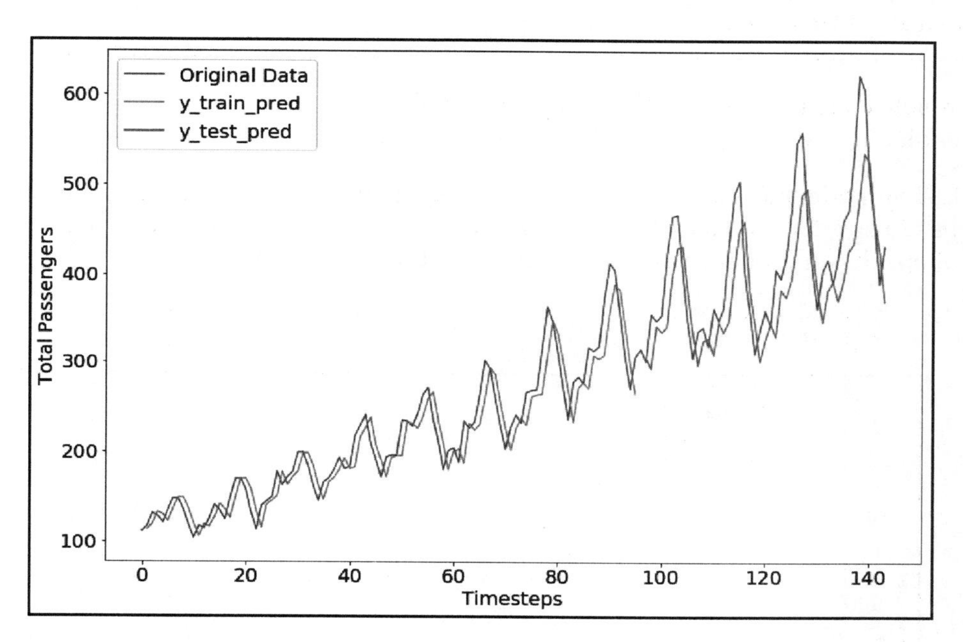

As we can see, this is not as perfect a fit as the ones we got in the TensorFlow section; however, this difference is because of hyperparameter values. We leave it to you to try different hyperparameter values to tune this Keras model to get even better results.

LSTM with Keras

Creating an LSTM model is only a matter of adding the LSTM layer instead of the SimpleRNN layer, as follows:

```
model.add(LSTM(units=4, input_shape=(X_train.shape[1], X_train.shape[2])))
```

The model structure appears as the following:

```
_________________________________________________________________
Layer (type)                 Output Shape              Param #
=================================================================
lstm_1 (LSTM)                (None, 4)                 96
_________________________________________________________________
```

```
dense_1 (Dense)              (None, 1)                 5
=================================================================
Total params: 101
Trainable params: 101
Non-trainable params: 0
_________________________________________________________________
```

The complete code for the LSTM model is provided in notebook `ch-07b_RNN_TimeSeries_Keras`.

As the LSTM model has more parameters that need to be trained, for the same number of iterations (20 epochs), we get a higher error score. We leave it to you to explore various values of epochs and other hyperparameters to get better results:

```
Train Score: 32.21 RMSE
Test Score: 84.68 RMSE
```

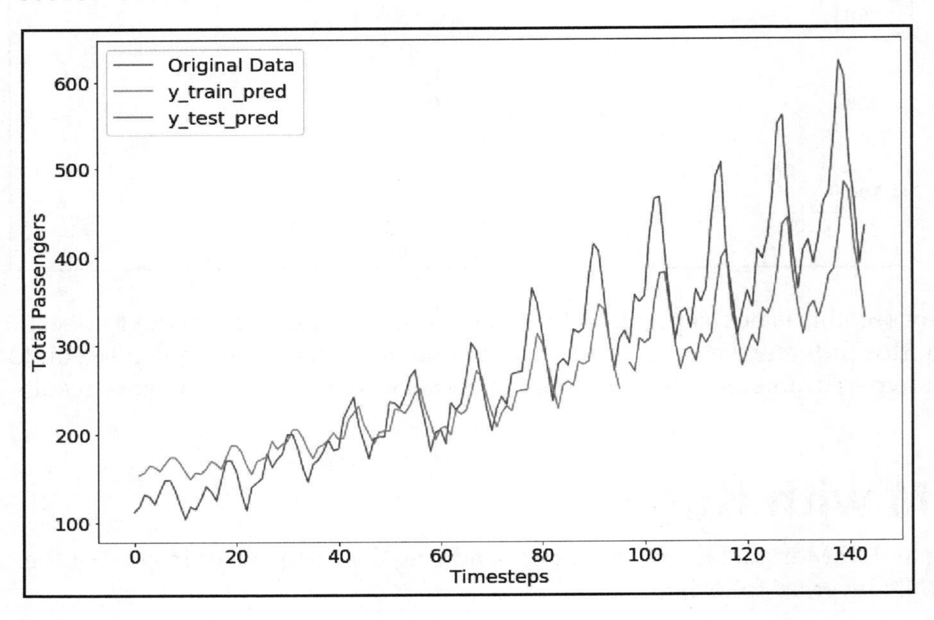

GRU with Keras

An advantage of using TensorFlow and Keras is that they make it easy to create models. Just like LSTM, creating a GRU model is only a matter of adding the GRU layer instead of LSTM or SimpleRNN layer, as follows:

```
model.add(GRU(units=4, input_shape=(X_train.shape[1], X_train.shape[2])))
```

The model structure is as follows:

```
Layer (type)                  Output Shape              Param #
=================================================================
gru_1 (GRU)                   (None, 4)                 72
_________________________________________________________________
dense_1 (Dense)               (None, 1)                 5
=================================================================
Total params: 77
Trainable params: 77
Non-trainable params: 0
```

The complete code for the GRU model is provided in notebook `ch-07b_RNN_TimeSeries_Keras`.

As expected, the GRU model shows almost the same performance as LSTM, and we leave it to you to try different values of hyperparameters to optimize this model:

```
Train Score: 31.49 RMSE
Test Score: 92.75 RMSE
```

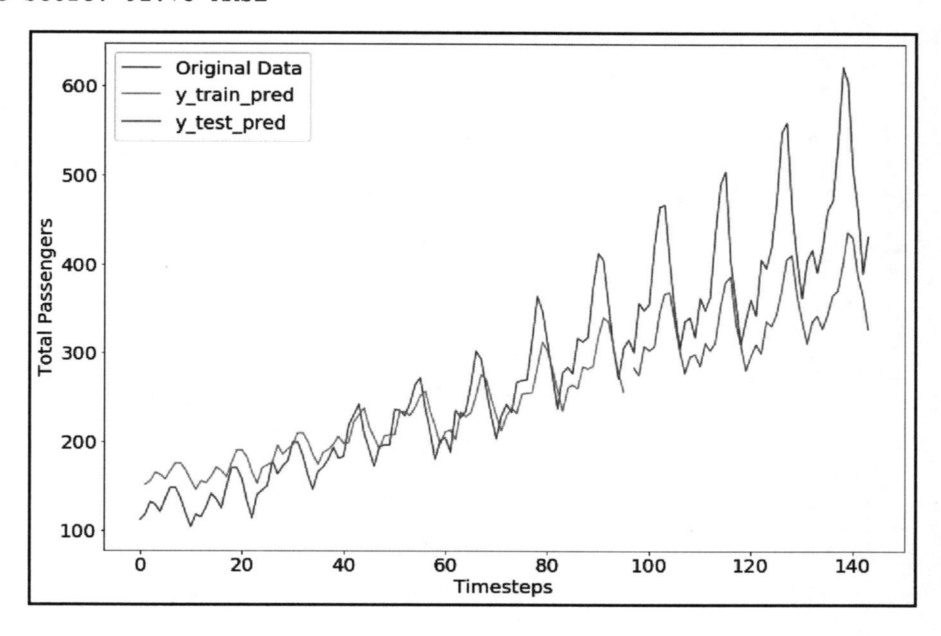

Summary

Time series data is sequence-based data, thus RNN models are a relevant architecture to learn from time series data. In this chapter, you learned how to create different kinds of RNN models with TensorFlow, a low-level library, and Keras, a high-level library. We only covered SimpleRNN, LSTM, and GRU, but you should explore many other RNN variants that can be created using TensorFlow and Keras.

In the next chapter, we will use the foundation built in the current chapter and the previous chapters to create RNN models for text data for various **Natural Language Processing** (**NLP**) tasks.

8
RNN for Text Data with TensorFlow and Keras

Text data can be viewed as a sequence of characters, words, sentences or paragraphs. **Recurrent neural networks** (**RNN**) have proven highly useful neural network architecture for sequences. For the purpose of applying neural network models to **Natural Language Processing** (**NLP**) tasks, the text is viewed as a sequence of words. This has proven highly successful for NLP tasks such as:

- Question answering
- Conversational agents or chatbots
- Document classification
- Sentiment analysis
- Image caption or description text generation
- Named entity recognition
- Speech recognition and tagging

NLP with TensorFlow deep learning techniques is a vast area and difficult to capture in one chapter. Hence, we have attempted to equip you with the most prevalent and important examples in this area using Tensorflow and Keras. Do not forget to explore and experiment with other areas of NLP once you have mastered the content in this chapter.

In this chapter, we shall learn about the following topics:

- Word vector representations
- Preparing data for word2vec models
- skip-gram model in TensorFlow and Keras
- Visualizing the word embeddings using t-SNE
- Text generation example with LSTM models in TensorFlow and Keras

Word vector representations

In order to learn the parameters of our neural network model from textual data, first, we have to convert the text or natural language data into a format that can be ingested by the neural networks. The neural networks generally ingest the text in the form of numeric vectors. The algorithms that convert raw text data into numeric vectors are known as word embedding algorithms.

One of the popular methods of word embedding is the **one-hot encoding** that we saw in MNIST image classification. Let's say our text dataset is made up of 60,000 dictionary words. Then each word can be represented by a one-hot encoded vector with 60,000 elements where all other elements have the value zero except the one element that represents this word which has the value one.

However, the one-hot encoding method has its drawbacks. Firstly, for vocabularies with a large number of words, the dimensionality of the one-hot word vector becomes very large. Secondly, one cannot find out word similarity with one-hot encoded vectors. For example, let's say cat and kitten have vectors as `[1 0 0 0 0 0]` and `[0 0 0 0 0 1]` respectively. There is no similarity in these vectors.

There are other corpus-based methods for converting text-based corpus to numeric vectors such as:

- Term Frequency-InverseDocument Frequency (TF-IDF)
- Latent Semantic Analysis (LSA)
- Topic Modeling

Recently, the focus of representing words with numerical vectors has shifted to the methods based on distributional hypothesis, which implies that words with similar semantic meaning have the tendency to appear in a similar context.

Two of the most widely used methods are called word2vec and GloVe. We shall practice with word2vec in this chapter. As we learned in the preceding paragraph, one-hot encoding gives a dimensionality of the size of the total number of words in the dictionary of a corpus. The dimensionality of word vectors created using word2vec is much lower.

The word2vec family of models is built using two architectures:

- **Continuous Bag of Words**: The model is trained to learn the probability distributions of the center words given the context words. Thus, given a set of context words, the model predicts the center words in a fill-in-the-blanks fashion that you did in your high-school language class. The CBOW architecture works best for datasets with smaller vocabularies.
- **Skip-gram**: The model is trained to learn the probability distributions of the context words given the center word. Thus, given a center word, the model predicts the context words in a complete-the-sentence fashion that you did in your high-school language class.

For example, let's consider the sentence:

> *Vets2data.org is a non-profit for educating the US Military Veterans Community on Artificial Intelligence and Data Science.*

In CBOW architecture, given the words *Military* and *Community*, the model learns the probability of the word *Veterans* and in skip-gram architecture, given the words *Veterans*, the model learns the probability of the words *Military* and *Community*.

The word2vec models learn word vectors from a corpus of text in an unsupervised fashion. The corpus of text is divided into pairs of context words and target words. While these pairs are the true pairs, the false pairs are generated with randomly pairing context words and context words, thus creating noise in the data. A classifier is trained to learn the parameters for distinguishing the true pairs from false pairs. The parameters of this classifier become the word2vec model or word vectors.

More information on the mathematics and theory behind word2vec theory can be learned from the following papers:

Mikolov, T., I. Sutskever, K. Chen, G. Corrado, and J. Dean. Distributed Representations of Words and Phrases and Their Compositionality. *Advances in Neural Information Processing Systems*, 2013, pp. 3111–3119.

Mikolov, T., K. Chen, G. Corrado, and J. Dean. Efficient Estimation of Word Representations in Vector Space. *arXiv*, 2013, pp. 1–12.

Rong, X. word2vec Parameter Learning Explained. *arXiv:1411.2738*, 2014, pp. 1–19.

Baroni, M., G. Dinu, and G. Kruszewski. Don't Count, Predict! A Systematic Comparison of Context-Counting vs. Context-Predicting Semantic Vectors. 2014.

You should practice with GloVe along with word2vec and apply the method that works for your text data.

More information on the GLoVe algorithm can be learned from the following paper:

Pennington, J., R. Socher, and C. Manning. GloVe: Global Vectors for Word Representation. 2014.

Let's understand the word2vec models by creating word vectors in TensorFlow and Keras.

You can follow along with the code for the next few sections in the Jupyter notebook `ch-08a_Embeddings_in_TensorFlow_and_Keras`.

Preparing the data for word2vec models

We shall use the popular PTB and text8 datasets for our demonstrations.

The **Penn Treebank (PTB)** dataset is a by-product of Penn Treebank project carried out at UPenn (`https://catalog.ldc.upenn.edu/ldc99t42`). The PTB project team extracted about one million words from the three years of Wall Street Journal stories and annotated them in Treebank II style. The PTB dataset comes in two flavors: Basic Examples, that are about 35 MB in size, and Advanced Examples, that are about 235 MB in size. We shall use the simple dataset that consists of 929K words for training, 73K words for validation, and 82K words for testing. You are encouraged to explore the advanced dataset. Further details on the PTB dataset are available at the following link: `http://www.fit.vutbr.cz/~imikolov/rnnlm/simple-examples.tgz`.

The PTB dataset can be downloaded from the following link: `http://www.fit.vutbr.cz/~imikolov/rnnlm/rnn-rt07-example.tar.gz`.

The **text8** dataset is a shorter and cleaned up version of larger Wikipedia data dump of about 1 GB in size. The process of how the text8 dataset was created is explained at the following link: `http://mattmahoney.net/dc/textdata.html`.

The text8 dataset can be downloaded from the following link: `http://mattmahoney.net/dc/text8.zip`.

The datasets are loaded using the `load_data` code in our custom library `datasetslib`:

The `load_data()` function does the following things:

1. It downloads the data archive from the dataset's URL if it is not available locally.
2. Since the *PTB* data comes in three files, it reads the text from the training file first, while for *text8* it reads the first file from the archive.
3. It converts the words in training files into a vocabulary, and assigns each vocabulary word a unique number, the word-id, stores it in collection `word2id`, and also prepares the reverse dictionary, so we can look up the word from the ID, and stores it in collection `id2word`.

4. It uses the collection `word2id` to convert the text files into a sequence of IDs.
5. Thus, at the end of the `load_data` we have a sequence of numbers in the training dataset and an ID to word mapping in collection `id2word`.

Let's see a glimpse of the data loaded from text8 and PTB datasets:

Loading and preparing the PTB dataset

Start by importing the modules and loading the data as follows::

```
from datasetslib.ptb import PTBSimple
ptb = PTBSimple()
# downloads data, converts words to ids, converts files to a list of ids
ptb.load_data()
print('Train :',ptb.part['train'][0:5])
print('Test: ',ptb.part['test'][0:5])
print('Valid: ',ptb.part['valid'][0:5])
print('Vocabulary Length = ',ptb.vocab_len)
```

The first five elements of each of the datasets along with the vocabulary length are printed as follows:

```
Train : [9970, 9971, 9972, 9974, 9975]
Test:  [102, 14, 24, 32, 752]
Valid:  [1132, 93, 358, 5, 329]
Vocabulary Length =  10000
```

We set the window of context to two words and get the CBOW pairs:

```
ptb.skip_window=2
ptb.reset_index_in_epoch()
# in CBOW input is the context word and output is the target word
y_batch, x_batch = ptb.next_batch_cbow()

print('The CBOW pairs : context,target')
for i in range(5 * ptb.skip_window):
    print('(', [ptb.id2word[x_i] for x_i in x_batch[i]],
          ',', y_batch[i], ptb.id2word[y_batch[i]], ')')
```

And the output is:

```
The CBOW pairs : context,target
( ['aer', 'banknote', 'calloway', 'centrust'] , 9972 berlitz )
( ['banknote', 'berlitz', 'centrust', 'cluett'] , 9974 calloway )
( ['berlitz', 'calloway', 'cluett', 'fromstein'] , 9975 centrust )
( ['calloway', 'centrust', 'fromstein', 'gitano'] , 9976 cluett )
( ['centrust', 'cluett', 'gitano', 'guterman'] , 9980 fromstein )
( ['cluett', 'fromstein', 'guterman', 'hydro-quebec'] , 9981 gitano )
( ['fromstein', 'gitano', 'hydro-quebec', 'ipo'] , 9982 guterman )
( ['gitano', 'guterman', 'ipo', 'kia'] , 9983 hydro-quebec )
( ['guterman', 'hydro-quebec', 'kia', 'memotec'] , 9984 ipo )
( ['hydro-quebec', 'ipo', 'memotec', 'mlx'] , 9986 kia )
```

Now let's see the skip-gram pairs:

```
ptb.skip_window=2
ptb.reset_index_in_epoch()
# in skip-gram input is the target word and output is the context word
x_batch, y_batch = ptb.next_batch()

print('The skip-gram pairs : target,context')
for i in range(5 * ptb.skip_window):
    print('(',x_batch[i], ptb.id2word[x_batch[i]],
        ',', y_batch[i], ptb.id2word[y_batch[i]],')')
```

And the output is:

```
The skip-gram pairs : target,context
( 9972 berlitz , 9970 aer )
( 9972 berlitz , 9971 banknote )
( 9972 berlitz , 9974 calloway )
( 9972 berlitz , 9975 centrust )
( 9974 calloway , 9971 banknote )
( 9974 calloway , 9972 berlitz )
( 9974 calloway , 9975 centrust )
( 9974 calloway , 9976 cluett )
( 9975 centrust , 9972 berlitz )
( 9975 centrust , 9974 calloway )
```

Loading and preparing the text8 dataset

Now we do the same loading and preprocessing steps with the text8 dataset:

```
from datasetslib.text8 import Text8
text8 = Text8()
text8.load_data()
# downloads data, converts words to ids, converts files to a list of ids
print('Train:', text8.part['train'][0:5])
print('Vocabulary Length = ',text8.vocab_len)
```

We find the vocabulary length is about 254,000 words:

```
Train: [5233, 3083, 11, 5, 194]
Vocabulary Length =  253854
```

Some of the tutorials manipulate this data by finding the most common words or truncate the vocabulary size to 10,000 words. However, we have used the complete dataset and complete vocabulary from the first file of the text8 dataset.

Preparing the CBOW pairs:

```
text8.skip_window=2
text8.reset_index_in_epoch()
# in CBOW input is the context word and output is the target word
y_batch, x_batch = text8.next_batch_cbow()

print('The CBOW pairs : context,target')
for i in range(5 * text8.skip_window):
    print('(', [text8.id2word[x_i] for x_i in x_batch[i]],
          ',', y_batch[i], text8.id2word[y_batch[i]], ')')
```

And the output is:

```
The CBOW pairs : context,target
( ['anarchism', 'originated', 'a', 'term'] , 11 as )
( ['originated', 'as', 'term', 'of'] , 5 a )
( ['as', 'a', 'of', 'abuse'] , 194 term )
( ['a', 'term', 'abuse', 'first'] , 1 of )
( ['term', 'of', 'first', 'used'] , 3133 abuse )
( ['of', 'abuse', 'used', 'against'] , 45 first )
( ['abuse', 'first', 'against', 'early'] , 58 used )
( ['first', 'used', 'early', 'working'] , 155 against )
( ['used', 'against', 'working', 'class'] , 127 early )
( ['against', 'early', 'class', 'radicals'] , 741 working )
```

Preparing the skip-gram pairs:

```
text8.skip_window=2
text8.reset_index_in_epoch()
# in skip-gram input is the target word and output is the context word
x_batch, y_batch = text8.next_batch()

print('The skip-gram pairs : target,context')
for i in range(5 * text8.skip_window):
    print('(',x_batch[i], text8.id2word[x_batch[i]],
        ',', y_batch[i], text8.id2word[y_batch[i]],')')
```

And the output is:

```
The skip-gram pairs : target,context
( 11 as , 5233 anarchism )
( 11 as , 3083 originated )
( 11 as , 5 a )
( 11 as , 194 term )
( 5 a , 3083 originated )
( 5 a , 11 as )
( 5 a , 194 term )
( 5 a , 1 of )
( 194 term , 11 as )
( 194 term , 5 a )
```

Preparing the small validation set

For the purpose of demonstrating the example, we create a small validation set of 8 words, each word is randomly chosen from the word with a word-id between 0 to 10 x 8.

```
valid_size = 8
x_valid = np.random.choice(valid_size * 10, valid_size, replace=False)
print(x_valid)
```

As an example, we get the following as our validation set:

```
valid:  [64 58 59 4 69 53 31 77]
```

We shall use this validation set to demonstrate the result of the word embeddings by printing the five closest words.

skip-gram model with TensorFlow

Now that we have training and validation data prepared, let's create a skip-gram model in TensorFlow.

We start by defining the hyper-parameters:

```
batch_size = 128
embedding_size = 128
skip_window = 2
n_negative_samples = 64
ptb.skip_window=2
learning_rate = 1.0
```

- The `batch_size` is the number of pairs of target and context words to be fed into the algorithms in a single batch
- The `embedding_size` is the dimension of the word vector or embedding for each word
- The `ptb.skip_window` is the number of words to be considered in the context of the target words in both directions
- The `n_negative_samples` is the number of negative samples to be generated by the NCE loss function, explained further in this chapter

In some tutorials, including the one in the TensorFlow documentation, one more parameter `num_skips` is used. In such tutorials, the authors pick the `num_skips` number of pairs of (target, context). For example, if the `skip_window` is two, then the total number of pairs would be four, and if `num_skips` is set to two, then only two pairs will be randomly picked for the training. However, we considered all the pairs for keeping the training exercise simple.

Define the input and output placeholders for the training data and a tensor of validation data:

```
inputs = tf.placeholder(dtype=tf.int32, shape=[batch_size])
outputs = tf.placeholder(dtype=tf.int32, shape=[batch_size,1])
inputs_valid = tf.constant(x_valid, dtype=tf.int32)
```

Define an embedding matrix that has rows equal to the vocabulary length and columns equal to the embedding dimensions. Each row in this matrix would represent the word vector for one word in the vocabulary. Populate this embedding matrix with values uniformly sampled over -1.0 to 1.0.

```
# define embeddings matrix with vocab_len rows and embedding_size columns
# each row represents vectore representation or embedding of a word
# in the vocbulary

embed_dist = tf.random_uniform(shape=[ptb.vocab_len, embedding_size],
                               minval=-1.0,maxval=1.0)
embed_matrix = tf.Variable(embed_dist,name='embed_matrix')
```

Using this matrix, define an embedding lookup table implemented using the `tf.nn.embedding_lookup()`. The `tf.nn.embedding_lookup()` is given two parameters: the embedding matrix and the inputs placeholder. The lookup function returns the word vectors for the words in the `inputs` placeholder.

```
# define the embedding lookup table
# provides the embeddings of the word ids in the input tensor
embed_ltable = tf.nn.embedding_lookup(embed_matrix, inputs)
```

The `embed_ltable` can also be interpreted as the embedding layer on top of the input layer. Next, the output of the embedding layer is fed into softmax or the noise-contrastive estimation (NCE) layer. The NCE is based on a very simple idea of training the logistic regression-based binary classifier to learn the parameters from a mix of true and noisy data.

TensorFlow documentation describes the NCE in further detail: `https://www.tensorflow.org/tutorials/word2vec`.

In summary, the softmax loss-based models are computationally expensive because probability distribution is computed and normalized over the entire vocabulary. The NCE loss-based models reduce this to a binary classification problem, i.e. identifying true samples from noisy samples.

Underlying mathematical details of NCE can be found in the following NIPS paper: *Learning word embeddings efficiently with noise-contrastive estimation* by Andriy Mnih and Koray Kavukcuoglu. The paper is available at the following link: `http://papers.nips.cc/paper/5165-learning-word-embeddings-efficiently-with-noise-contrastive-estimation.pdf`.

The `tf.nn.nce_loss()` function automatically generates negative samples when it is evaluated to calculate the loss: the parameter `num_sampled`is set equal to the number of negative samples (`n_negative_samples`). This parameter specifies how many negative samples to draw.

```
# define noise-contrastive estimation (NCE) loss layer
nce_dist = tf.truncated_normal(shape=[ptb.vocab_len, embedding_size],
                               stddev=1.0 /
                               tf.sqrt(embedding_size * 1.0)
                               )
nce_w = tf.Variable(nce_dist)
nce_b = tf.Variable(tf.zeros(shape=[ptb.vocab_len]))

loss = tf.reduce_mean(tf.nn.nce_loss(weights=nce_w,
                                     biases=nce_b,
                                     inputs=embed_ltable,
                                     labels=outputs,
                                     num_sampled=n_negative_samples,
                                     num_classes=ptb.vocab_len
                                     )
                      )
```

Next, compute the cosine similarity between the samples in the validation set and the embedding matrix:

1. In order to compute the similarity score, first, compute the L2 norm for each word vector in the embedding matrix.

```
# Compute the cosine similarity between validation set samples
# and all embeddings.
norm = tf.sqrt(tf.reduce_sum(tf.square(embed_matrix), 1,
                             keep_dims=True))
normalized_embeddings = embed_matrix / norm
```

2. Look up the embeddings or word vectors for the samples in the validation set:

```
embed_valid = tf.nn.embedding_lookup(normalized_embeddings,
                                     inputs_valid)
```

3. Compute the similarity score by multiplying embeddings for the validation set with the embedding matrix.

```
similarity = tf.matmul(
    embed_valid, normalized_embeddings, transpose_b=True)
```

This gives a tensor that has the shape of (`valid_size`, `vocab_len`). Each row in the tensor refers to the similarity score between a validation word and the vocabulary words.

Next, define the SGD optimizer with a learning rate of 0.9 for 50 epochs.

```
n_epochs = 10
learning_rate = 0.9
n_batches = ptb.n_batches(batch_size)
optimizer = tf.train.GradientDescentOptimizer(learning_rate)
        .minimize(loss)
```

For each epoch:

1. Run the optimizer over the entire dataset, batch by batch.

```
ptb.reset_index_in_epoch()
for step in range(n_batches):
    x_batch, y_batch = ptb.next_batch()
    y_batch = dsu.to2d(y_batch,unit_axis=1)
    feed_dict = {inputs: x_batch, outputs: y_batch}
    _, batch_loss = tfs.run([optimizer, loss], feed_dict=feed_dict)
    epoch_loss += batch_loss
```

2. Calculate and print the average loss for the epoch.

```
epoch_loss = epoch_loss / n_batches
print('\n','Average loss after epoch ', epoch, ': ', epoch_loss)
```

3. At the end of the epoch, compute the similarity score.

```
similarity_scores = tfs.run(similarity)
```

4. For each word in the validation set, print the five words having the highest similarity score.

```
top_k = 5
for i in range(valid_size):
    similar_words = (-similarity_scores[i,:])
                    .argsort()[1:top_k + 1]
    similar_str = 'Similar to {0:}:'
```

```
                              .format(ptb.id2word[x_valid[i]])
            for k in range(top_k):
                similar_str = '{0:} {1:},'.format(similar_str,
                                    ptb.id2word[similar_words[k]])
            print(similar_str)
```

Finally, after all the epochs are done, compute the embeddings vector that can be utilized further in the learning process:

```
    final_embeddings = tfs.run(normalized_embeddings)
```

The complete training code is as follows:

```
n_epochs = 10
learning_rate = 0.9
n_batches = ptb.n_batches_wv()
optimizer = tf.train.GradientDescentOptimizer(learning_rate).minimize(loss)

with tf.Session() as tfs:
    tf.global_variables_initializer().run()
    for epoch in range(n_epochs):
        epoch_loss = 0
        ptb.reset_index()
        for step in range(n_batches):
            x_batch, y_batch = ptb.next_batch_sg()
            y_batch = nputil.to2d(y_batch, unit_axis=1)
            feed_dict = {inputs: x_batch, outputs: y_batch}
            _, batch_loss = tfs.run([optimizer, loss], feed_dict=feed_dict)
            epoch_loss += batch_loss
        epoch_loss = epoch_loss / n_batches
        print('\nAverage loss after epoch ', epoch, ': ', epoch_loss)

        # print closest words to validation set at end of every epoch
        similarity_scores = tfs.run(similarity)
        top_k = 5
        for i in range(valid_size):
            similar_words = (-similarity_scores[i, :]
                            ).argsort()[1:top_k + 1]
            similar_str = 'Similar to {0:}:'.format(
                ptb.id2word[x_valid[i]])
            for k in range(top_k):
                similar_str = '{0:} {1:},'.format(
                    similar_str, ptb.id2word[similar_words[k]])
            print(similar_str)
    final_embeddings = tfs.run(normalized_embeddings)
```

Here is the output we get after the 1st and 10th epoch respectively:

```
Average loss after epoch  0 :  115.644006802
Similar to we: types, downturn, internal, by, introduce,
Similar to been: said, funds, mcgraw-hill, street, have,
Similar to also: will, she, next, computer, 's,
Similar to of: was, and, milk, dollars, $,
Similar to last: be, october, acknowledging, requested, computer,
Similar to u.s.: plant, increase, many, down, recent,
Similar to an: commerce, you, some, american, a,
Similar to trading: increased, describes, state, companies, in,

Average loss after epoch  9 :  5.56538496033
Similar to we: types, downturn, introduce, internal, claims,
Similar to been: exxon, said, problem, mcgraw-hill, street,
Similar to also: will, she, ssangyong, audit, screens,
Similar to of: seasonal, dollars, motor, none, deaths,
Similar to last: acknowledging, allow, incorporated, joint, requested,
Similar to u.s.: undersecretary, typically, maxwell, recent, increase,
Similar to an: banking, officials, imbalances, americans, manager,
Similar to trading: describes, increased, owners, committee, else,
```

Finally, we run the model for 5,000 epochs and get the following result:

```
Average loss after epoch  4999 :  2.74216903135
Similar to we: matter, noted, here, classified, orders,
Similar to been: good, precedent, medium-sized, gradual, useful,
Similar to also: introduce, england, index, able, then,
Similar to of: indicator, cleveland, theory, the, load,
Similar to last: dec., office, chrysler, march, receiving,
Similar to u.s.: label, fannie, pressures, squeezed, reflection,
Similar to an: knowing, outlawed, milestones, doubled, base,
Similar to trading: associates, downturn, money, portfolios, go,
```

Try running it further, up to 50,000 epochs to get even better results.

Similarly, we get the following results with the text8 model after 50 epochs:

```
Average loss after epoch  49 :  5.74381046423
Similar to four: five, three, six, seven, eight,
Similar to all: many, both, some, various, these,
Similar to between: with, through, thus, among, within,
Similar to a: another, the, any, each, tpvgames,
Similar to that: which, however, although, but, when,
Similar to zero: five, three, six, eight, four,
Similar to is: was, are, has, being, busan,
Similar to no: any, only, the, another, trinomial,
```

Visualize the word embeddings using t-SNE

Let's visualize the word embeddings that we generated in the previous section. The t-SNE is the most popular method to display high-dimensional data in two-dimensional spaces. We shall use the method from the scikit-learn library and reuse the code given in TensorFlow documentation to draw a graph of the word embeddings we just learned.

The original code from the TensorFlow documentation is available at the following link: `https://github.com/tensorflow/tensorflow/blob/r1.3/tensorflow/examples/tutorials/word2vec/word2vec_basic.py`.

Here is how we implement the procedure:

1. Create the `tsne` model:

```
tsne = TSNE(perplexity=30, n_components=2,
            init='pca', n_iter=5000, method='exact')
```

2. Limit the number of embeddings to display to 500, otherwise, the graph becomes very unreadable:

```
n_embeddings = 500
```

3. Create the low-dimensional representation by calling the `fit_transform()` method on the `tsne` model and passing the first `n_embeddings` of `final_embeddings` as input.

```
low_dim_embeddings = tsne.fit_transform(
    final_embeddings[:n_embeddings, :])
```

4. Find the textual representation for the word vectors we selected for the graph:

```
labels = [ptb.id2word[i] for i in range(n_embeddings)]
```

5. Finally, plot the embeddings:

```
plot_with_labels(low_dim_embeddings, labels)
```

We get the following plot:

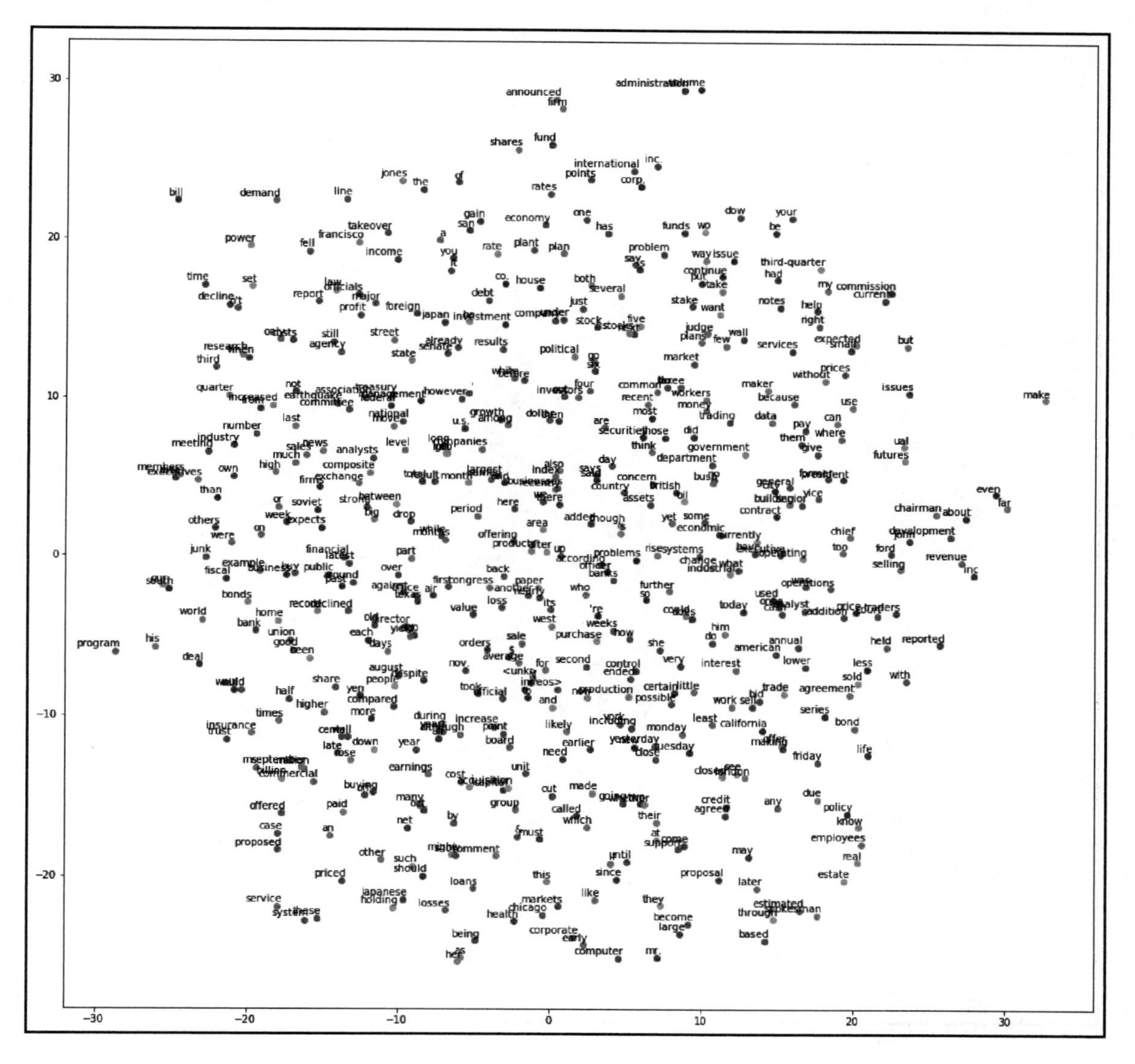

t-SNE visualization of embeddings for PTB data set

Similarly, from the text8 model, we get the following plot:

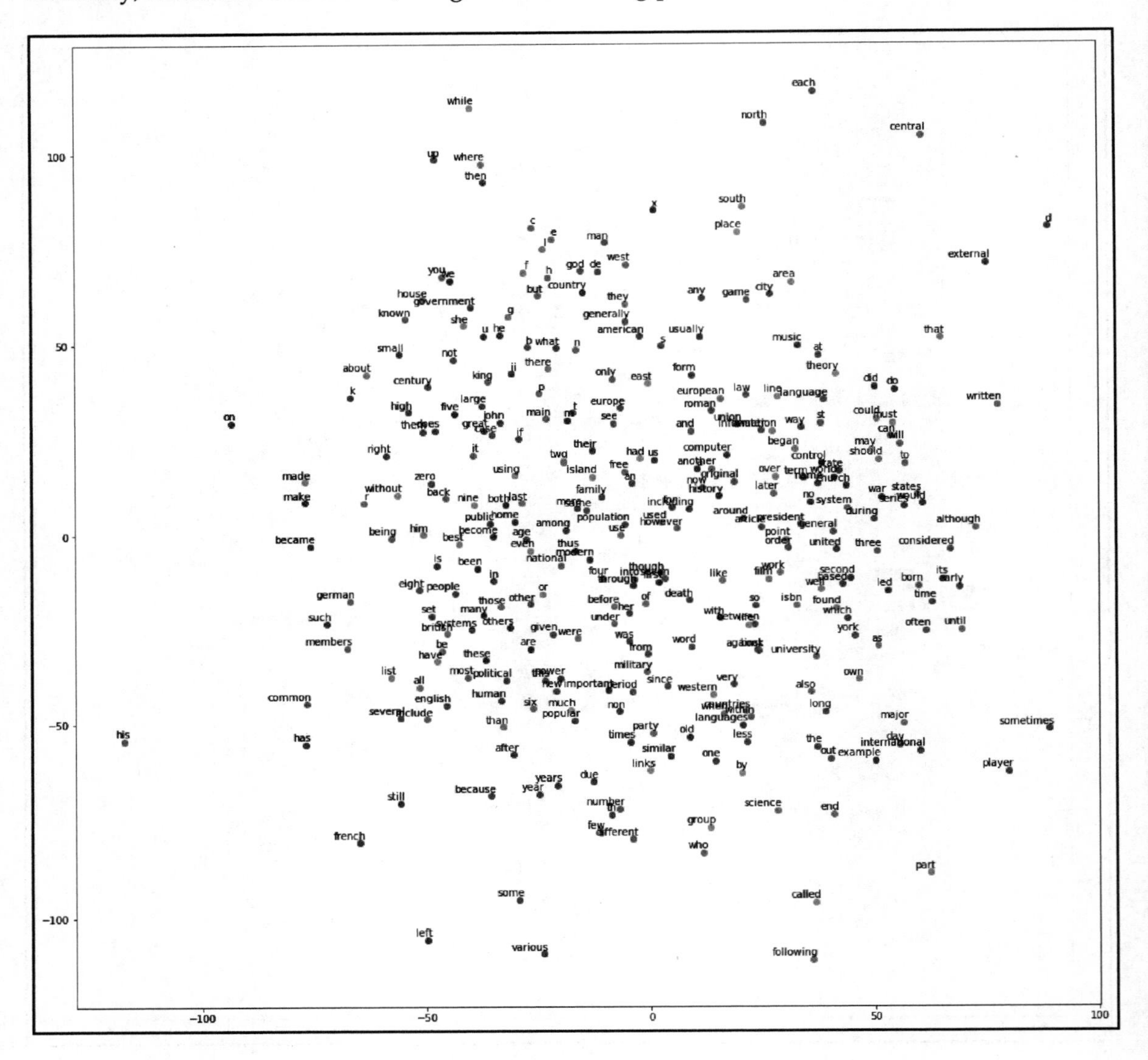

t-SNE visualization of embeddings for text8 data set

skip-gram model with Keras

The flow of the embedding model with Keras remains the same as TensorFlow.

- Create the network architecture in the Keras functional or sequential model
- Feed the true and false pairs of the target and context words to the network
- Look up the word vector for target and context words
- Perform a dot product of the word vectors to get the similarity score
- Pass the similarity score through a sigmoid layer to get the output as the true or false pair

Now let's implement these steps using the Keras functional API:

1. Import the required libraries:

```
from keras.models import Model
from keras.layers.embeddings import Embedding
from keras.preprocessing import sequence
from keras.preprocessing.sequence import skipgrams
from keras.layers import Input, Dense, Reshape, Dot, merge
import keras
```

   Reset graphs so that any after effects left from previous runs in Jupyter Notebook are cleared up:

```
# reset the jupyter buffers
tf.reset_default_graph()
keras.backend.clear_session()
```

2. Create a validation set that we will use to print the similar words that our model found at the end of the training:

```
valid_size = 8
x_valid = np.random.choice(valid_size * 10, valid_size,
replace=False)
print('valid: ',x_valid)
```

3. Define the required hyper-parameters:

```
batch_size = 1024
embedding_size = 512
n_negative_samples = 64
ptb.skip_window=2
```

4. Use the `make_sampling_table()` function from `keras.preprocessing.sequence` to create a sample table of the size equal to the length of our vocabulary. Next, use the function `skipgrams()` from `keras.preprocessing.sequence` to generate the pairs of context and target words along with the labels representing if they are true or fake pairs.

```
sample_table = sequence.make_sampling_table(ptb.vocab_len)
pairs, labels= sequence.skipgrams(ptb.part['train'],
        ptb.vocab_len,window_size=ptb.skip_window,
        sampling_table=sample_table)
```

5. Let's print some of the fake and true pairs that are generated with the following code:

```
print('The skip-gram pairs : target,context')
for i in range(5 * ptb.skip_window):
    print(['{} {}'.format(id,ptb.id2word[id]) \
        for id in pairs[i]],':',labels[i])
```

And the pairs are as follows:

```
The skip-gram pairs : target,context
['547 trying', '5 to'] : 1
['4845 bargain', '2 <eos>'] : 1
['1705 election', '198 during'] : 1
['4704 flows', '8117 gun'] : 0
['13 is', '37 company'] : 1
['625 above', '132 three'] : 1
['5768 pessimistic', '1934 immediate'] : 0
['637 china', '2 <eos>'] : 1
['258 five', '1345 pence'] : 1
['1956 chrysler', '8928 exercises'] : 0
```

6. Split the target and context words from the pairs generated above so they can be fed to the model. Convert the target and context words to two-dimensional arrays.

```
x,y=zip(*pairs)
x=np.array(x,dtype=np.int32)
x=dsu.to2d(x,unit_axis=1)
y=np.array(y,dtype=np.int32)
y=dsu.to2d(y,unit_axis=1)
labels=np.array(labels,dtype=np.int32)
labels=dsu.to2d(labels,unit_axis=1)
```

7. Define the architecture of the network. As we discussed, the target and context words have to be fed into the network and their vectors need to be looked up from the embedding layers. Thus, first we define the input, embedding, and reshape layers for the target and context words respectively:

```
# build the target word model
target_in = Input(shape=(1,),name='target_in')
target = Embedding(ptb.vocab_len,embedding_size,input_length=1,
            name='target_em')(target_in)
target = Reshape((embedding_size,1),name='target_re')(target)

# build the context word model
context_in = Input((1,),name='context_in')
context = Embedding(ptb.vocab_len,embedding_size,input_length=1,
            name='context_em')(context_in)
context = Reshape((embedding_size,1),name='context_re')(context)
```

8. Next, build the dot product of these two models, which feeds into the sigmoid layer to generate the output labels:

```
# merge the models with the dot product to check for
# similarity and add sigmoid layer
output = Dot(axes=1,name='output_dot')([target,context])
output = Reshape((1,),name='output_re')(output)
output = Dense(1, activation='sigmoid',name='output_sig')(output)
```

9. Build the functional model from the input and output models we just created:

```
# create the functional model for finding word vectors
model = Model(inputs=[target_in,context_in],outputs=output)
model.compile(loss='binary_crossentropy', optimizer='adam')
```

10. Additionally, build a model that will be used to predict the similarity to all the words, given the input target word:

```
# merge the models and create model to check for cosine similarity
similarity = Dot(axes=0,normalize=True,
            name='sim_dot')([target,context])
similarity_model = Model(inputs=[target_in,context_in],
            outputs=similarity)
```

Let's print the model summary:

```
_________________________________________________________________________________
Layer (type)                    Output Shape          Param #     Connected to
=================================================================================
target_in (InputLayer)          (None, 1)             0
_________________________________________________________________________________
context_in (InputLayer)         (None, 1)             0
_________________________________________________________________________________
target_em (Embedding)           (None, 1, 512)        5120000     target_in[0][0]
_________________________________________________________________________________
context_em (Embedding)          (None, 1, 512)        5120000     context_in[0][0]
_________________________________________________________________________________
target_re (Reshape)             (None, 512, 1)        0           target_em[0][0]
_________________________________________________________________________________
context_re (Reshape)            (None, 512, 1)        0           context_em[0][0]
_________________________________________________________________________________
output_dot (Dot)                (None, 1, 1)          0           target_re[0][0]
                                                                  context_re[0][0]
_________________________________________________________________________________
output_re (Reshape)             (None, 1)             0           output_dot[0][0]
_________________________________________________________________________________
output_sig (Dense)              (None, 1)             2           output_re[0][0]
=================================================================================
Total params: 10,240,002
Trainable params: 10,240,002
Non-trainable params: 0
_________________________________________________________________________________
```

11. Next, train the model. We only trained it for 5 epochs, but you should try it with more epochs, at least 1,000 or 10,000 epochs.

Remember, it will take several hours since this is not the most optimized code. You are welcome to optimize the code further with the tips and tricks from this book and from other sources.

```
n_epochs = 5
batch_size = 1024
model.fit([x,y],labels,batch_size=batch_size, epochs=n_epochs)
```

Let's print the similarity of the words based on the word vectors discovered by this model:

```
# print closest words to validation set at end of training
top_k = 5
y_val = np.arange(ptb.vocab_len, dtype=np.int32)
y_val = dsu.to2d(y_val,unit_axis=1)
for i in range(valid_size):
    x_val = np.full(shape=(ptb.vocab_len,1),fill_value=x_valid[i],
            dtype=np.int32)
    similarity_scores = similarity_model.predict([x_val,y_val])
    similarity_scores=similarity_scores.flatten()
    similar_words = (-similarity_scores).argsort()[1:top_k + 1]
    similar_str = 'Similar to {0:}:'.format(ptb.id2word[x_valid[i]])
    for k in range(top_k):
        similar_str = '{0:} {1:},'.format(similar_str,
                                ptb.id2word[similar_words[k]])
    print(similar_str)
```

We get the following output:

```
Similar to we: rake, kia, sim, ssangyong, memotec,
Similar to been: nahb, sim, rake, punts, rubens,
Similar to also: photography, snack-food, rubens, nahb, ssangyong,
Similar to of: isi, rake, memotec, kia, mlx,
Similar to last: rubens, punts, memotec, sim, photography,
Similar to u.s.: mlx, memotec, punts, rubens, kia,
Similar to an: memotec, isi, ssangyong, rake, sim,
Similar to trading: rake, rubens, swapo, mlx, nahb,
```

So far we have seen how to create word vector or embeddings using TensorFlow and its high-level library, Keras. Now let's see how to use TensorFlow and Keras to learn the models and apply the model to the prediction of some of the NLP-related tasks.

Text generation with RNN models in TensorFlow and Keras

Text generation is one of the major applications of RNN models in NLP. An RNN model is trained on the sequences of text and then used to generate the sequences of text by providing a seed text as input. Let's try that on the text8 dataset.

Let's load the text8 dataset and print the first 100 words:

```
from datasetslib.text8 import Text8
text8 = Text8()
# downloads data, converts words to ids, converts files to a list of ids
text8.load_data()
print(' '.join([text8.id2word[x_i] for x_i in text8.part['train'][0:100]]))
```

We get the following output:

```
anarchism originated as a term of abuse first used against early working
class radicals including the diggers of the english revolution and the sans
culottes of the french revolution whilst the term is still used in a
pejorative way to describe any act that used violent means to destroy the
organization of society it has also been taken up as a positive label by
self defined anarchists the word anarchism is derived from the greek
without archons ruler chief king anarchism as a political philosophy is the
belief that rulers are unnecessary and should be abolished although there
are differing
```

In our notebook examples, we clip the data load at 5,000 words of text as the larger text requires advanced techniques such as distributed or batched processing and we wanted to keep the example simple.

```
from datasetslib.text8 import Text8
text8 = Text8()
text8.load_data(clip_at=5000)
print('Train:', text8.part['train'][0:5])
print('Vocabulary Length = ',text8.vocab_len)
```

We see that the vocabulary size is reduced to 1,457 words now.

```
Train: [  8 497   7   5 116]
Vocabulary Length =  1457
```

In our example, we construct a very simple one-layer LSTM. For training the model, we use the 5 words as input to learn the parameters for the sixth word. The input layer is the 5 words, and the hidden layer is the LSTM cell with 128 units and the final layer is a fully connected layer with outputs equal to the vocabulary size. Since we are demonstrating the example, we did not use word vectors but a very simple one-hot encoded output vector.

Once the model is trained, we test it with 2 different strings as the seed for generating further characters:

- `random5`: String generated from randomly selecting 5 words.
- `first5`: String generated from first 5 words of the text.

```
random5 = np.random.choice(n_x * 50, n_x, replace=False)
print('Random 5 words: ',id2string(random5))
first5 = text8.part['train'][0:n_x].copy()
print('First 5 words: ',id2string(first5))
```

And we see that the seed strings are:

```
Random 5 words:  free bolshevik be n another
First 5 words:  anarchism originated as a term
```

For your executions, the random seed string may be different.

Now let's create the LSTM model in TensorFlow first.

Text generation LSTM in TensorFlow

You can follow along with the code for this section in the Jupyter notebook `ch-08b_RNN_Text_TensorFlow`.

We implement text generation LSTM in TensorFlow with the following steps:

1. Let's define the parameters and placeholders for `x` and `y`:

```
batch_size = 128
n_x = 5 # number of input words
n_y = 1 # number of output words
n_x_vars = 1 # in case of our text, there is only 1 variable at
each timestep
n_y_vars = text8.vocab_len
```

```
state_size = 128
learning_rate = 0.001
x_p = tf.placeholder(tf.float32, [None, n_x, n_x_vars], name='x_p')
y_p = tf.placeholder(tf.float32, [None, n_y_vars], name='y_p')
```

For input, we are using the integer representations of the words, hence `n_x_vars` is 1. For output we are using one-hot encoded values hence the number of outputs is equal to the vocabulary length.

2. Next, create a list of tensors of length `n_x`:

```
x_in = tf.unstack(x_p,axis=1,name='x_in')
```

3. Next, create an LSTM cell and a static RNN network from the inputs and the cell:

```
cell = tf.nn.rnn_cell.LSTMCell(state_size)
rnn_outputs, final_states = tf.nn.static_rnn(cell,
x_in,dtype=tf.float32)
```

4. Next, we define the weight, biases, and formulae for the final layer. The final layer only needs to pick the output for the sixth word, hence we apply the following formulae to grab only the last output:

```
# output node parameters
w = tf.get_variable('w', [state_size, n_y_vars], initializer=
tf.random_normal_initializer)
b = tf.get_variable('b', [n_y_vars],
initializer=tf.constant_initializer(0.0))
y_out = tf.matmul(rnn_outputs[-1], w) + b
```

5. Next, create a loss function and optimizer:

```
loss = tf.reduce_mean(tf.nn.softmax_cross_entropy_with_logits(
        logits=y_out, labels=y_p))
optimizer = tf.train.AdamOptimizer(learning_rate=learning_rate)
            .minimize(loss)
```

6. Create the accuracy function that we can run in the session block to check the accuracy of trained mode:

```
n_correct_pred = tf.equal(tf.argmax(y_out,1), tf.argmax(y_p,1))
accuracy = tf.reduce_mean(tf.cast(n_correct_pred, tf.float32))
```

7. Finally, we train the model for 1,000 epochs and print the results every 100 epochs. Also every 100 epochs, we print the generated text from both the seed strings we described above.

LSTM and RNN networks require training on a large dataset for a large number of epochs for better results. Please try loading the complete dataset and running for 50,000 or 80,000 epochs on your computer and play with other hyper-parameters to improve the results.

```
n_epochs = 1000
learning_rate = 0.001
text8.reset_index_in_epoch()
n_batches = text8.n_batches_seq(batch_size=batch_size,n_tx=n_x,n_ty=n_y)
n_epochs_display = 100

with tf.Session() as tfs:
    tf.global_variables_initializer().run()

    for epoch in range(n_epochs):
        epoch_loss = 0
        epoch_accuracy = 0
        for step in range(n_batches):
            x_batch, y_batch = text8.next_batch_seq(batch_size=batch_size,
                                        n_tx=n_x,n_ty=n_y)
            y_batch = dsu.to2d(y_batch,unit_axis=1)
            y_onehot = np.zeros(shape=[batch_size,text8.vocab_len],
                        dtype=np.float32)
            for i in range(batch_size):
                y_onehot[i,y_batch[i]]=1

            feed_dict = {x_p: x_batch.reshape(-1, n_x, n_x_vars),
                         y_p: y_onehot}
            _, batch_accuracy, batch_loss = tfs.run([optimizer,accuracy,
                                    loss],feed_dict=feed_dict)
            epoch_loss += batch_loss
            epoch_accuracy += batch_accuracy

        if (epoch+1) % (n_epochs_display) == 0:
            epoch_loss = epoch_loss / n_batches
            epoch_accuracy = epoch_accuracy / n_batches
            print('\nEpoch {0:}, Average loss:{1:}, Average accuracy:{2:}'.
                    format(epoch,epoch_loss,epoch_accuracy ))

            y_pred_r5 = np.empty([10])
            y_pred_f5 = np.empty([10])
```

```
        x_test_r5 = random5.copy()
        x_test_f5 = first5.copy()
        # let us generate text of 10 words after feeding 5 words
        for i in range(10):
            for x,y in zip([x_test_r5,x_test_f5],
                           [y_pred_r5,y_pred_f5]):
                x_input = x.copy()
                feed_dict = {x_p: x_input.reshape(-1, n_x, n_x_vars)}
                y_pred = tfs.run(y_out, feed_dict=feed_dict)
                y_pred_id = int(tf.argmax(y_pred, 1).eval())
                y[i]=y_pred_id
                x[:-1] = x[1:]
                x[-1] = y_pred_id
        print(' Random 5 prediction:',id2string(y_pred_r5))
        print(' First 5 prediction:',id2string(y_pred_f5))
```

Here are the results:

```
Epoch 99, Average loss:1.3972469369570415, Average
accuracy:0.8489583333333334
  Random 5 prediction: labor warren together strongly profits strongly
supported supported co without
  First 5 prediction: market own self free together strongly profits
strongly supported supported

Epoch 199, Average loss:0.7894854595263799, Average
accuracy:0.9186197916666666
  Random 5 prediction: syndicalists spanish class movements also also
anarcho anarcho anarchist was
  First 5 prediction: five civil association class movements also anarcho
anarcho anarcho anarcho

Epoch 299, Average loss:1.360412875811259, Average accuracy:0.865234375
  Random 5 prediction: anarchistic beginnings influenced true tolstoy
tolstoy tolstoy tolstoy tolstoy tolstoy
  First 5 prediction: early civil movement be for was two most most most

Epoch 399, Average loss:1.1692512730757396, Average
accuracy:0.8645833333333334
  Random 5 prediction: including war than than revolutionary than than war
than than
  First 5 prediction: left including including including other other other
other other other

Epoch 499, Average loss:0.5921860883633295, Average accuracy:0.923828125
  Random 5 prediction: ever edited interested interested variety variety
variety variety variety variety
  First 5 prediction: english market herbert strongly price interested
```

```
variety variety variety variety

Epoch 599, Average loss:0.8356450994809469, Average
accuracy:0.8958333333333334
  Random 5 prediction: management allow trabajo trabajo national national
mag mag ricardo ricardo
  First 5 prediction: spain prior am working n war war war self self

Epoch 699, Average loss:0.7057955612738928, Average
accuracy:0.8971354166666666
  Random 5 prediction: teachings can directive tend resist obey
christianity author christianity christianity
  First 5 prediction: early early called social called social social social
social social

Epoch 799, Average loss:0.772875706354777, Average accuracy:0.90234375
  Random 5 prediction: associated war than revolutionary revolutionary
revolutionary than than revolutionary revolutionary
  First 5 prediction: political been hierarchy war than see anti anti anti
anti

Epoch 899, Average loss:0.43675946692625683, Average accuracy:0.9375
  Random 5 prediction: individualist which which individualist warren
warren tucker benjamin how tucker
  First 5 prediction: four at warren individualist warren published
considered considered considered considered

Epoch 999, Average loss:0.23202441136042276, Average
accuracy:0.9602864583333334
  Random 5 prediction: allow allow trabajo you you you you you you you
  First 5 prediction: labour spanish they they they movement movement
anarcho anarcho two
```

The repetitive words in the generated text are the common occurrence and should go away with better training of the model. Although the accuracy of the model improved to 96%, still it was not enough to generate the legible text. Try increasing the number of LSTM cells/hidden layers along with running the model for a large number of epochs on a larger dataset.

Now let's build the same model in Keras:

Text generation LSTM in Keras

You can follow along with the code for this section in the Jupyter notebook `ch-08b_RNN_Text_Keras`.

We implement text generation LSTM in Keras with the following steps:

1. First, we convert all of the data into two tensors, tensor `x` with five columns since we input five words at a time, and tensor `y` with just one column of output. We convert the `y` or label tensor into one-hot encoded representation.

Remember that in practice for large datasets you would use word2vec embeddings instead of one-hot representations.

```
# get the data
x_train, y_train =
text8.seq_to_xy(seq=text8.part['train'],n_tx=n_x,n_ty=n_y)
# reshape input to be [samples, time steps, features]
x_train = x_train.reshape(x_train.shape[0], x_train.shape[1],1)
y_onehot =
np.zeros(shape=[y_train.shape[0],text8.vocab_len],dtype=np.float32)
for i in range(y_train.shape[0]):
    y_onehot[i,y_train[i]]=1
```

2. Next, define the LSTM model with only one hidden LSTM Layer. Since our output is not a sequence, we also set the `return_sequences` to `False`:

```
n_epochs = 1000
batch_size=128
state_size=128
n_epochs_display=100

# create and fit the LSTM model
model = Sequential()
model.add(LSTM(units=state_size,
               input_shape=(x_train.shape[1], x_train.shape[2]),
               return_sequences=False
               )
          )
model.add(Dense(text8.vocab_len))
model.add(Activation('softmax'))
```

```
model.compile(loss='categorical_crossentropy', optimizer='adam')
model.summary()
```

The model looks like the following:

```
Layer (type)                 Output Shape              Param #
=================================================================
lstm_1 (LSTM)                (None, 128)               66560
_________________________________________________________________
dense_1 (Dense)              (None, 1457)              187953
_________________________________________________________________
activation_1 (Activation)    (None, 1457)              0
=================================================================
Total params: 254,513
Trainable params: 254,513
Non-trainable params: 0
_________________________________________________________________
```

3. For Keras, we run a loop to run the 10 times, within each iteration train the model for 100 epochs and print the results of text generation. Here is the full code for training the model and generating the text:

```
for j in range(n_epochs // n_epochs_display):
    model.fit(x_train, y_onehot, epochs=n_epochs_display,
                    batch_size=batch_size,verbose=0)
    # generate text
    y_pred_r5 = np.empty([10])
    y_pred_f5 = np.empty([10])
    x_test_r5 = random5.copy()
    x_test_f5 = first5.copy()
    # let us generate text of 10 words after feeding 5 words
    for i in range(10):
        for x,y in zip([x_test_r5,x_test_f5],
                    [y_pred_r5,y_pred_f5]):
            x_input = x.copy()
            x_input = x_input.reshape(-1, n_x, n_x_vars)
            y_pred = model.predict(x_input)[0]
            y_pred_id = np.argmax(y_pred)
            y[i]=y_pred_id
            x[:-1] = x[1:]
            x[-1] = y_pred_id
    print('Epoch: ',((j+1) * n_epochs_display)-1)
    print(' Random5 prediction:',id2string(y_pred_r5))
    print(' First5 prediction:',id2string(y_pred_f5))
```

4. The output is no surprise, from repeating the words, the model improves a little bit, but could improve further with more LSTM Layers, more data, more training iterations and other hyper-parameter tunings.

```
Random 5 words: free bolshevik be n another
First 5 words: anarchism originated as a term
```

The output of prediction is as follows:

```
Epoch: 99
    Random5 prediction: anarchistic anarchistic wrote wrote wrote
wrote wrote wrote wrote wrote
    First5 prediction: right philosophy than than than than than
than than than

Epoch: 199
    Random5 prediction: anarchistic anarchistic wrote wrote wrote
wrote wrote wrote wrote wrote
    First5 prediction: term i revolutionary than war war french
french french french

Epoch: 299
    Random5 prediction: anarchistic anarchistic wrote wrote wrote
wrote wrote wrote wrote wrote
    First5 prediction: term i revolutionary revolutionary
revolutionary revolutionary revolutionary revolutionary
revolutionary revolutionary

Epoch: 399
    Random5 prediction: anarchistic anarchistic wrote wrote wrote
wrote wrote wrote wrote wrote
    First5 prediction: term i revolutionary labor had had french
french french french

Epoch: 499
    Random5 prediction: anarchistic anarchistic amongst wrote wrote
wrote wrote wrote wrote wrote
    First5 prediction: term i revolutionary labor individualist had
had french french french

Epoch: 599
    Random5 prediction: tolstoy wrote tolstoy wrote wrote wrote
wrote wrote wrote wrote     First5 prediction: term i revolutionary
labor individualist had had had had had

Epoch: 699
    Random5 prediction: tolstoy wrote tolstoy wrote wrote wrote
```

```
wrote wrote wrote wrote      First5 prediction: term i revolutionary
labor individualist had had had had had

Epoch: 799
    Random5 prediction: tolstoy wrote tolstoy tolstoy tolstoy
tolstoy tolstoy tolstoy tolstoy tolstoy
    First5 prediction: term i revolutionary labor individualist had
had had had had

Epoch: 899
    Random5 prediction: tolstoy wrote tolstoy tolstoy tolstoy
tolstoy tolstoy tolstoy tolstoy tolstoy
    First5 prediction: term i revolutionary labor should warren
warren warren warren warren

Epoch: 999
    Random5 prediction: tolstoy wrote tolstoy tolstoy tolstoy
tolstoy tolstoy tolstoy tolstoy tolstoy
    First5 prediction: term i individualist labor should warren
warren warren warren warren
```

If you noticed that we got repetitive words in the output from LSTM models for text generation. Although hyper-parameter and network tuning can eliminate some of the repetitions, there are other methods to solve this problem. The reason we get repetitive words is that the model is always picking the word with the highest probability from a probability distribution of words. This can be changed to pick the words such as to introduce greater variability between consecutive words.

Summary

In this chapter, we learned methods for word embedding to find better representations of textual data elements. As neural networks and deep learning ingest larger amounts of text data, one-hot encoding and other methods of word representation become inefficient. We also learned how to visualize word embedding using t-SNE plots. We used a simple LSTM model to generate the text in TensorFlow and Keras. Similar concepts can be applied to various other tasks, such as sentiment analysis, question answering, and neural machine translation.

Before we dive deeper into advanced TensorFlow features such as Transfer Learning, Reinforcement Learning, Generative Networks, and Distributed TensorFlow, in the next chapter, we shall see how to take TensorFlow models into production.

9

CNN with TensorFlow and Keras

Convolutional Neural Network (**CNN**) is a special kind of feed-forward neural network that includes convolutional and pooling layers in its architecture. Also known as ConvNets, the general pattern for the CNN architecture is to have these layers in the following sequence:

1. Fully connected input layer
2. Multiple combinations of convolutional, pooling, and fully connected layers
3. Fully connected output layer with softmax activation

CNN architectures have proven to be highly successful in solving problems that involve learning from images, such as image recognition and object identification.

In this chapter, we shall learn the following topics related to ConvNets:

- Understanding Convolution
- Understanding Pooling
- CNN architecture pattern-LeNet
- LeNet for MNIST dataset
    - LeNet for MNIST with TensorFlow
    - LeNet for MNIST with Keras
- LeNet for CIFAR dataset
    - LeNet CNN for CIFAR10 with TensorFlow
    - LeNet CNN for CIFAR10 with Keras

Let us start by learning the core concepts behind the ConvNets.

Understanding convolution

Convolution is the central concept behind the CNN architecture. In simple terms, convolution is a mathematical operation that combines information from two sources to produce a new set of information. Specifically, it applies a special matrix known as the *kernel* to the input tensor to produce a set of matrices known as the *feature maps*. The kernel can be applied to the input tensor using any of the popular algorithms.

The most commonly used algorithm to produce the convolved matrix is as follows:

```
N_STRIDES = [1,1]
1. Overlap the kernel with the top-left cells of the image matrix.
2. Repeat while the kernel overlaps the image matrix:
    2.1 c_col = 0
    2.2 Repeat while the kernel overlaps the image matrix:
        2.1.1 set c_row = 0
        2.1.2 convolved_scalar = scalar_prod(kernel, overlapped cells)
        2.1.3 convolved_matrix(c_row,c_col) = convolved_scalar
        2.1.4 Slide the kernel down by N_STRIDES[0] rows.
        2.1.5 c_row = c_row  + 1
    2.3 Slide the kernel to (topmost row, N_STRIDES[1] columns right)
    2.4 c_col = c_col + 1
```

For example, let us assume the kernel matrix is a 2 x 2 matrix, and the input image is a 3 x 3 matrix. The following diagrams show the above algorithm step by step:

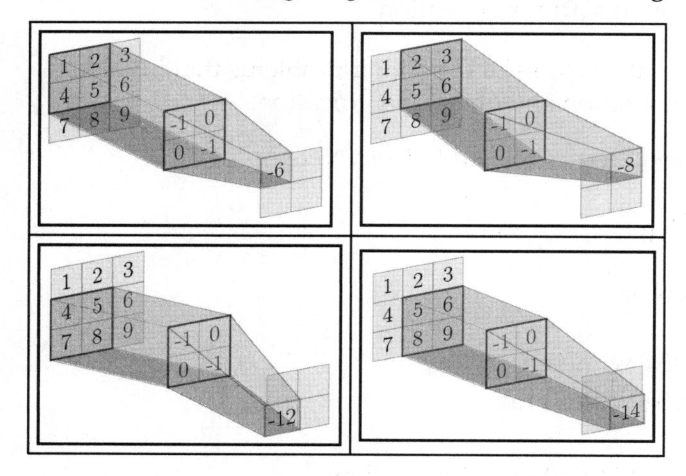

At the end of the convolution operation we get the following feature map:

-6	-8
-12	-14

In the example above the resulting feature map is smaller in size as compared to the original input to the convolution. Generally, the size of the feature maps gets reduced by (kernel size-1). Thus the size of feature map is :

$$size_{feature\_map} = size_{features} - size_{kernel} + 1$$

The 3-D Tensor

For 3-D tensors with an additional depth dimension, you can think of the preceding algorithm being applied to each layer in the depth dimension. The output of applying convolution to a 3D tensor is also a 2D tensor as convolution operation adds the three channels.

The Strides

The ***strides*** in array N_STRIDES is the number the rows or columns by which you want to slide the kernel across. In our example, we used a stride of 1. If we use a higher number of strides, then the size of the feature map gets reduced further as per the following equation:

$$size_{feature\_map} = \frac{size_{features} - size_{kernel}}{n_{strides}} + 1$$

The Padding

If we do not wish to reduce the size of the feature map, then we can use padding on all sides of the input such that the size of features is increased by double of the padding size. With padding, the size of the feature map can be calculated as follows:

$$size_{feature\_map} = \frac{size_{features} + 2 * size_{padding} - size_{kernel}}{n_{strides}} + 1$$

TensorFlow allows two kinds of padding: SAME or VALID. The SAME padding means to add a padding such that the output feature map has the same size as input features. VALID padding means no padding.

The result of applying the previously-mentioned convolution algorithm is the feature map which is the filtered version of the original tensor. For example, the feature map could have only the outlines filtered from the original image. Hence, the kernel is also known as the filter. For each kernel, you get a separate 2D feature map.

Depending on which features you want the network to learn, you have to apply the appropriate filters to emphasize the required features. However, with CNN, the model can automatically learn which kernels work best in the convolution layer.

Convolution Operation in TensorFlow

TensorFlow provides the convolutional layers that implement the convolution algorithm. For example, the `tf.nn.conv2d()` operation with the following signature:

```
tf.nn.conv2d(
  input,
  filter,
  strides,
  padding,
  use_cudnn_on_gpu=None,
  data_format=None,
  name=None
)
```

`input` and `filter` represent the data tensor of the shape `[batch_size, input_height, input_width, input_depth]` and kernel tensor of the shape `[filter_height, filter_width, input_depth, output_depth]`. The `output_depth` in he kernel tensor represents the number of kernels that should be applied to the input. The `strides` tensor represents the number of cells to slide in each dimension. The `padding` is VALID or SAME as described above.

You can find more information on convolution operations available in TensorFlow at the following link: `https://www.tensorflow.org/api_guides/python/nn#Convolution`

You can find more information on convolution layers available in Keras at the following link: `https://keras.io/layers/convolutional/`

The following links provide a detailed mathematical explanation of convolution:
`http://colah.github.io/posts/2014-07-Understanding-Convolutions/`
`http://ufldl.stanford.edu/tutorial/supervised/FeatureExtractionUsingConvolution/`

The convolution layer or operation connects the input values or neurons to the next hidden layer neurons. Each hidden layer neuron is connected to the same number of input neurons as the number of elements in the kernel. So in our previous example, the kernel has 4 elements, thus the hidden layer neuron is connected to 4 neurons (out of the 3 x 3 neurons) of the input layer. This area of 4 neurons of the input layer in our example is known as the **receptive field** in CNN theory.

The convolution layer has the separate weights and bias parameters for each kernel. The number of weight parameters is equal to the number of elements in the kernel, and only one bias parameter. All connections for the kernel share the same weights and bias parameters. Thus in our example, there would be 4 weight parameters and 1 bias parameter, but if we use 5 kernels in our convolution layer, then there would be total of 5 x 4 weight parameters and 5 x 1 bias parameters, a set of (4 weights, 1 bias) parameters for each feature map.

Understanding pooling

Generally, in the convolution operation several different kernels are applied that result in generation of several feature maps. Thus, the convolution operation results in generating a large sized dataset.

As an example, applying a kernel of shape 3 x 3 x 1 to an MNIST dataset that has images of shape 28 x 28 x 1 pixels, produces a feature map of shape 26 x 26 x 1. If we apply 32 such filters in a convolutional layer, then the output will be of shape 32 x 26 x 26 x 1, that is, 32 feature maps of shape 26 x 26 x 1.

This is a huge dataset as compared to the original dataset of shape 28 x 28 x 1. Thus, to simplify the learning for the next layer, we apply the concept of *pooling*.

Pooling refers to calculating the aggregate statistic over the regions of the convolved feature space. Two most popular aggregate statistics are the maximum and the average. The output of applying max-pooling is the maximum of the region selected, while the output of applying the average-pooling is the mean of the numbers in the region.

As an example, let us say the feature map is of shape 3 x 3 and the pooling region of shape 2 x 2. The following images show the max pool operation applied with a stride of [1,1]:

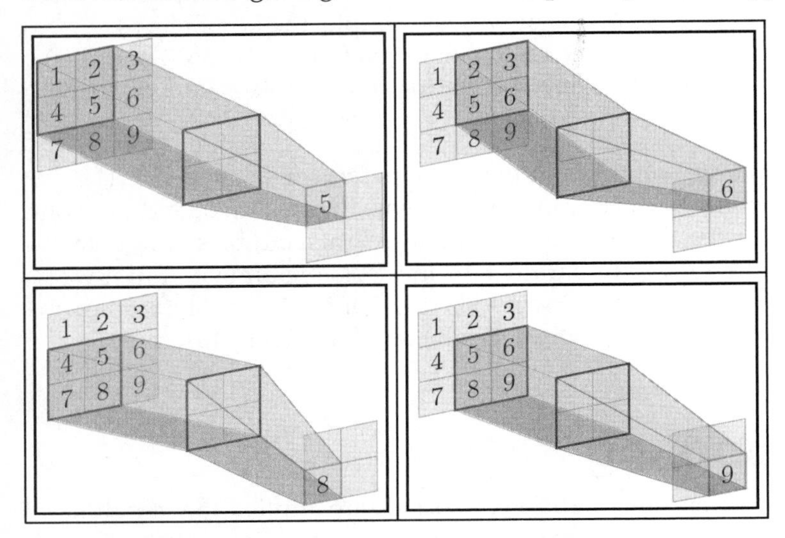

At the end of the max pool operation we get the following matrix:

5	6
8	9

Generally, the pooling operation is applied with non-overlapping regions, thus the stride tensor and the region tensor are set to the same values.

As an example, TensorFlow has the `max_pooling` operation with the following signature:

```
max_pool(
  value,
  ksize,
  strides,
  padding,
  data_format='NHWC',
  name=None
)
```

`value` represents the input tensor of the shape `[batch_size, input_height, input_width, input_depth]`. The pooling operation is performed on rectangular regions of shape `ksize`. These regions are offset by the shape `strides`.

You can find more information on the pooling operations available in TensorFlow at the following link: https://www.tensorflow.org/api_guides/python/nn#Pooling

Find more information on the pooling layers available in Keras at the following link: https://keras.io/layers/pooling/

The following link provides a detailed mathematical explanation of the pooling: http://ufldl.stanford.edu/tutorial/supervised/Pooling/

CNN architecture pattern - LeNet

LeNet is a popular architectural pattern for implementing CNN. In this chapter, we shall learn to build CNN model based on LeNet pattern by creating the layers in the following sequence:

1. The input layer
2. The convolutional layer 1 that produces a set of feature maps, with ReLU activation
3. The pooling layer 1 that produces a set of statistically aggregated feature maps
4. The convolutional layer 2 that produces a set of feature maps, with ReLU activation
5. The pooling layer 2 that produces a set of statistically aggregated feature maps
6. The fully connected layer that flattens the feature maps, with ReLU activation
7. The output layer that produces the output by applying simple linear activation

LeNet family of models were introduced by Yann LeCun and his fellow researchers. More details on the LeNet family of models can be found at the following link: `http://yann.lecun.com/exdb/publis/pdf/lecun-01a.pdf`.

Yann LeCun maintains a list of the LeNet family of models at the following link: `http://yann.lecun.com/exdb/lenet/index.html`.

LeNet for MNIST data

You can follow along with the code in the Jupyter notebook `ch-09a_CNN_MNIST_TF_and_Keras`.

Prepare the MNIST data into test and train sets:

```
from tensorflow.examples.tutorials.mnist import input_data
mnist = input_data.read_data_sets(os.path.join('.','mnist'), one_hot=True)
X_train = mnist.train.images
X_test = mnist.test.images
Y_train = mnist.train.labels
Y_test = mnist.test.labels
```

LeNet CNN for MNIST with TensorFlow

In TensorFlow, apply the following steps to build the LeNet based CNN models for MNIST data:

1. Define the hyper-parameters, and the placeholders for x and y (input images and output labels):

```
n_classes = 10 # 0-9 digits
n_width = 28
n_height = 28
n_depth = 1
n_inputs = n_height * n_width * n_depth # total pixels
learning_rate = 0.001
n_epochs = 10
batch_size = 100
n_batches = int(mnist.train.num_examples/batch_size)
```

```
# input images shape: (n_samples,n_pixels)
x = tf.placeholder(dtype=tf.float32, name="x", shape=[None,
n_inputs])
# output labels
y = tf.placeholder(dtype=tf.float32, name="y", shape=[None,
n_classes])
```

Reshape the input x to shape (`n_samples`, `n_width`, `n_height`, `n_depth`):

```
x_ = tf.reshape(x, shape=[-1, n_width, n_height, n_depth])
```

2. Define the first convolutional layer with 32 kernels of shape 4 x 4, thus producing 32 feature maps.

- First, define the weights and biases for the first convolutional layer. We populate the parameters with the normal distribution:

```
layer1_w = tf.Variable(tf.random_normal(shape=[4,4,n_depth,32],
             stddev=0.1),name='l1_w')
layer1_b = tf.Variable(tf.random_normal([32]),name='l1_b')
```

- Next, define the convolutional layer with the `tf.nn.conv2d` function. The function argument `stride` defines the elements by which the kernel tensor should slide in each dimension. The dimension order is determined by `data_format`, which could be either `'NHWC'` or `'NCHW'` (by default, `'NHWC'`). Generally, the first and last element in `stride` is set to '1'. The function argument `padding` could be `SAME` or `VALID`. `SAME padding` means that the input would be padded with zeroes such that after convolution the output is of the same shape as the input. Add the `relu` activation using the `tf.nn.relu()` function:

```
layer1_conv = tf.nn.relu(tf.nn.conv2d(x_,layer1_w,
                                       strides=[1,1,1,1],
                                       padding='SAME'
                                      ) +
                         layer1_b
                        )
```

- Define the first pooling layer with the `tf.nn.max_pool()` function. The argument `ksize` represents the pooling operation using 2 x 2 x 1 regions, and the argument `stride` represents to slide the regions by 2 x 2 x 1 pixels. Thus the regions do not overlap with each other. Since we use `max_pool`, pooling operation selects the maximum in 2 x 2 x 1 regions:

```
layer1_pool = tf.nn.max_pool(layer1_conv,ksize=[1,2,2,1],
                    strides=[1,2,2,1],padding='SAME')
```

The first convolution layer produces 32 feature maps of size 28 x 28 x 1, which are then pooled into data of shape 32 x 14 x 14 x 1.

3. Define the second convolutional layer that takes this data as input and produces 64 feature maps.

- First, define the weights and biases for the second convolutional layer. We populate the parameters with a normal distribution:

```
layer2_w = tf.Variable(tf.random_normal(shape=[4,4,32,64],
           stddev=0.1),name='l2_w')
layer2_b = tf.Variable(tf.random_normal([64]),name='l2_b')
```

- Next, define the convolutional layer with the `tf.nn.conv2d` function:

```
layer2_conv = tf.nn.relu(tf.nn.conv2d(layer1_pool,
                                      layer2_w,
                                      strides=[1,1,1,1],
                                      padding='SAME'
                                     ) +
                         layer2_b
                        )
```

- Define the second pooling layer with the `tf.nn.max_pool` function:

```
layer2_pool = tf.nn.max_pool(layer2_conv,
                             ksize=[1,2,2,1],
                             strides=[1,2,2,1],
                             padding='SAME'
                            )
```

The output of the second convolution layer is of shape 64 x 14 x 14 x 1, which then gets pooled into an output of shape 64 x 7 x 7 x 1.

4. Reshape this output before feeding into the fully connected layer of 1024 neurons to produce a flattened output of size 1024:

```
layer3_w = tf.Variable(tf.random_normal(shape=[64*7*7*1,1024],
                       stddev=0.1),name='l3_w')
layer3_b = tf.Variable(tf.random_normal([1024]),name='l3_b')
layer3_fc = tf.nn.relu(tf.matmul(tf.reshape(layer2_pool,
            [-1, 64*7*7*1]),layer3_w) + layer3_b)
```

5. The output of the fully connected layer is fed into a linear output layer with 10 outputs. We did not use softmax in this layer because our loss function automatically applies the softmax to the output:

```
layer4_w = tf.Variable(tf.random_normal(shape=[1024, n_classes],
                                  stddev=0.1),name='l)
layer4_b = tf.Variable(tf.random_normal([n_classes]),name='l4_b')
layer4_out = tf.matmul(layer3_fc,layer4_w)+layer4_b
```

This creates our first CNN model that we save in the variable `model`:

```
model = layer4_out
```

The reader is encouraged to explore different convolutional and pooling operators available in TensorFlow with different hyper-parameter values.

For defining the loss, we use the `tf.nn.softmax_cross_entropy_with_logits` function, and for optimizer, we use the `AdamOptimizer` function. You should try to explore the different optimizer functions available in TensorFlow.

```
entropy = tf.nn.softmax_cross_entropy_with_logits(logits=model, labels=y)
loss = tf.reduce_mean(entropy)
optimizer = tf.train.AdamOptimizer(learning_rate).minimize(loss)
```

Finally, we train the model by iterating over `n_epochs`, and within each epoch train over `n_batches`, each batch of the size of `batch_size`:

```
with tf.Session() as tfs:
    tf.global_variables_initializer().run()
    for epoch in range(n_epochs):
        total_loss = 0.0
        for batch in range(n_batches):
            batch_x,batch_y = mnist.train.next_batch(batch_size)
            feed_dict={x:batch_x, y: batch_y}
            batch_loss,_ = tfs.run([loss, optimizer],
```

```
                                         feed_dict=feed_dict)
            total_loss += batch_loss
        average_loss = total_loss / n_batches
        print("Epoch: {0:04d} loss = {1:0.6f}".format(epoch,average_loss))
    print("Model Trained.")

    predictions_check = tf.equal(tf.argmax(model,1),tf.argmax(y,1))
    accuracy = tf.reduce_mean(tf.cast(predictions_check, tf.float32))
    feed_dict = {x:mnist.test.images, y:mnist.test.labels}
    print("Accuracy:", accuracy.eval(feed_dict=feed_dict))
```

We get the following output:

```
Epoch: 0000   loss = 1.418295
Epoch: 0001   loss = 0.088259
Epoch: 0002   loss = 0.055410
Epoch: 0003   loss = 0.042798
Epoch: 0004   loss = 0.030471
Epoch: 0005   loss = 0.023837
Epoch: 0006   loss = 0.019800
Epoch: 0007   loss = 0.015900
Epoch: 0008   loss = 0.012918
Epoch: 0009   loss = 0.010322
Model Trained.
Accuracy: 0.9884
```

Now that is some pretty good accuracy as compared to the approaches we have seen in the previous chapters so far. Aren't CNN models almost magical when it comes to learning from the image data?

LeNet CNN for MNIST with Keras

Let us revisit the same LeNet architecture with the same dataset to build and train the CNN model in Keras:

1. Import the required Keras modules:

```
import keras
from keras.models import Sequential
from keras.layers import Conv2D,MaxPooling2D, Dense, Flatten,
Reshape
from keras.optimizers import SGD
```

2. Define the number of filters for each layer:

```
n_filters=[32,64]
```

3. Define other hyper-parameters:

```
learning_rate = 0.01
n_epochs = 10
batch_size = 100
```

4. Define the sequential model and add the layer to reshape the input data to shape `(n_width,n_height,n_depth)`:

```
model = Sequential()
model.add(Reshape(target_shape=(n_width,n_height,n_depth),
                  input_shape=(n_inputs,))
         )
```

5. Add the first convolutional layer with 4 x 4 kernel filter, `SAME` padding and `relu` activation:

```
model.add(Conv2D(filters=n_filters[0],kernel_size=4,
                 padding='SAME',activation='relu')
         )
```

6. Add the pooling layer with region size of 2 x 2 and stride of 2 x 2:

```
model.add(MaxPooling2D(pool_size=(2,2),strides=(2,2)))
```

7. Add the second convolutional and pooling layer in the same way as we added the first layer:

```
model.add(Conv2D(filters=n_filters[1],kernel_size=4,
                 padding='SAME',activation='relu')
         )
model.add(MaxPooling2D(pool_size=(2,2),strides=(2,2)))
```

8. Add a layer to flatten the output of the second pooling layer and a fully connected layer of 1024 neurons to handle the flattened output:

```
model.add(Flatten())
model.add(Dense(units=1024, activation='relu'))
```

9. Add the final output layer with the `softmax` activation:

```
model.add(Dense(units=n_outputs, activation='softmax'))
```

10. See the model summary with the following code:

```
model.summary()
```

The model is described as follows:

```
Layer (type)                 Output Shape              Param #
=================================================================
reshape_1 (Reshape)          (None, 28, 28, 1)         0
_________________________________________________________________
conv2d_1 (Conv2D)            (None, 28, 28, 32)        544
_________________________________________________________________
max_pooling2d_1 (MaxPooling2 (None, 14, 14, 32)        0
_________________________________________________________________
conv2d_2 (Conv2D)            (None, 14, 14, 64)        32832
_________________________________________________________________
max_pooling2d_2 (MaxPooling2 (None, 7, 7, 64)          0
_________________________________________________________________
flatten_1 (Flatten)          (None, 3136)              0
_________________________________________________________________
dense_1 (Dense)              (None, 1024)              3212288
_________________________________________________________________
dense_2 (Dense)              (None, 10)                10250
=================================================================
Total params: 3,255,914
Trainable params: 3,255,914
Non-trainable params: 0
_________________________________________________________________
```

11. Compile, train, and evaluate the model:

```
model.compile(loss='categorical_crossentropy',
              optimizer=SGD(lr=learning_rate),
              metrics=['accuracy'])
model.fit(X_train, Y_train,batch_size=batch_size,
          epochs=n_epochs)
score = model.evaluate(X_test, Y_test)
print('\nTest loss:', score[0])
print('Test accuracy:', score[1])
```

We get the following output:

```
Epoch 1/10
55000/55000 [====================] - 267s - loss: 0.8854 - acc: 0.7631
Epoch 2/10
55000/55000 [====================] - 272s - loss: 0.2406 - acc: 0.9272
Epoch 3/10
55000/55000 [====================] - 267s - loss: 0.1712 - acc: 0.9488
Epoch 4/10
55000/55000 [====================] - 295s - loss: 0.1339 - acc: 0.9604
Epoch 5/10
55000/55000 [====================] - 278s - loss: 0.1112 - acc: 0.9667
Epoch 6/10
55000/55000 [====================] - 279s - loss: 0.0957 - acc: 0.9714
Epoch 7/10
55000/55000 [====================] - 316s - loss: 0.0842 - acc: 0.9744
Epoch 8/10
55000/55000 [====================] - 317s - loss: 0.0758 - acc: 0.9773
Epoch 9/10
55000/55000 [====================] - 285s - loss: 0.0693 - acc: 0.9790
Epoch 10/10
55000/55000 [====================] - 217s - loss: 0.0630 - acc: 0.9804
Test loss: 0.0628845927377
Test accuracy: 0.9785
```

The difference in accuracy could be attributed to the fact that we used SGD optimizer here, which does not implement some of the advanced features provided by AdamOptimizer we used for the TensorFlow model.

LeNet for CIFAR10 Data

Now that we have learned to build and train the CNN model using MNIST data set with TensorFlow and Keras, let us repeat the exercise with CIFAR10 dataset.

The CIFAR-10 dataset consists of 60,000 RGB color images of the shape 32x32 pixels. The images are equally divided into 10 different categories or classes: airplane, automobile, bird, cat, deer, dog, frog, horse, ship, and truck. CIFAR-10 and CIFAR-100 are subsets of a large image dataset comprising of 80 million images. The CIFAR data sets were collected and labelled by Alex Krizhevsky, Vinod Nair, and Geoffrey Hinton. The numbers 10 and 100 represent the number of classes of images.

More details about the CIFAR dataset are available at the following links: `http://www.cs.toronto.edu/~kriz/cifar.html` and `http://www.cs.toronto.edu/~kriz/learning-features-2009-TR.pdf`.

We picked CIFAR 10, since it has 3 channels, i.e. the depth of the images is 3, while the MNIST data set had only one channel. For the sake of brevity, we leave out the details to download and split the data into training and test set and provide the code in the datasetslib package in the code bundle for this book.

You can follow along with the code in the Jupyter notebook `ch-09b_CNN_CIFAR10_TF_and_Keras`.

We load and preprocess the CIFAR10 data using the following code:

```
from datasetslib.cifar import cifar10
from datasetslib import imutil
dataset = cifar10()
dataset.x_layout=imutil.LAYOUT_NHWC
dataset.load_data()
dataset.scaleX()
```

The data is loaded such that the images are in the `'NHWC'` format, that makes the data variable of shape (`number_of_samples, image_height, image_width, image_channels`). We refer to image channels as image depth. Each pixel in the images is a number from 0 to 255. The dataset is scaled using MinMax scaling to normalize the images by dividing all pixel values with 255.

The loaded and pre-processed data becomes available in the dataset object variables as `dataset.X_train`, `dataset.Y_train`, `dataset.X_test`, and `dataset.Y_test`.

ConvNets for CIFAR10 with TensorFlow

We keep the layers, filters, and their sizes the same as in the MNIST examples earlier, with one new addition of regularization layer. Since this data set is complex as compared to the MNIST, we add additional dropout layers for the purpose of regularization:

```
tf.nn.dropout(layer1_pool, keep_prob)
```

The placeholder `keep_prob` is set to 1 during prediction and evaluation. That way we can reuse the same model for training as well as prediction and evaluation.

The complete code for LeNet model for CIFAR10 data is provided in the notebook `ch-09b_CNN_CIFAR10_TF_and_Keras`.

On running the model we get the following output:

```
Epoch: 0000   loss = 2.115784
Epoch: 0001   loss = 1.620117
Epoch: 0002   loss = 1.417657
Epoch: 0003   loss = 1.284346
Epoch: 0004   loss = 1.164068
Epoch: 0005   loss = 1.058837
Epoch: 0006   loss = 0.953583
Epoch: 0007   loss = 0.853759
Epoch: 0008   loss = 0.758431
Epoch: 0009   loss = 0.663844
Epoch: 0010   loss = 0.574547
Epoch: 0011   loss = 0.489902
Epoch: 0012   loss = 0.410211
Epoch: 0013   loss = 0.342640
Epoch: 0014   loss = 0.280877
Epoch: 0015   loss = 0.234057
Epoch: 0016   loss = 0.195667
Epoch: 0017   loss = 0.161439
Epoch: 0018   loss = 0.140618
Epoch: 0019   loss = 0.126363
Model Trained.
Accuracy: 0.6361
```

We did not get good accuracy as compared to the accuracy we achieved on the MNIST data. It is possible to achieve much better accuracy by tuning the different hyper-parameters and varying the combinations of convolutional and pooling layers. We leave it as a challenge for the reader to explore and try different variations of the LeNet architecture and hyper-parameters to achieve better accuracy.

ConvNets for CIFAR10 with Keras

Let us repeat the LeNet CNN model building and training for CIFAR10 data in Keras. We keep the architecture same as previous examples in order to explain the concepts easily. In Keras, the dropout layer is added as follows:

```
model.add(Dropout(0.2))
```

The complete code in Keras for CIFAR10 CNN model is provided in the notebook `ch-09b_CNN_CIFAR10_TF_and_Keras`.

On running the model we get the following model description:

```
_________________________________________________________________
Layer (type)                 Output Shape              Param #
=================================================================
conv2d_1 (Conv2D)            (None, 32, 32, 32)        1568
_________________________________________________________________
max_pooling2d_1 (MaxPooling2 (None, 16, 16, 32)        0
_________________________________________________________________
dropout_1 (Dropout)          (None, 16, 16, 32)        0
_________________________________________________________________
conv2d_2 (Conv2D)            (None, 16, 16, 64)        32832
_________________________________________________________________
max_pooling2d_2 (MaxPooling2 (None, 8, 8, 64)          0
_________________________________________________________________
dropout_2 (Dropout)          (None, 8, 8, 64)          0
_________________________________________________________________
flatten_1 (Flatten)          (None, 4096)              0
_________________________________________________________________
dense_1 (Dense)              (None, 1024)              4195328
_________________________________________________________________
dropout_3 (Dropout)          (None, 1024)              0
_________________________________________________________________
dense_2 (Dense)              (None, 10)                10250
=================================================================
Total params: 4,239,978
Trainable params: 4,239,978
Non-trainable params: 0
_________________________________________________________________
```

We get the following training and evaluation output:

```
Epoch 1/10
50000/50000 [====================] - 191s - loss: 1.5847 - acc: 0.4364
Epoch 2/10
50000/50000 [====================] - 202s - loss: 1.1491 - acc: 0.5973
Epoch 3/10
50000/50000 [====================] - 223s - loss: 0.9838 - acc: 0.6582
Epoch 4/10
50000/50000 [====================] - 223s - loss: 0.8612 - acc: 0.7009
Epoch 5/10
50000/50000 [====================] - 224s - loss: 0.7564 - acc: 0.7394
Epoch 6/10
50000/50000 [====================] - 217s - loss: 0.6690 - acc: 0.7710
Epoch 7/10
50000/50000 [====================] - 222s - loss: 0.5925 - acc: 0.7945
Epoch 8/10
50000/50000 [====================] - 221s - loss: 0.5263 - acc: 0.8191
Epoch 9/10
50000/50000 [====================] - 237s - loss: 0.4692 - acc: 0.8387
Epoch 10/10
50000/50000 [====================] - 230s - loss: 0.4320 - acc: 0.8528
Test loss: 0.849927025414
Test accuracy: 0.7414
```

Once again, we leave it as a challenge for the reader to explore and try different variations of the LeNet architecture and hyper-parameters to achieve better accuracy.

Summary

In this chapter, we learned how to create convolutional neural networks with TensorFlow and Keras. We learned the core concepts of convolution and pooling, that lay the foundation of CNN. We learned the LeNet family of architectures and created, trained, and evaluated the LeNet family model for MNIST and CIFAR datasets. TensorFlow and Keras offer many convolutional and pooling layers and operations. The reader is encouraged to explore the layers and operations that were not covered in this chapter.

In the next chapter, we shall continue our journey to learn how to apply TensorFlow on image data with the AutoEncoder architecture.

10
Autoencoder with TensorFlow and Keras

Autoencoder is a neural network architecture that is often associated with unsupervised learning, dimensionality reduction, and data compression. Autoencoders learn to produce the same output as given to the input layer by using lesser number of neurons in the hidden layers. This allows hidden layers to learn the features of input with lesser number of parameters. This process of using lesser number of neurons to learn the features of the input data, in turn, reduces the dimensionality of the input dataset.

An autoencoder architecture has two stages: encoder and decoder. In the encoder stage, the model learns to represent the input to a compressed vector with lesser dimensions, and in the decoder stage, the model learns to represent the compressed vector to an output vector. The loss is calculated as entropy distance between the output and input, thus by minimizing the loss, we learn parameters that encode the input into a representation that is capable of producing the input back, with yet another set of learned parameters.

In this chapter, you will learn how to use TensorFlow and Keras to create autoencoder architectures in the following topics:

- Autoencoder types
- Stacked autoencoder in TensorFlow and Keras
- Denoising autoencoder in TensorFlow and Keras
- Variational autoencoder in TensorFlow and Keras

Autoencoder types

Autoencoder architectures can be found in a variety of configurations such as simple autoencoders, sparse autoencoders, denoising autoencoders, and convolutional autoencoders.

- **Simple autoencoder:** In simple autoencoder, the hidden layers have lesser number of nodes or neurons as compared to the input. For example, in the MNIST dataset, an input of 784 features can be connected to the hidden layer of 512 nodes or 256 nodes, which is connected to the 784-feature output layer. Thus, during training, the 784 features would be learned by only 256 nodes. Simple autoencoders are also known as *undercomplete* autoencoders.

  Simple autoencoder could be single-layer or multi-layer. Generally, single-layer autoencoder does not perform very good in production. Multi-layer autoencoder has more than one hidden layer, divided into encoder and decoder groupings. Encoder layers encode a large number of features into a smaller number of neurons, and decoder layers then decode the learned compressed features back into the original or a reduced number of features. Multi-layer autoencoder is known as the **stacked autoencoder**.
- **Sparse autoencoder**: In sparse autoencoder, a regularization term is added as the penalty and hence, the representation becomes more sparse as compared to simple autoencoders.
- **Denoising autoencoder** (DAE): In the DAE architecture, the input is introduced with stochastic noise. The DAE recreates the input and attempts to remove noise. The loss function in the DAE compares the denoised recreated output to the original uncorrupted input.
- **Convolutional autoencoder** (CAE): The autoencoders discussed previously use fully-connected layers, a pattern similar to the multilayer perceptron models. We can also use convolutional layers instead of fully connected or dense layers. When we use convolutional layers to create an autoencoder, it is known as a convolutional autoencoder. As an example, we could have the following layers for the CAE:

  input -> convolution -> pooling -> convolution -> pooling -> output

  The first set of convolution and pooling layers acts as the encoder, reducing the high-dimensional input feature space to low dimensional feature space. The second set of convolutional and pooling layers acts as the decoder, converting it back to high-dimensional feature space.

- **Variational autoencoder** (VAE): The variational autoencoder architecture is the latest development in the field of autoencoders. VAE is a kind of generative model, that is, it produces parameters of the probability distribution from which the original data or the data very similar to original data can be generated.

  In a VAE, the encoder turns the input samples into parameters in latent space using which the latent points are sampled. The decoder then uses the latent points to regenerate the original input data. Hence, the focus of learning in VAE shifts to maximizing the probability of the input data in place of trying to recreate the output from input.

Now let's build autoencoders in TensorFlow and Keras in the following sections. We will use the MNIST dataset to build the autoencoders. The autoencoder will learn to represent the handwritten digits of MNIST dataset with a lesser number of neurons or features.

You can follow along with the code in the Jupyter notebook `ch-10_AutoEncoders_TF_and_Keras`.

As usual, we first read the MNIST dataset with the following code:

```
from tensorflow.examples.tutorials.mnist.input_data import input_data
dataset_home = os.path.join(datasetslib.datasets_root,'mnist')
mnist = input_data.read_data_sets(dataset_home,one_hot=False)

X_train = mnist.train.images
X_test = mnist.test.images
Y_train = mnist.train.labels
Y_test = mnist.test.labels

pixel_size = 28
```

We extract four distinct images and their respective labels from training and test datasets:

```
while True:
     train_images,train_labels = mnist.train.next_batch(4)
     if len(set(train_labels))==4:
        break
while True:
     test_images,test_labels = mnist.test.next_batch(4)
     if len(set(test_labels))==4:
        break
```

Now let's look at the code to build an autoencoder using the MNIST dataset.

You can follow along with the code in the Jupyter notebook `ch-10_AutoEncoders_TF_and_Keras`.

Stacked autoencoder in TensorFlow

The steps to build a stacked autoencoder model in TensorFlow are as follows:

1. First, define the hyper-parameters as follows:

```
learning_rate = 0.001
n_epochs = 20
batch_size = 100
n_batches = int(mnist.train.num_examples/batch_size)
```

2. Define the number of inputs (that is, features) and outputs (that is, targets). The number of outputs will be the same as the number of inputs:

```
# number of pixels in the MNIST image as number of inputs
n_inputs = 784
n_outputs = n_inputs
```

3. Define the placeholders for input and output images:

```
x = tf.placeholder(dtype=tf.float32, name="x", shape=[None,
n_inputs])
y = tf.placeholder(dtype=tf.float32, name="y", shape=[None,
n_outputs])
```

4. Add the number of neurons for encoder and decoder layers as `[512,256,256,512]`:

```
# number of hidden layers
n_layers = 2
# neurons in each hidden layer
n_neurons = [512,256]
# add number of decoder layers:
n_neurons.extend(list(reversed(n_neurons)))
n_layers = n_layers * 2
```

5. Define the *w* and *b* parameters:

```
w=[]
b=[]

for i in range(n_layers):
    w.append(tf.Variable(tf.random_normal([n_inputs \
                      if i==0 else n_neurons[i-1],n_neurons[i]]),
                                 name="w_{0:04d}".format(i)
                                )
                  )
    b.append(tf.Variable(tf.zeros([n_neurons[i]]),
                                 name="b_{0:04d}".format(i)
                                )
                  )
w.append(tf.Variable(tf.random_normal([n_neurons[n_layers-1] \
                          if n_layers > 0 else n_inputs,n_outputs]),
                          name="w_out"
                         )
            )
b.append(tf.Variable(tf.zeros([n_outputs]),name="b_out"))
```

6. Build the network and use the sigmoid activation function for each layer:

```
# x is input layer
layer = x
# add hidden layers
for i in range(n_layers):
layer = tf.nn.sigmoid(tf.matmul(layer, w[i]) + b[i])
# add output layer
layer = tf.nn.sigmoid(tf.matmul(layer, w[n_layers]) + b[n_layers])
model = layer
```

7. Define the `loss` function using `mean_squared_error` and the `optimizer` function using `AdamOptimizer`:

```
mse = tf.losses.mean_squared_error
loss = mse(predictions=model, labels=y)
optimizer = tf.train.AdamOptimizer(learning_rate=learning_rate)
optimizer = optimizer.minimize(loss)
```

8. Train the model and predict the images for the `train` as well as `test` sets:

```
with tf.Session() as tfs:
    tf.global_variables_initializer().run()
    for epoch in range(n_epochs):
        epoch_loss = 0.0
        for batch in range(n_batches):
            X_batch, _ = mnist.train.next_batch(batch_size)
            feed_dict={x: X_batch,y: X_batch}
            _,batch_loss = tfs.run([optimizer,loss], feed_dict)
            epoch_loss += batch_loss
        if (epoch%10==9) or (epoch==0):
            average_loss = epoch_loss / n_batches
            print('epoch: {0:04d} loss = {1:0.6f}'
                  .format(epoch,average_loss))
    # predict images using trained autoencoder model
    Y_train_pred = tfs.run(model, feed_dict={x: train_images})
    Y_test_pred = tfs.run(model, feed_dict={x: test_images})
```

9. We see the following output as the loss reduces significantly after 20 epochs:

```
epoch: 0000   loss = 0.156696
epoch: 0009   loss = 0.091367
epoch: 0019   loss = 0.078550
```

10. Now that the model is trained, let's display the predicted images from the trained model. We wrote a helper function `display_images` to help us display images:

```
import random

# Function to display the images and labels
# images should be in NHW or NHWC format
def display_images(images, labels, count=0, one_hot=False):
    # if number of images to display is not provided, then display
all the images
    if (count==0):
        count = images.shape[0]

    idx_list = random.sample(range(len(labels)),count)
    for i in range(count):
        plt.subplot(4, 4, i+1)
        plt.title(labels[i])
        plt.imshow(images[i])
        plt.axis('off')
    plt.tight_layout()
    plt.show()
```

Using this function, we first display the four images from the training set and the images predicted by the autoencoder.

The first row indicates the actual images and second row indicates the generated images:

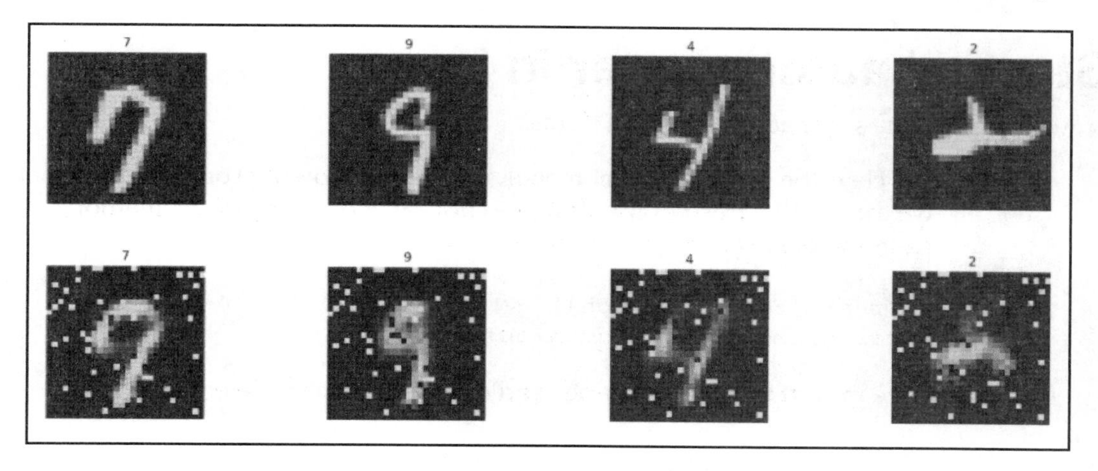

The images generated have a little bit of noise that can be removed with more training and hyper-parameter tuning. Now predicting the training set images is not magical as we trained the autoencoder on those images, hence it knows about them. Let's see the result of predicting the test set images. The first row indicates the actual images and second row indicates the generated images:

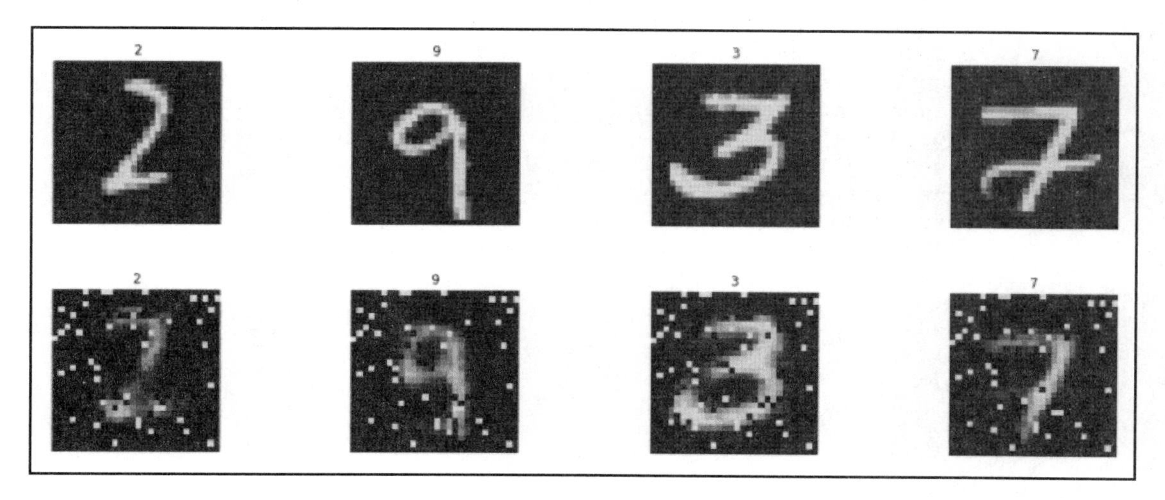

Wow! The trained autoencoder was able to generate the same digits with just 256 features that it learned out of 768. The noise in generated images can be improved by hyper-parameter tuning and more training.

Stacked autoencoder in Keras

Now let's build the same autoencoder in Keras.

We clear the graph in the notebook using the following commands so that we can build a fresh graph that does not carry over any of the memory from the previous session or graph:

```
tf.reset_default_graph()
keras.backend.clear_session()
```

1. First, we import the keras libraries and define hyperparameters and layers:

```
import keras
from keras.layers import Dense
from keras.models import Sequential

learning_rate = 0.001
n_epochs = 20
batch_size = 100
n_batches = int(mnist.train.num_examples/batch_sizee
# number of pixels in the MNIST image as number of inputs
n_inputs = 784
n_outputs = n_i
# number of hidden layers
n_layers = 2
# neurons in each hidden layer
n_neurons = [512,256]
# add decoder layers:
n_neurons.extend(list(reversed(n_neurons)))
n_layers = n_layers * 2
```

2. Next, we build a sequential model and add dense layers to it. For a change, we use `relu` activation for the hidden layers and `linear` activation for the final layer:

```
model = Sequential()

# add input to first layer
model.add(Dense(units=n_neurons[0], activation='relu',
    input_shape=(n_inputs,)))

for i in range(1,n_layers):
    model.add(Dense(units=n_neurons[i], activation='relu'))

# add last layer as output layer
model.add(Dense(units=n_outputs, activation='linear'))
```

3. Now let's display the model summary to see how the model looks:

```
model.summary()
```

The model has a total of 1,132,816 parameters in five dense layers:

```
_________________________________________________________________
Layer (type)                 Output Shape              Param #
=================================================================
dense_1 (Dense)              (None, 512)               401920
_________________________________________________________________
dense_2 (Dense)              (None, 256)               131328
_________________________________________________________________
dense_3 (Dense)              (None, 256)               65792
_________________________________________________________________
dense_4 (Dense)              (None, 512)               131584
_________________________________________________________________
dense_5 (Dense)              (None, 784)               402192
=================================================================
Total params: 1,132,816
Trainable params: 1,132,816
Non-trainable params: 0
_________________________________________________________________
```

4. Let's compile the model with the mean squared loss as in the previous example:

```
model.compile(loss='mse',
    optimizer=keras.optimizers.Adam(lr=learning_rate),
    metrics=['accuracy'])

model.fit(X_train, X_train,batch_size=batch_size,
    epochs=n_epochs)
```

Just in 20 epochs, we are able to get a loss of 0.0046 as compared to 0.078550 that we got before:

```
Epoch 1/20
55000/55000 [==========================] - 18s - loss: 0.0193 - acc: 0.0117
Epoch 2/20
55000/55000 [==========================] - 18s - loss: 0.0087 - acc: 0.0139
...
...
...
Epoch 20/20
55000/55000 [==========================] - 16s - loss: 0.0046 - acc: 0.0171
```

Now let's predict and display the train and test images generated by the model. The first row indicates the actual images and second row indicates the generated images. The following are the train set images:

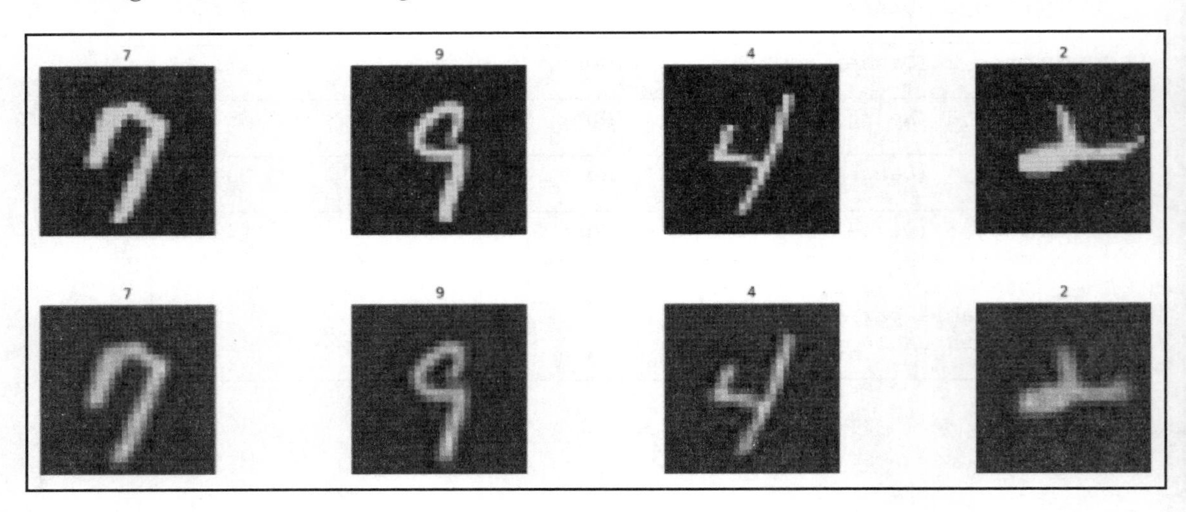

The following are the test set images:

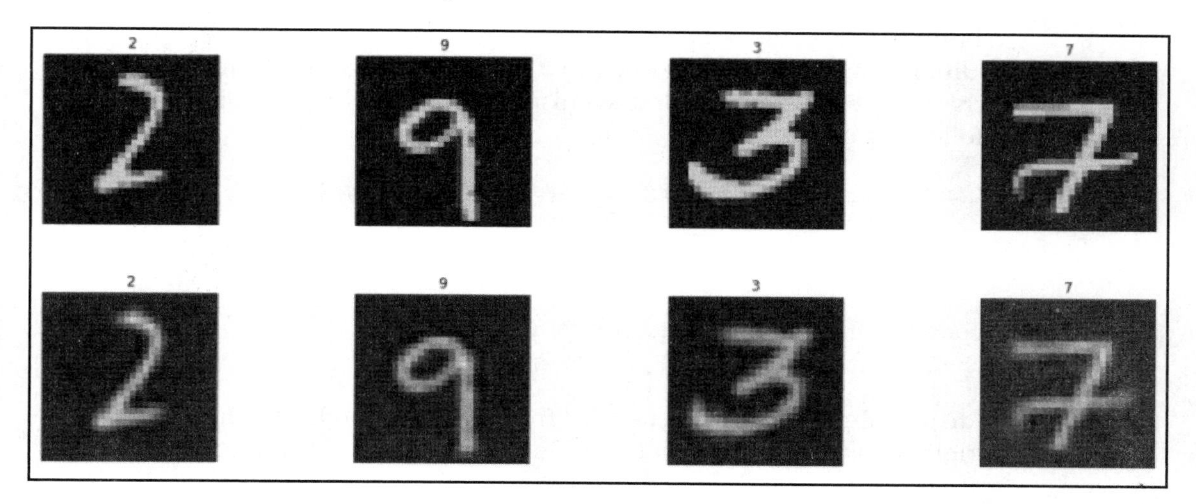

This is pretty good accuracy that we achieved in being able to generate the images from just 256 features.

Denoising autoencoder in TensorFlow

As you learned in the first section of this chapter, denoising autoencoders can be used to train the models such that they are able to remove the noise from the images input to the trained model:

1. For the purpose of this example, we write the following helper function to help us add noise to the images:

```
def add_noise(X):
    return X + 0.5 * np.random.randn(X.shape[0],X.shape[1])
```

2. Then we add noise to test images and store it in a separate list:

```
test_images_noisy = add_noise(test_images)
```

We will use these test images to test the output from our denoising model examples.

3. We build and train the denoising autoencoder as in the preceding example, with one difference: While training, we input the noisy images to the input layer and we check the reconstruction and denoising error with the non-noisy images, as the following code shows:

```
X_batch, _ = mnist.train.next_batch(batch_size)
X_batch_noisy = add_noise(X_batch)
feed_dict={x: X_batch_noisy, y: X_batch}
_,batch_loss = tfs.run([optimizer,loss], feed_dict=feed_dict)
```

The complete code for the denoising autoencoder is provided in the notebook `ch-10_AutoEncoders_TF_and_Keras`.

Now let's first display the test images generated from the DAE model; the first row indicates the original non-noisy test images and second row indicates the generated test images:

```
display_images(test_images.reshape(-1,pixel_size,pixel_size),test_labels)
display_images(Y_test_pred1.reshape(-1,pixel_size,pixel_size),test_labels)
```

The result of the preceding code is as follows:

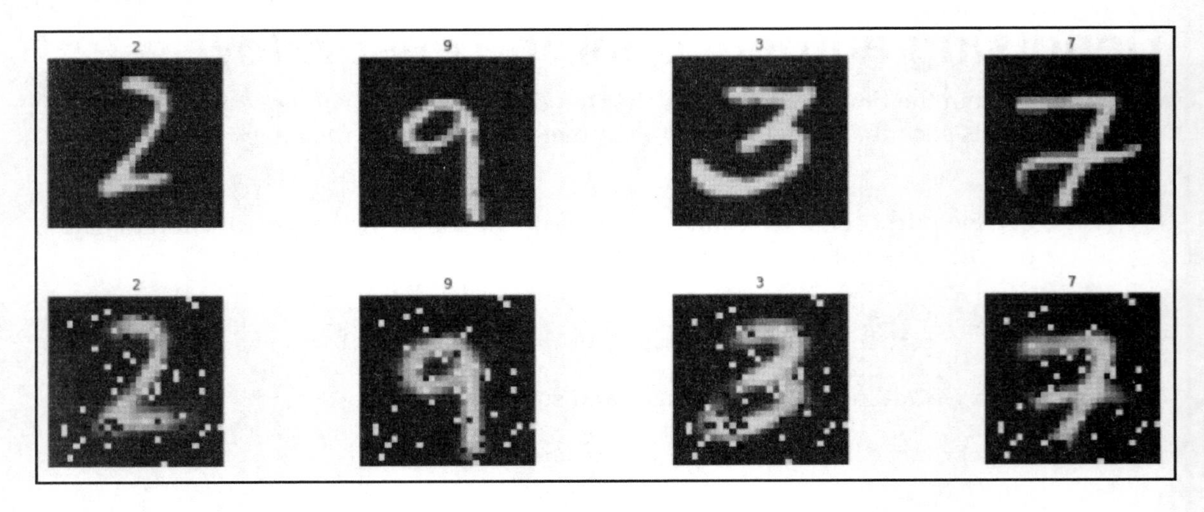

Next, we display the generated images when we input the noisy test images:

```
display_images(test_images_noisy.reshape(-1,pixel_size,pixel_size),
    test_labels)
display_images(Y_test_pred2.reshape(-1,pixel_size,pixel_size),test_labels)
```

The result of the preceding code is as follows:

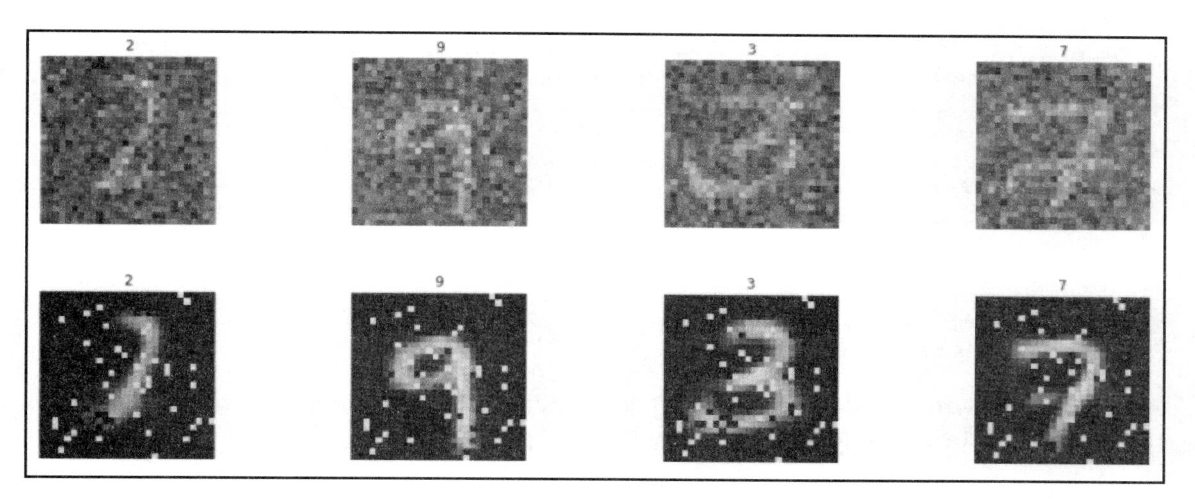

That is super cool!! The model learned the images and generated almost correct images even from a very noisy set. The quality of regeneration can be further improved with proper hyperparameter tuning.

Denoising autoencoder in Keras

Now let's build the same denoising autoencoder in Keras.

As Keras takes care of feeding the training set by batch size, we create a noisy training set to feed as input for our model:

```
X_train_noisy = add_noise(X_train)
```

The complete code for the DAE in Keras is provided in the notebook `ch-10_AutoEncoders_TF_and_Keras`.

The DAE Keras model looks like the following:

```
Layer (type)                 Output Shape              Param #
=================================================================
dense_1 (Dense)              (None, 512)               401920
_________________________________________________________________
dense_2 (Dense)              (None, 256)               131328
_________________________________________________________________
dense_3 (Dense)              (None, 256)               65792
_________________________________________________________________
dense_4 (Dense)              (None, 512)               131584
_________________________________________________________________
dense_5 (Dense)              (None, 784)               402192
=================================================================
Total params: 1,132,816
Trainable params: 1,132,816
Non-trainable params: 0
```

As DAE models are complex, for the purpose of demonstration, we had to increase the number of epochs to 100 to train the model:

```
n_epochs=100

model.fit(x=X_train_noisy, y=X_train,
          batch_size=batch_size,
          epochs=n_epochs,
          verbose=0)

Y_test_pred1 = model.predict(test_images)
Y_test_pred2 = model.predict(test_images_noisy)
```

Print the resulting images:

```
display_images(test_images.reshape(-1,pixel_size,pixel_size),test_labels)
display_images(Y_test_pred1.reshape(-1,pixel_size,pixel_size),test_labels)
```

The first row is the original test images and the second row is the generated test images:

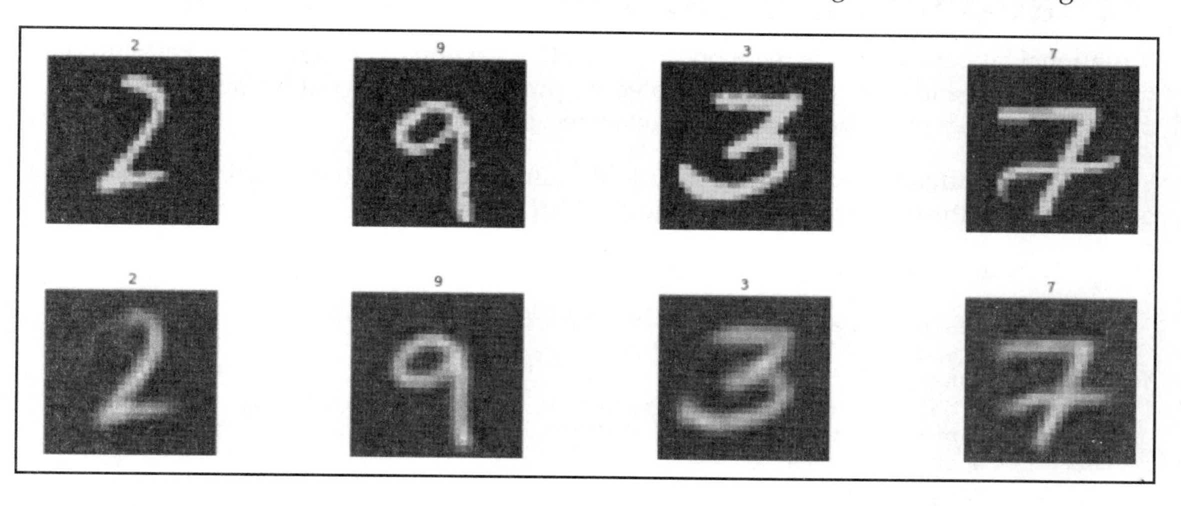

```
display_images(test_images_noisy.reshape(-1,pixel_size,pixel_size),
    test_labels)
display_images(Y_test_pred2.reshape(-1,pixel_size,pixel_size),test_labels)
```

The first row is the noisy test images and the second row is the generated test images:

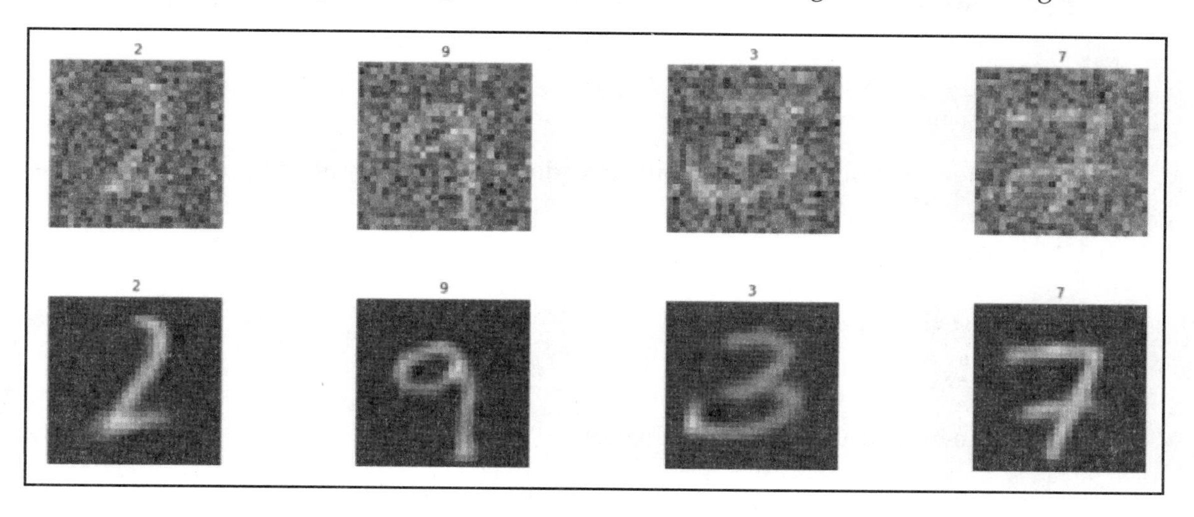

As we can see, the denoising autoencoder does a pretty good job of generating the images from the noisy version of the images.

Variational autoencoder in TensorFlow

Variational autoencoders are the modern generative version of autoencoders. Let's build a variational autoencoder for the same preceding problem. We will test the autoencoder by providing images from the original and noisy test set.

We will use a different coding style to build this autoencoder for the purpose of demonstrating the different styles of coding with TensorFlow:

1. Start by defining the hyper-parameters:

```
learning_rate = 0.001
n_epochs = 20
batch_size = 100
n_batches = int(mnist.train.num_examples/batch_size)
# number of pixels in the MNIST image as number of inputs
n_inputs = 784
n_outputs = n_inputs
```

2. Next, define a parameter dictionary to hold the weight and bias parameters:

```
params={}
```

3. Define the number of hidden layers in each of the encoder and decoder:

```
n_layers = 2
# neurons in each hidden layer
n_neurons = [512,256]
```

4. The new addition in a variational encoder is that we define the dimensions of the latent variable `z`:

```
n_neurons_z = 128 # the dimensions of latent variables
```

5. We use the activation `tanh`:

```
activation = tf.nn.tanh
```

6. Define input and output placeholders:

```
x = tf.placeholder(dtype=tf.float32, name="x",
                   shape=[None, n_inputs])
y = tf.placeholder(dtype=tf.float32, name="y",
                   shape=[None, n_outputs])
```

7. Define the input layer:

```
# x is input layer
layer = x
```

8. Define the biases and weights for the encoder network and add layers. The encoder network for variational autoencoders is also known as recognition network or inference network or probabilistic encoder network:

```
for i in range(0,n_layers):
    name="w_e_{0:04d}".format(i)
    params[name] = tf.get_variable(name=name,
        shape=[n_inputs if i==0 else n_neurons[i-1],
        n_neurons[i]],
        initializer=tf.glorot_uniform_initializer()
        )
    name="b_e_{0:04d}".format(i)
    params[name] = tf.Variable(tf.zeros([n_neurons[i]]),
        name=name
        )
    layer = activation(tf.matmul(layer,
        params["w_e_{0:04d}".format(i)]
        ) + params["b_e_{0:04d}".format(i)]
        )
```

9. Next, add the layers for mean and variance of the latent variables:

```
name="w_e_z_mean"
params[name] = tf.get_variable(name=name,
    shape=[n_neurons[n_layers-1], n_neurons_z],
    initializer=tf.glorot_uniform_initializer()
    )
name="b_e_z_mean"
params[name] = tf.Variable(tf.zeros([n_neurons_z]),
    name=name
    )
z_mean = tf.matmul(layer, params["w_e_z_mean"]) +
            params["b_e_z_mean"]
name="w_e_z_log_var"
params[name] = tf.get_variable(name=name,
    shape=[n_neurons[n_layers-1], n_neurons_z],
    initializer=tf.glorot_uniform_initializer()
    )
name="b_e_z_log_var"
params[name] = tf.Variable(tf.zeros([n_neurons_z]),
    name="b_e_z_log_var"
    )
```

```
z_log_var = tf.matmul(layer, params["w_e_z_log_var"]) +
            params["b_e_z_log_var"]
```

10. Next, define the epsilon variable representing the noise distribution of the same shape as the variable holding the variance of z:

```
epsilon = tf.random_normal(tf.shape(z_log_var),
    mean=0,
    stddev=1.0,
    dtype=tf.float32,
    name='epsilon'
    )
```

11. Define a posterior distribution based on the mean, log variance, and noise:

```
z = z_mean + tf.exp(z_log_var * 0.5) * epsilon
```

12. Next, define the weights and biases for the decoder network and add the decoder layers. The decoder network in variational autoencoder is also known as probabilistic decoder or generator network.

```
# add generator / probablistic decoder network parameters and
layers
layer = z

for i in range(n_layers-1,-1,-1):
name="w_d_{0:04d}".format(i)
    params[name] = tf.get_variable(name=name,
    shape=[n_neurons_z if i==n_layers-1 else n_neurons[i+1],
    n_neurons[i]],
    initializer=tf.glorot_uniform_initializer()
    )
name="b_d_{0:04d}".format(i)
params[name] = tf.Variable(tf.zeros([n_neurons[i]]),
    name=name
    )
layer = activation(tf.matmul(layer,
params["w_d_{0:04d}".format(i)]) +
    params["b_d_{0:04d}".format(i)])
```

13. Finally, define the output layer:

```
name="w_d_z_mean"
params[name] = tf.get_variable(name=name,
     shape=[n_neurons[0],n_outputs],
     initializer=tf.glorot_uniform_initializer()
     )
name="b_d_z_mean"
     params[name] = tf.Variable(tf.zeros([n_outputs]),
     name=name
     )
name="w_d_z_log_var"
params[name] = tf.Variable(tf.random_normal([n_neurons[0],
     n_outputs]),
     name=name
     )
name="b_d_z_log_var"
params[name] = tf.Variable(tf.zeros([n_outputs]),
     name=name
     )
layer = tf.nn.sigmoid(tf.matmul(layer, params["w_d_z_mean"]) +
     params["b_d_z_mean"])

model = layer
```

14. In variation autoencoders, we have the reconstruction loss and the regularization loss. Define the loss function as the sum of reconstruction loss and regularization loss:

```
rec_loss = -tf.reduce_sum(y * tf.log(1e-10 + model) + (1-y)
                          * tf.log(1e-10 + 1 - model), 1)
reg_loss = -0.5*tf.reduce_sum(1 + z_log_var - tf.square(z_mean)
                              - tf.exp(z_log_var), 1)
loss = tf.reduce_mean(rec_loss+reg_loss)
```

15. Define the optimizer function based on `AdapOptimizer`:

```
optimizer = tf.train.AdamOptimizer(learning_rate=learning_rate)
            .minimize(loss)
```

16. Now let's train the model and generate the images from non-noisy and noisy test images:

```
with tf.Session() as tfs:
    tf.global_variables_initializer().run()
    for epoch in range(n_epochs):
        epoch_loss = 0.0
        for batch in range(n_batches):
            X_batch, _ = mnist.train.next_batch(batch_size)
            feed_dict={x: X_batch,y: X_batch}
            _,batch_loss = tfs.run([optimizer,loss],
                                    feed_dict=feed_dict)
            epoch_loss += batch_loss
        if (epoch%10==9) or (epoch==0):
            average_loss = epoch_loss / n_batches
            print("epoch: {0:04d} loss = {1:0.6f}"
                  .format(epoch,average_loss))

    # predict images using autoencoder model trained
    Y_test_pred1 = tfs.run(model, feed_dict={x: test_images})
    Y_test_pred2 = tfs.run(model, feed_dict={x: test_images_noisy})
```

We get the following output:

```
epoch: 0000   loss = 180.444682
epoch: 0009   loss = 106.817749
epoch: 0019   loss = 102.580904
```

Now let's display the images:

```
display_images(test_images.reshape(-1,pixel_size,pixel_size),test_labels)
display_images(Y_test_pred1.reshape(-1,pixel_size,pixel_size),test_labels)
```

The result is as follows:

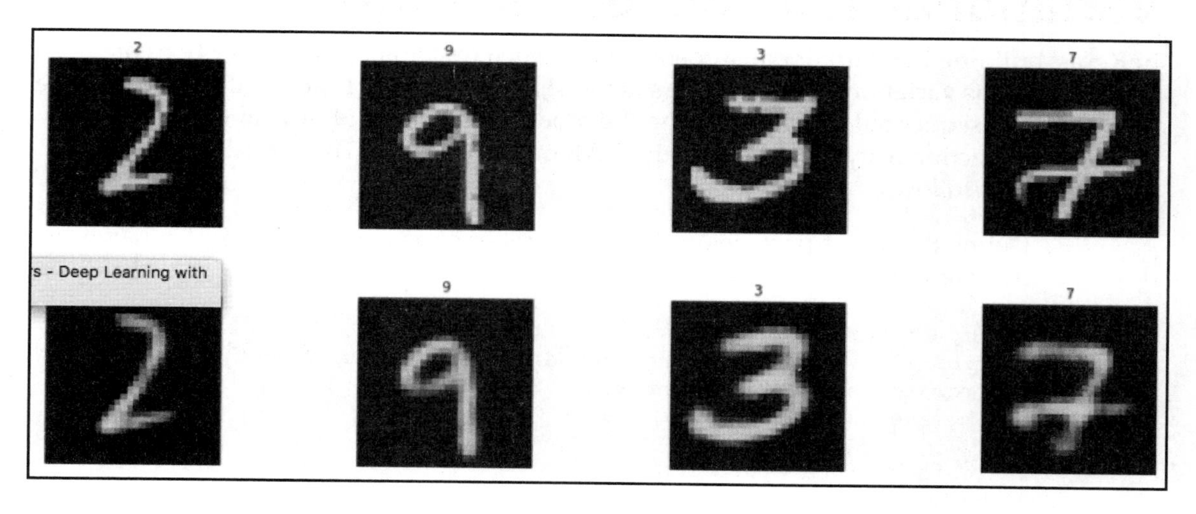


```
display_images(test_images_noisy.reshape(-1,pixel_size,pixel_size),
    test_labels)
display_images(Y_test_pred2.reshape(-1,pixel_size,pixel_size),test_labels)
```

The result is as follows:

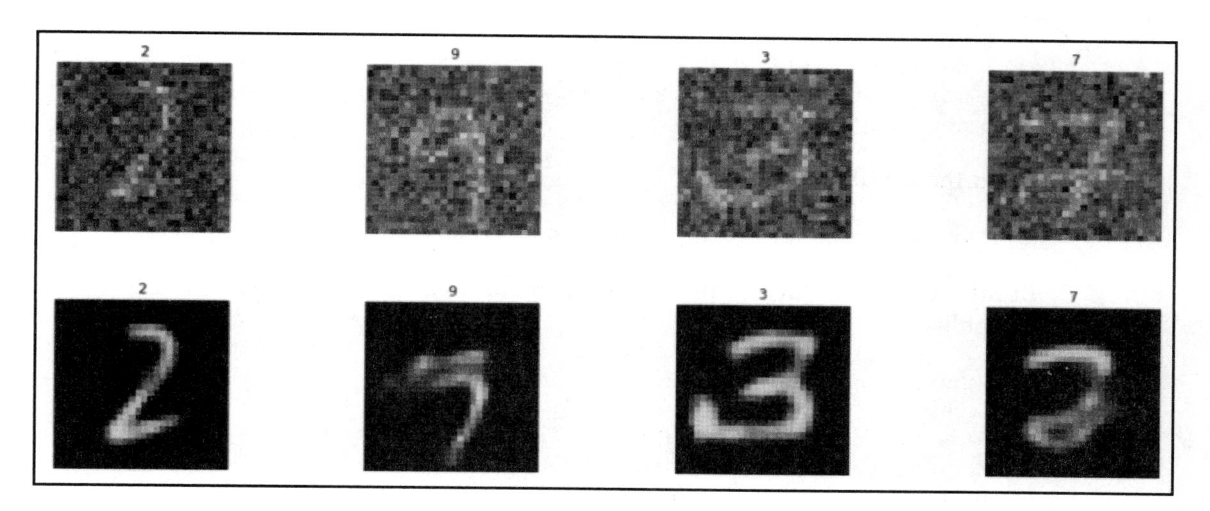

Again, the results can be improved with hyperparameter tuning and increasing the amount of learning.

Variational autoencoder in Keras

In Keras, building the variational autoencoder is much easier and with lesser lines of code. The Keras variational autoencoders are best built using the functional style. So far we have used the sequential style of building the models in Keras, and now in this example, we will see the functional style of building the VAE model in Keras. The steps to build a VAE in Keras are as follows:

1. Define the hyper-parameters and the number of neurons in the hidden layers and the latent variables layer:

```
import keras
from keras.layers import Lambda, Dense, Input, Layer
from keras.models import Model
from keras import backend as K

learning_rate = 0.001
batch_size = 100
n_batches = int(mnist.train.num_examples/batch_size)
# number of pixels in the MNIST image as number of inputs
n_inputs = 784
n_outputs = n_inputs
# number of hidden layers
n_layers = 2
# neurons in each hidden layer
n_neurons = [512,256]
# the dimensions of latent variables
n_neurons_z = 128
```

2. Build the input layer:

```
x = Input(shape=(n_inputs,), name='input')
```

3. Build the encoder layers, along with mean and variance layers for the latent variables:

```
# build encoder
layer = x
for i in range(n_layers):
    layer = Dense(units=n_neurons[i],
activation='relu',name='enc_{0}'.format(i))(layer)

z_mean = Dense(units=n_neurons_z,name='z_mean')(layer)
z_log_var = Dense(units=n_neurons_z,name='z_log_v')(layer)
```

4. Create the noise and posterior distributions:

```
# noise distribution
epsilon = K.random_normal(shape=K.shape(z_log_var),
               mean=0,stddev=1.0)

# posterior distribution
z = Lambda(lambda zargs: zargs[0] + K.exp(zargs[1] * 0.5) *
epsilon,
    name='z')([z_mean,z_log_var])
```

5. Add the decoder layers:

```
# add generator / probablistic decoder network layers
layer = z
for i in range(n_layers-1,-1,-1):
    layer = Dense(units=n_neurons[i], activation='relu',
            name='dec_{0}'.format(i))(layer)
```

6. Define the final output layer:

```
y_hat = Dense(units=n_outputs, activation='sigmoid',
        name='output')(layer)
```

7. Finally, define the model from the input layer and the output layer and display the model summary:

```
model = Model(x,y_hat)
model.summary()
```

We see the following summary:

```
Layer (type)            Output Shape          Param #     Connected to
==========================================================================
input (InputLayer)      (None, 784)           0
__________________________________________________________________________
enc_0 (Dense)           (None, 512)           401920      input[0][0]
__________________________________________________________________________
enc_1 (Dense)           (None, 256)           131328      enc_0[0][0]
__________________________________________________________________________
z_mean (Dense)          (None, 128)           32896       enc_1[0][0]
__________________________________________________________________________
z_log_v (Dense)         (None, 128)           32896       enc_1[0][0]
__________________________________________________________________________
z (Lambda)              (None, 128)           0           z_mean[0][0]
                                                          z_log_v[0][0]
```

```
____________________________________________________________________________
dec_1 (Dense)              (None, 256)          33024        z[0][0]
____________________________________________________________________________
dec_0 (Dense)              (None, 512)          131584       dec_1[0][0]
____________________________________________________________________________
output (Dense)             (None, 784)          402192       dec_0[0][0]
============================================================================
Total params: 1,165,840
Trainable params: 1,165,840
Non-trainable params: 0
____________________________________________________________________________
```

8. Define a function that calculates the sum of reconstruction and regularization loss:

```
def vae_loss(y, y_hat):
    rec_loss = -K.sum(y * K.log(1e-10 + y_hat) + (1-y) *
               K.log(1e-10 + 1 - y_hat), axis=-1)
    reg_loss = -0.5 * K.sum(1 + z_log_var - K.square(z_mean) -
               K.exp(z_log_var), axis=-1)
    loss = K.mean(rec_loss+reg_loss)
    return loss
```

9. Use this loss function to compile the model:

```
model.compile(loss=vae_loss,
    optimizer=keras.optimizers.Adam(lr=learning_rate))
```

10. Let's train the model for 50 epochs and predict the images, as we have done in previous sections:

```
n_epochs=50
model.fit(x=X_train_noisy,y=X_train,batch_size=batch_size,
    epochs=n_epochs,verbose=0)
Y_test_pred1 = model.predict(test_images)
Y_test_pred2 = model.predict(test_images_noisy)
```

Let's display the resulting images:

```
display_images(test_images.reshape(-1,pixel_size,pixel_size),test_labels)
display_images(Y_test_pred1.reshape(-1,pixel_size,pixel_size),test_labels)
```

We get the result as follows:

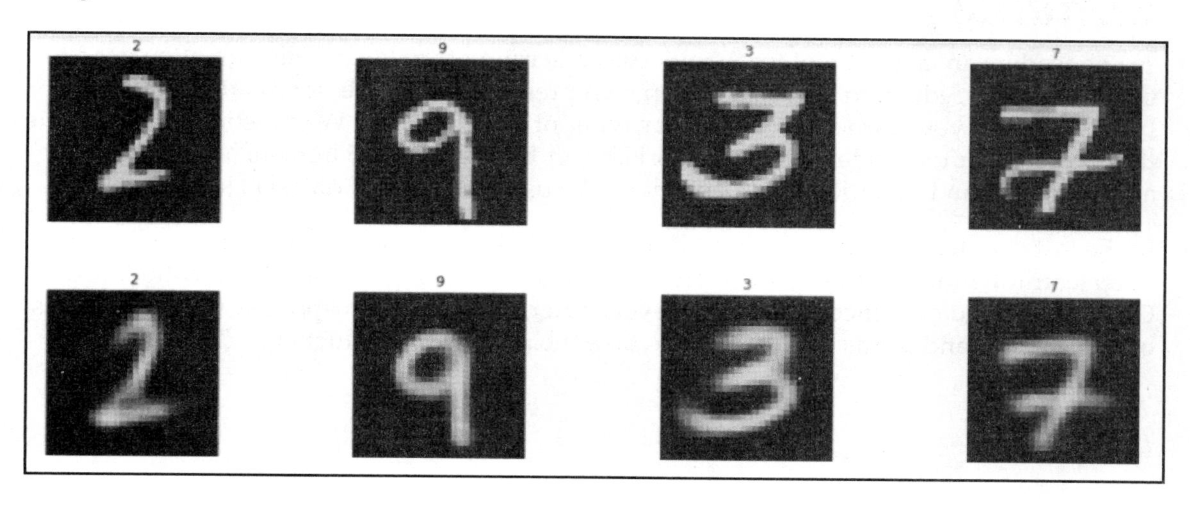


```
display_images(test_images_noisy.reshape(-1,pixel_size,pixel_size),
    test_labels)
display_images(Y_test_pred2.reshape(-1,pixel_size,pixel_size),test_labels)
```

We get the following results:

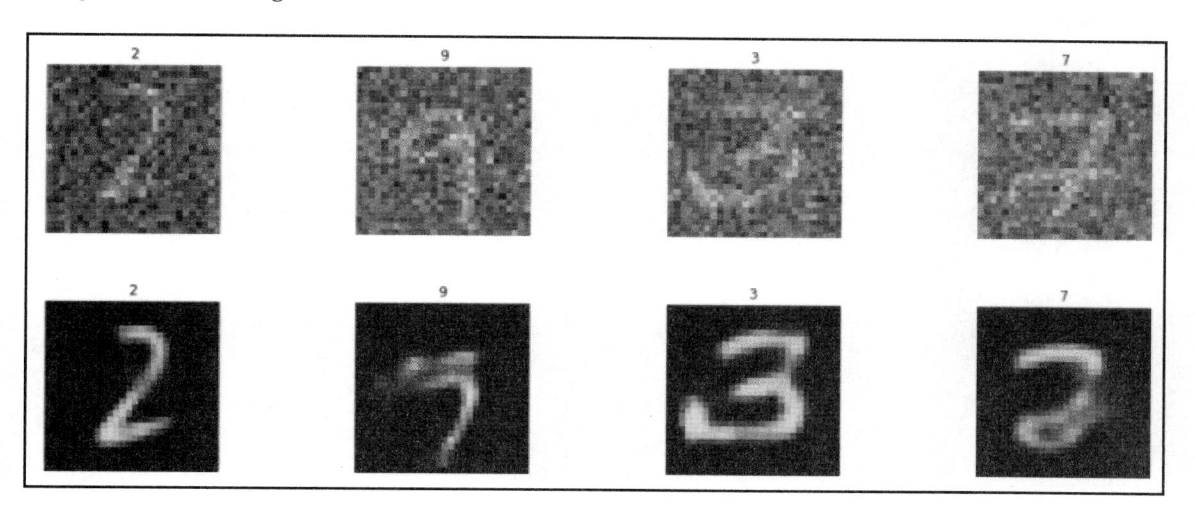

This is great!! The resulting generated images are much clearer and sharper.

Summary

Autoencoders are a great tool for unsupervised learning from data. They are often used for dimensionality reduction so that data can be represented by the lesser number of features. In this chapter, you learned about various types of autoencoders. We practiced building the three types of autoencoders using TensorFlow and Keras: stacked autoencoders, denoising autoencoders, and variational autoencoders. We used the MNIST dataset as an example.

In the last chapters, you have learned how to build various kinds of machine learning and deep learning models with TensorFlow and Keras, such as regression, classification, MLP, CNN, RNN, and autoencoders. In the next chapter, you will learn about advanced features of TensorFlow and Keras that allow us to take the models to production.

11

TensorFlow Models in Production with TF Serving

The TensorFlow models are trained and validated in the development environment. Once released, they need to be hosted somewhere to be made available to application engineers and software engineers to integrate into various applications. TensorFlow provides a high-performance server for this purpose, known as TensorFlow Serving.

For serving TensorFlow models in production, one would need to save them after training offline and then restore the trained models in the production environment. A TensorFlow model consists of the following files when saved:

- **meta-graph**: The meta-graph represents the protocol buffer definition of the graph. The meta-graph is saved in files with the `.meta` extension.
- **checkpoint**: The checkpoint represents the values of various variables. The checkpoint is saved in two files: one with the `.index` extension and one with the `.data-00000-of-00001` extension.

In this chapter, we shall learn various ways to save and restore models and how to serve them using the TF Serving. We will use the MNIST example to keep things simple and cover the following topics:

- Saving and restoring Models in TensorFlow with the `Saver` class
- Saving and restoring Keras models
- TensorFlow Serving
- Installing TF Serving
- Saving the model for TF Serving

- Serving the model with TF Serving
- TF Serving in the Docker container
- TF Serving on the Kubernetes

Saving and Restoring models in TensorFlow

You can save and restore the models and the variables in TensorFlow by one of the following two methods:

- A saver object created from the `tf.train.Saver` class
- A `SavedModel` format based object created from the `tf.saved_model_builder.SavedModelBuilder` class

Let us see both the methods in action.

You can follow along with the code in the Jupyter notebook `ch-11a_Saving_and_Restoring_TF_Models`.

Saving and restoring all graph variables with the saver class

We proceed as follows:

1. To use the `saver` class, first an object of this class is created:

```
saver = tf.train.Saver()
```

2. The simplest way to save all the variables in a graph is to call the `save()` method with the following two parameters: the session object and the path to the file on the disk where the variables will be saved:

```
with tf.Session() as tfs:
    ...
    saver.save(tfs,"saved-models/model.ckpt")
```

3. To restore the variables, the `restore()` method is called:

```
with tf.Session() as tfs:
    saver.restore(tfs,"saved-models/model.ckpt")
    ...
```

4. Let us revisit the example from `Chapter 1`, *TensorFlow 101*, the code to save the variables in the graph in the simple example is as follows:

```
# Assume Linear Model y = w * x + b
# Define model parameters
w = tf.Variable([.3], tf.float32)
b = tf.Variable([-.3], tf.float32)
# Define model input and output
x = tf.placeholder(tf.float32)
y = w * x + b
output = 0

# create saver object
saver = tf.train.Saver()

with tf.Session() as tfs:
    # initialize and print the variable y
    tfs.run(tf.global_variables_initializer())
    output = tfs.run(y,{x:[1,2,3,4]})
    saved_model_file = saver.save(tfs,
        'saved-models/full-graph-save-example.ckpt')
    print('Model saved in {}'.format(saved_model_file))
    print('Values of variables w,b: {}{}'
        .format(w.eval(),b.eval()))
    print('output={}'.format(output))
```

We get the following output:

```
Model saved in saved-models/full-graph-save-example.ckpt
Values of variables w,b: [ 0.30000001][-0.30000001]
output=[ 0.          0.30000001  0.60000002  0.90000004]
```

5. Now let us restore the variables from the checkpoint file we just created:

```
# Assume Linear Model y = w * x + b
# Define model parameters
w = tf.Variable([0], dtype=tf.float32)
b = tf.Variable([0], dtype=tf.float32)
# Define model input and output
x = tf.placeholder(dtype=tf.float32)
y = w * x + b
```

```
    output = 0

    # create saver object
    saver = tf.train.Saver()

    with tf.Session() as tfs:
        saved_model_file = saver.restore(tfs,
            'saved-models/full-graph-save-example.ckpt')
        print('Values of variables w,b: {}{}'
            .format(w.eval(),b.eval()))
        output = tfs.run(y,{x:[1,2,3,4]})
        print('output={}'.format(output))
```

You will notice that in the restore code we did not call `tf.global_variables_initializer()`, because there is no need to initialize the variables since they will be restored from the file. We get the following output, which is calculated from the restored variables:

```
INFO:tensorflow:Restoring parameters from saved-models/full-graph-save-
example.ckpt
Values of variables w,b: [ 0.30000001][-0.30000001]
output=[ 0.          0.30000001  0.60000002  0.90000004]
```

Saving and restoring selected variables with the saver class

By default, the `Saver()` class saves all the variables in a graph, but you can select which variables to save by passing the list of variables to the constructor of the `Saver()` class:

```
# create saver object
saver = tf.train.Saver({'weights': w})
```

The variables names can be passed either as a list or as a dictionary. In case the variable names are passed as a list then each variable in the list will be saved with its own name. The variables can also be passed as a dictionary consisting of key-value pairs where the key is the name to be used for saving and the value is the name of the variable to be saved.

The following is the code for the example we just saw, but this time we are only saving weights from the `w` variable; name it `weights` when you save it:

```
# Saving selected variables in a graph in TensorFlow

# Assume Linear Model y = w * x + b
# Define model parameters
```

```
w = tf.Variable([.3], tf.float32)
b = tf.Variable([-.3], tf.float32)
# Define model input and output
x = tf.placeholder(tf.float32)
y = w * x + b
output = 0

# create saver object
saver = tf.train.Saver({'weights': w})

with tf.Session() as tfs:
    # initialize and print the variable y
    tfs.run(tf.global_variables_initializer())
    output = tfs.run(y,{x:[1,2,3,4]})
    saved_model_file = saver.save(tfs,
        'saved-models/weights-save-example.ckpt')
    print('Model saved in {}'.format(saved_model_file))
    print('Values of variables w,b: {}{}'
        .format(w.eval(),b.eval()))
    print('output={}'.format(output))
```

We get the following output:

```
Model saved in saved-models/weights-save-example.ckpt
Values of variables w,b: [ 0.30000001][-0.30000001]
output=[ 0.          0.30000001  0.60000002  0.90000004]
```

The checkpoint file has only saved weights and not the biases. Now let us initialize the biases and weights to zero, and restore the weights. The code is given here for this example:

```
# Restoring selected variables in a graph in TensorFlow
tf.reset_default_graph()
# Assume Linear Model y = w * x + b
# Define model parameters
w = tf.Variable([0], dtype=tf.float32)
b = tf.Variable([0], dtype=tf.float32)
# Define model input and output
x = tf.placeholder(dtype=tf.float32)
y = w * x + b
output = 0

# create saver object
saver = tf.train.Saver({'weights': w})

with tf.Session() as tfs:
    b.initializer.run()
    saved_model_file = saver.restore(tfs,
        'saved-models/weights-save-example.ckpt')
```

```
    print('Values of variables w,b: {}{}'
        .format(w.eval(),b.eval()))
    output = tfs.run(y,{x:[1,2,3,4]})
    print('output={}'.format(output))
```

As you can see, this time we had to initialize biases by using `b.initializer.run()`. We do not use `tfs.run(tf.global_variables_initializer())` since that would initialize all the variables, and there is no need to initialize the weights since they would be restored from the checkpoint file.

We get the following output, as the calculations use only the restored weights while the biases are set to zero:

```
INFO:tensorflow:Restoring parameters from saved-models/weights-save-
example.ckpt
Values of variables w,b: [ 0.30000001][ 0.]
output=[ 0.30000001  0.60000002  0.90000004  1.20000005]
```

Saving and restoring Keras models

In Keras, saving and restoring models is very simple. Keras provides three options:

- Save the complete model with its network architecture, weights (parameters), training configuration, and optimizer state.
- Save only the architecture.
- Save only the weights.

For saving the complete model, use the `model.save(filepath)` function. This will save the complete model in an HDF5 file. The saved model can be loaded back using the `keras.models.load_model(filepath)` function. This function loads everything back, and then also compiles the model.

For saving the architecture of a model, use either the `model.to_json()` or `model.to_yaml()` function. These functions return a string that can be written to the disk file. While restoring the architecture, the string can be read back and the model architecture restored using the `keras.models.model_from_json(json_string)` or the `keras.models.model_from_yaml(yaml_string)` function. Both these functions return a model instance.

For saving the weights of a model, use the `model.save_weights(path_to_h5_file)` function. The weights can be restored using the `model.load_weights(path_to_h5_file)` function.

TensorFlow Serving

TensorFlow Serving (TFS) is a high-performance server architecture for serving the machine learning models in production. It offers out-of-the-box integration with the models built using TensorFlow.

In TFS, a **model** is composed of one or more **servables**. A servable is used to perform computation, for example:

- A lookup table for embedding lookups
- A single model returning predictions
- A tuple of models returning a tuple of predictions
- A shard of lookup tables or models

The ***manager*** component manages the full lifecycle for the *servables* including loading/unloading a *servable* and serving the *servable*.

The internal architecture and workflow of TensorFlow Serving is described at the following link: `https://www.tensorflow.org/serving/architecture_overview`.

Installing TF Serving

Follow the instructions in this section to install the TensorFlow ModelServer on Ubuntu using `aptitude`.

1. First, add TensorFlow Serving distribution URI as a package source (one-time setup) with the following command at shell prompt:

```
$ echo "deb [arch=amd64]
http://storage.googleapis.com/tensorflow-serving-apt stable
tensorflow-model-server tensorflow-model-server-universal" | sudo
tee /etc/apt/sources.list.d/tensorflow-serving.list

$ curl
```

```
https://storage.googleapis.com/tensorflow-serving-apt/tensorflow-se
rving.release.pub.gpg | sudo apt-key add -
```

2. Install and update TensorFlow ModelServer with the following command at shell prompt:

```
$ sudo apt-get update && sudo apt-get install tensorflow-model-
server
```

This installs the version of ModelServer that uses platform-specific compiler optimizations, such as utilizing the SSE4 and AVX instructions. However, if the optimized version install does not work on older machines, then you can install the universal version:

```
$ sudo apt-get remove tensorflow-model-server
$ sudo apt-get update && sudo apt-get install tensorflow-model-
server-universal
```

For other operating systems and for installing from source, refer to the following link: `https://www.tensorflow.org/serving/setup`

When the new versions of ModelServer are released, you can upgrade to newer versions using the following command:

```
$ sudo apt-get update && sudo apt-get upgrade tensorflow-model-
server
```

3. Now that ModelServer is installed, run the server with the following command:

```
$ tensorflow-model-server
```

4. To connect to `tensorflow-model-server`, install the python client package with pip:

```
$ sudo pip2 install tensorflow-serving-api
```

The TF Serving API is only available for Python 2 and not yet available for Python 3.

Saving models for TF Serving

In order to serve the models, they need to be saved first. In this section, we demonstrate a slightly modified version of the MNIST example from the official TensorFlow documentation, available at the following link: `https://www.tensorflow.org/serving/serving_basic`.

The TensorFlow team recommends using SavedModel for saving and restoring models built and trained in TensorFlow. According to the TensorFlow documentation:

> *SavedModel is a language-neutral, recoverable, hermetic serialization format. SavedModel enables higher-level systems and tools to produce, consume, and transform TensorFlow models.*

You can follow along with the code in the Jupyter notebook `ch-11b_Saving_TF_Models_with_SavedModel_for_TF_Serving`.

We proceed with saving models as follows:

1. Define the model variables:

```
model_name = 'mnist'
model_version = '1'
model_dir = os.path.join(models_root,model_name,model_version)
```

2. Get the MNIST data as we did in chapter 4 - MLP models:

```
from tensorflow.examples.tutorials.mnist import input_data
dataset_home = os.path.join('.','mnist')
mnist = input_data.read_data_sets(dataset_home, one_hot=True)
x_train = mnist.train.images
x_test = mnist.test.images
y_train = mnist.train.labels
y_test = mnist.test.labels
pixel_size = 28
num_outputs = 10 # 0-9 digits
num_inputs = 784 # total pixels
```

3. Define an MLP function that would build and return the model:

```
def mlp(x, num_inputs, num_outputs,num_layers,num_neurons):
    w=[]
    b=[]
```

```
        for i in range(num_layers):
            w.append(tf.Variable(tf.random_normal(
                [num_inputs if i==0 else num_neurons[i-1],
                num_neurons[i]]),name="w_{0:04d}".format(i)
                 )
                )
            b.append(tf.Variable(tf.random_normal(
                  [num_neurons[i]]),
                  name="b_{0:04d}".format(i)
                 )
                )
        w.append(tf.Variable(tf.random_normal(
            [num_neurons[num_layers-1] if num_layers > 0 \
            else num_inputs, num_outputs]),name="w_out"))
        b.append(tf.Variable(tf.random_normal([num_outputs]),
                name="b_out"))

        # x is input layer
        layer = x
        # add hidden layers
        for i in range(num_layers):
            layer = tf.nn.relu(tf.matmul(layer, w[i]) + b[i])
        # add output layer
        layer = tf.matmul(layer, w[num_layers]) + b[num_layers]
        model = layer
        probs = tf.nn.softmax(model)

        return model,probs
```

The `mlp()` function described above returns model and probabilities. Probabilities are the softmax activation applied to the model.

4. Define the `x_p` and `y_p` placeholders for image input and target output:

```
    # input images
    serialized_tf_example = tf.placeholder(tf.string,
            name='tf_example')
    feature_configs = {'x': tf.FixedLenFeature(shape=[784],
            dtype=tf.float32),}
    tf_example = tf.parse_example(serialized_tf_example,
            feature_configs)
    # use tf.identity() to assign name
    x_p = tf.identity(tf_example['x'], name='x_p')
    # target output
    y_p = tf.placeholder(dtype=tf.float32, name="y_p",
            shape=[None, num_outputs])
```

5. Create the model, along with the loss, optimizer, accuracy, and training functions:

```
num_layers = 2
num_neurons = []
for i in range(num_layers):
   num_neurons.append(256)

learning_rate = 0.01
n_epochs = 50
batch_size = 100
n_batches = mnist.train.num_examples//batch_size

model,probs = mlp(x=x_p,
    num_inputs=num_inputs,
    num_outputs=num_outputs,
    num_layers=num_layers,
    num_neurons=num_neurons)

loss_op = tf.nn.softmax_cross_entropy_with_logits
loss = tf.reduce_mean(loss_op(logits=model, labels=y_p))
optimizer = tf.train.GradientDescentOptimizer(learning_rate)
train_op = optimizer.minimize(loss)

pred_check = tf.equal(tf.argmax(probs,1), tf.argmax(y_p,1))
accuracy_op = tf.reduce_mean(tf.cast(pred_check, tf.float32))

values, indices = tf.nn.top_k(probs, 10)
table = tf.contrib.lookup.index_to_string_table_from_tensor(
        tf.constant([str(i) for i in range(10)]))
prediction_classes = table.lookup(tf.to_int64(indices))
```

6. In a TensorFlow session, train the model as we did before, but use the builder object to save the model:

```
from tf.saved_model.signature_constants import \
        CLASSIFY_INPUTS
from tf.saved_model.signature_constants import \
        CLASSIFY_OUTPUT_CLASSES
from tf.saved_model.signature_constants import \
        CLASSIFY_OUTPUT_SCORES
from tf.saved_model.signature_constants import \
        CLASSIFY_METHOD_NAME
from tf.saved_model.signature_constants import \
        PREDICT_METHOD_NAME
from tf.saved_model.signature_constants import \
        DEFAULT_SERVING_SIGNATURE_DEF_KEY
```

```
with tf.Session() as tfs:
    tfs.run(tf.global_variables_initializer())
    for epoch in range(n_epochs):
        epoch_loss = 0.0
        for batch in range(n_batches):
            x_batch, y_batch = mnist.train.next_batch(batch_size)
            feed_dict = {x_p: x_batch, y_p: y_batch}
            _,batch_loss = tfs.run([train_op,loss],
                               feed_dict=feed_dict)
            epoch_loss += batch_loss
        average_loss = epoch_loss / n_batches
        print("epoch: {0:04d}   loss = {1:0.6f}"
             .format(epoch,average_loss))
    feed_dict={x_p: x_test, y_p: y_test}
    accuracy_score = tfs.run(accuracy_op, feed_dict=feed_dict)
    print("accuracy={0:.8f}".format(accuracy_score))

    # save the model

    # definitions for saving the models
    builder = tf.saved_model.builder.SavedModelBuilder(model_dir)
    # build signature_def_map
    bti_op = tf.saved_model.utils.build_tensor_info
    bsd_op = tf.saved_model.utils.build_signature_def

    classification_inputs = bti_op(serialized_tf_example)
    classification_outputs_classes = bti_op(prediction_classes)
    classification_outputs_scores = bti_op(values)
    classification_signature = (bsd_op(
        inputs={CLASSIFY_INPUTS: classification_inputs},
        outputs={CLASSIFY_OUTPUT_CLASSES:
                    classification_outputs_classes,
                 CLASSIFY_OUTPUT_SCORES:
                    classification_outputs_scores
                },
        method_name=CLASSIFY_METHOD_NAME))

    tensor_info_x = bti_op(x_p)
    tensor_info_y = bti_op(probs)

    prediction_signature = (bsd_op(
            inputs={'inputs': tensor_info_x},
            outputs={'outputs': tensor_info_y},
            method_name=PREDICT_METHOD_NAME))

    legacy_init_op = tf.group(tf.tables_initializer(),
        name='legacy_init_op')
    builder.add_meta_graph_and_variables(
```

```
                tfs, [tf.saved_model.tag_constants.SERVING],
                signature_def_map={
                    'predict_images':prediction_signature,
                    DEFAULT_SERVING_SIGNATURE_DEF_KEY:
                        classification_signature,
                },
                legacy_init_op=legacy_init_op)

            builder.save()
```

The model is saved once we see the following output:

```
accuracy=0.92979997
INFO:tensorflow:No assets to save.
INFO:tensorflow:No assets to write.
INFO:tensorflow:SavedModel written to:
b'/home/armando/models/mnist/1/saved_model.pb'
```

Next, we run the ModelServer and serve the model we just saved.

Serving models with TF Serving

To run the ModelServer, execute the following command:

```
$ tensorflow_model_server --model_name=mnist --
model_base_path=/home/armando/models/mnist
```

The server starts serving the model on port 8500:

```
I tensorflow_serving/model_servers/main.cc:147] Building single TensorFlow
model file config: model_name: mnist model_base_path:
/home/armando/models/mnist
I tensorflow_serving/model_servers/server_core.cc:441] Adding/updating
models.
I tensorflow_serving/model_servers/server_core.cc:492] (Re-)adding model:
mnist
I tensorflow_serving/core/basic_manager.cc:705] Successfully reserved
resources to load servable {name: mnist version: 1}
I tensorflow_serving/core/loader_harness.cc:66] Approving load for servable
version {name: mnist version: 1}
I tensorflow_serving/core/loader_harness.cc:74] Loading servable version
{name: mnist version: 1}
I
external/org_tensorflow/tensorflow/contrib/session_bundle/bundle_shim.cc:36
0] Attempting to load native SavedModelBundle in bundle-shim from:
/home/armando/models/mnist/1
```

```
I external/org_tensorflow/tensorflow/cc/saved_model/loader.cc:236] Loading
SavedModel from: /home/armando/models/mnist/1
I
external/org_tensorflow/tensorflow/core/platform/cpu_feature_guard.cc:137]
Your CPU supports instructions that this TensorFlow binary was not compiled
to use: AVX2 FMA
I external/org_tensorflow/tensorflow/cc/saved_model/loader.cc:155]
Restoring SavedModel bundle.
I external/org_tensorflow/tensorflow/cc/saved_model/loader.cc:190] Running
LegacyInitOp on SavedModel bundle.
I external/org_tensorflow/tensorflow/cc/saved_model/loader.cc:284] Loading
SavedModel: success. Took 29853 microseconds.
I tensorflow_serving/core/loader_harness.cc:86] Successfully loaded
servable version {name: mnist version: 1}
E1121 ev_epoll1_linux.c:1051] grpc epoll fd: 3
I tensorflow_serving/model_servers/main.cc:288] Running ModelServer at
0.0.0.0:8500 ...
```

To test the server by calling the model to classify images, follow along with notebook `ch-11c_TF_Serving_MNIST`.

The first two cells of the notebook provide the test client functions from TensorFlow official examples in the serving repository. We have modified the example to send the `'input'` and receive the `'output'` in the function signature for calling the ModelServer.

Call the test client functions in the third cell of the notebook with the following code:

```
error_rate = do_inference(hostport='0.0.0.0:8500',
                          work_dir='/home/armando/datasets/mnist',
                          concurrency=1,
                          num_tests=100)
print('\nInference error rate: %s%%' % (error_rate * 100))
```

We get an almost 7% error rate! (you might get a different value):

```
Extracting /home/armando/datasets/mnist/train-images-idx3-ubyte.gz
Extracting /home/armando/datasets/mnist/train-labels-idx1-ubyte.gz
Extracting /home/armando/datasets/mnist/t10k-images-idx3-ubyte.gz
Extracting /home/armando/datasets/mnist/t10k-labels-idx1-ubyte.gz

..................................................
..................................................
Inference error rate: 7.0%
```

TF Serving in the Docker containers

Docker is a platform for packaging and deploying the application in containers. If you do not already know about the Docker containers, then visit the tutorials and information at the following link: `https://www.docker.com/what-container`.

We can also install and run the TensorFlow Serving in the Docker containers. The instructions for Ubuntu 16.04 provided in this section are derived from the links on TensorFlow's official website:

- `https://www.tensorflow.org/serving/serving_inception`
- `https://www.tensorflow.org/serving/serving_basic`

Let us dive right in!

Installing Docker

We install Docker as follows:

1. First, remove the previous installations of Docker:

```
$ sudo apt-get remove docker docker-engine docker.io
```

2. Install the pre-requisite software:

```
$ sudo apt-get install \
    apt-transport-https \
    ca-certificates \
    curl \
    software-properties-common
```

3. Add the GPG key for Docker repositories:

```
$ curl -fsSL https://download.docker.com/linux/ubuntu/gpg | sudo
apt-key add -
```

4. Add the Docker repository:

```
$ sudo add-apt-repository \
   "deb [arch=amd64] https://download.docker.com/linux/ubuntu \
   $(lsb_release -cs) \
   stable"
```

5. Install the Docker community edition:

```
$ sudo apt-get update && sudo apt-get install docker-ce
```

6. Once the installation is successfully completed, add the Docker as a system service:

```
$ sudo systemctl enable docker
```

7. To run Docker as a non-root user or without `sudo`, add the `docker` group:

```
$ sudo groupadd docker
```

8. Add your user to the `docker` group:

```
$ sudo usermod -aG docker $USER
```

9. Now log off and log in again so that group membership takes effect. Once logged in, run the following command to test Docker installation:

```
$ docker run --name hello-world-container hello-world
```

You should see output similar to the following:

```
Unable to find image 'hello-world:latest' locally
latest: Pulling from library/hello-world
ca4f61b1923c: Already exists
Digest:
sha256:be0cd392e45be79ffeffa6b05338b98ebb16c87b255f48e297ec7f98e123905c
Status: Downloaded newer image for hello-world:latest

Hello from Docker!
This message shows that your installation appears to be working correctly.

To generate this message, Docker took the following steps:
 1. The Docker client contacted the Docker daemon.
 2. The Docker daemon pulled the "hello-world" image from the Docker Hub.
 (amd64)
 3. The Docker daemon created a new container from that image which runs
the
 executable that produces the output you are currently reading.
 4. The Docker daemon streamed that output to the Docker client, which sent
it
 to your terminal.

To try something more ambitious, you can run an Ubuntu container with:
 $ docker run -it ubuntu bash
```

```
Share images, automate workflows, and more with a free Docker ID:
 https://cloud.docker.com/

For more examples and ideas, visit:
 https://docs.docker.com/engine/userguide/
```

Docker is installed successfully. Now let us build a Docker image for TensorFlow Serving.

Building a Docker image for TF serving

We proceed with the Docker image for serving as follows:

1. Create the file named `dockerfile` with the following content:

```
FROM ubuntu:16.04
MAINTAINER Armando Fandango <armando@geekysalsero.com>

RUN apt-get update && apt-get install -y \
 build-essential \
 curl \
 git \
 libfreetype6-dev \
 libpng12-dev \
 libzmq3-dev \
 mlocate \
 pkg-config \
 python-dev \
 python-numpy \
 python-pip \
 software-properties-common \
 swig \
 zip \
 zlib1g-dev \
 libcurl3-dev \
 openjdk-8-jdk\
 openjdk-8-jre-headless \
 wget \
 && \
 apt-get clean && \
 rm -rf /var/lib/apt/lists/*

RUN echo "deb [arch=amd64]
http://storage.googleapis.com/tensorflow-serving-apt stable
tensorflow-model-server tensorflow-model-server-universal" \
 | tee /etc/apt/sources.list.d/tensorflow-serving.list
```

```
RUN curl
https://storage.googleapis.com/tensorflow-serving-apt/tensorflow-se
rving.release.pub.gpg \
  | apt-key add -

RUN apt-get update && apt-get install -y \
 tensorflow-model-server

RUN pip install --upgrade pip
RUN pip install mock grpcio tensorflow tensorflow-serving-api

CMD ["/bin/bash"]
```

2. Run the following command to build the Docker image from this `dockerfile`:

```
$ docker build --pull -t $USER/tensorflow_serving -f dockerfile .
```

3. It will take a while to create the image. When you see something similar to the following, then you know the image is built:

```
Removing intermediate container 1d8e757d96e0
Successfully built 0f95ddba4362
Successfully tagged armando/tensorflow_serving:latest
```

4. Run the following command to start the container:

```
$ docker run --name=mnist_container -it $USER/tensorflow_serving
```

5. You will be logged into the container when you see the following prompt:

```
root@244ea14efb8f:/#
```

6. Give the `cd` command to go to the home folder.
7. In the home folder, give the following command to check that TensorFlow is serving code. We will be using examples from this code to demonstrate, but you can check out your own Git repository to run your own models:

```
$ git clone --recurse-submodules
https://github.com/tensorflow/serving
```

Once the repository is cloned, we are ready to build, train, and save the MNIST model.

8. Remove the temp folder, if you haven't already, with the following command:

```
$ rm -rf /tmp/mnist_model
```

9. Run the following command to build, train, and save the MNIST model.

```
$ python serving/tensorflow_serving/example/mnist_saved_model.py
/tmp/mnist_model
```

You will see something similar to the following:

```
Training model...
Successfully downloaded train-images-idx3-ubyte.gz 9912422 bytes.
Extracting /tmp/train-images-idx3-ubyte.gz
Successfully downloaded train-labels-idx1-ubyte.gz 28881 bytes.
Extracting /tmp/train-labels-idx1-ubyte.gz
Successfully downloaded t10k-images-idx3-ubyte.gz 1648877 bytes.
Extracting /tmp/t10k-images-idx3-ubyte.gz
Successfully downloaded t10k-labels-idx1-ubyte.gz 4542 bytes.
Extracting /tmp/t10k-labels-idx1-ubyte.gz
2017-11-22 01:09:38.165391: I
tensorflow/core/platform/cpu_feature_guard.cc:137] Your CPU
supports instructions that this TensorFlow binary was not compiled
to use: SSE4.1 SSE4.2 AVX AVX2 FMA
training accuracy 0.9092
Done training!
Exporting trained model to /tmp/mnist_model/1
Done exporting!
```

10. Detach from the Docker image by pressing *Ctrl+P* and *Ctrl+Q*.
11. Commit the changes to the new image and stop the container with the following commands:

```
$ docker commit mnist_container $USER/mnist_serving
$ docker stop mnist_container
```

12. Now you can run this container at any time by giving the following command:

```
$ docker run --name=mnist_container -it $USER/mnist_serving
```

13. Remove the temporary MNIST container we built to save the image:

```
$ docker rm mnist_container
```

Serving the model in the Docker container

To serve the model in the container, the instructions are as follows:

1. Start the MNIST container built in the previous section:

```
$ docker run --name=mnist_container -it $USER/mnist_serving
```

2. Give the `cd` command to go to the home folder.
3. Run the ModelServer with the following command:

```
$ tensorflow_model_server  --model_name=mnist --
model_base_path=/tmp/mnist_model/ &> mnist_log &
```

4. Check the prediction from the model with the sample client:

```
$ python serving/tensorflow_serving/example/mnist_client.py --
num_tests=100 --server=localhost:8500
```

5. We see the error rate to be 7%, just like our previous notebook example execution:

```
Extracting /tmp/train-images-idx3-ubyte.gz
Extracting /tmp/train-labels-idx1-ubyte.gz
Extracting /tmp/t10k-images-idx3-ubyte.gz
Extracting /tmp/t10k-labels-idx1-ubyte.gz
.............................................................
................................
Inference error rate: 7.0%
```

That's it! You have built a Docker image and served your model in the Docker image. Issue the `exit` command to get out of the container.

TensorFlow Serving on Kubernetes

According to `https://kubernets.io`:

> *Kubernetes is an open-source system for automating deployment, scaling, and management of containerized applications.*

TensorFlow models can be scaled to be served from hundreds or thousands of `TF Serving` services using Kubernetes clusters in the production environment. Kubernetes clusters can be run on all popular public clouds, such as GCP, AWS, Azure, as well as in your on-premises private cloud. So let us dive right in to learn to install Kubernetes and then deploy the MNIST model on Kubernetes Cluster.

Installing Kubernetes

We installed Kubernetes on Ubuntu 16.04 in a single-node local cluster mode as per the following steps:

1. Install LXD and Docker, which are prerequisites to install Kubernetes locally. LXD is the container manager that works with linux containers. We already learned how to install Docker in the previous section. To install LXD, run the following command:

```
$ sudo snap install lxd
lxd 2.19 from 'canonical' installed
```

2. Initialize `lxd` and create the virtual network:

```
$ sudo /snap/bin/lxd init --auto
LXD has been successfully configured.

$ sudo /snap/bin/lxc network create lxdbr0 ipv4.address=auto
ipv4.nat=true ipv6.address=none ipv6.nat=false
If this is your first time using LXD, you should also run: lxd init
 To start your first container, try: lxc launch ubuntu:16.04

Network lxdbr0 created
```

3. Add your user to the `lxd` group:

```
$ sudo usermod -a -G lxd $(whoami)
```

4. Install `conjure-up` and restart the machine:

```
$ sudo snap install conjure-up --classic
conjure-up 2.3.1 from 'canonical' installed
```

5. Fire up `conjure-up` to install Kubernetes:

```
$ conjure-up kubernetes
```

6. From the list of spells select **Kubernetes Core**.
7. From the list of available clouds select **localhost.**
8. From the list of networks select **lxbr0 bridge.**
9. Provide the sudo password for the option: **Download the kubectl and kubefed client programs to your local host**.
10. In the next screen, it would ask to select the apps to install. Install all five of the remaining apps.

You know the Kubernetes cluster is ready to brew when the final screen during install looks like this:

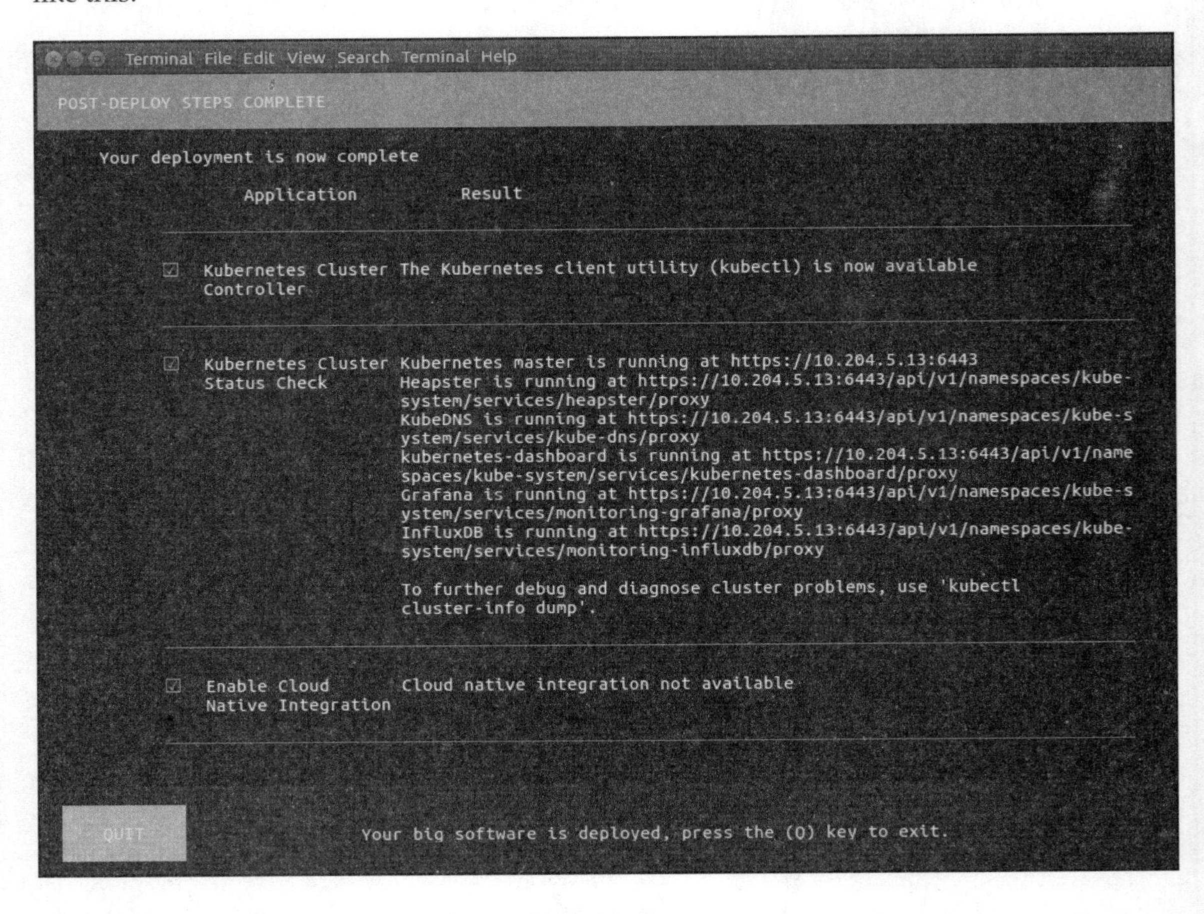

If you are having problems with the installation, please search for help on the internet, starting with the documentation at the following links:

https://kubernetes.io/docs/getting-started-guides/ubuntu/local/
https://kubernetes.io/docs/getting-started-guides/ubuntu/
https://tutorials.ubuntu.com/tutorial/install-kubernetes-with-conjure-up

Uploading the Docker image to the dockerhub

The steps to upload the Docker image to the dockerhub are as follows:

1. Create an account on the dockerhub if you haven't already.
2. Log in to the dockerhub account with the following command:

```
$ docker login --username=<username>
```

3. Tag the MNIST image with a repo that you have created on dockerhub. For example, we created `neurasights/mnist-serving`:

```
$ docker tag $USER/mnist_serving neurasights/mnist-serving
```

4. Push the tagged image to the dockerhub account.

```
$ docker push neurasights/mnist-serving
```

Deploying in Kubernetes

We proceed with deployment in Kubernotes as follows:

1. Create the `mnist.yaml` file with the following content:

```
apiVersion: extensions/v1beta1
kind: Deployment
metadata:
  name: mnist-deployment
spec:
  replicas: 3
  template:
    metadata:
      labels:
        app: mnist-server
    spec:
```

```
        containers:
        - name: mnist-container
          image: neurasights/mnist-serving
          command:
          - /bin/sh
          args:
          - -c
          - tensorflow_model_server --model_name=mnist --
model_base_path=/tmp/mnist_model
          ports:
          - containerPort: 8500
---
apiVersion: v1
kind: Service
metadata:
  labels:
    run: mnist-service
  name: mnist-service
spec:
  ports:
  - port: 8500
    targetPort: 8500
  selector:
    app: mnist-server
#  type: LoadBalancer
```

If you are running it in AWS or GCP clouds then uncomment the `LoadBalancer` line from the preceding file. Since we are running the whole cluster locally on a single node, we do not have external LoadBalancer.

2. Create the Kubernetes deployment and service:

```
$ kubectl create -f mnist.yaml
deployment "mnist-deployment" created
service "mnist-service" created
```

3. Check the deployments, pods, and services:

```
$ kubectl get deployments
NAME               DESIRED   CURRENT   UP-TO-DATE   AVAILABLE   AGE
mnist-deployment   3         3         3            0           1m

$ kubectl get pods
NAME                                READY   STATUS
RESTARTS   AGE
default-http-backend-bbchw          1/1     Running              3
```

```
9d
mnist-deployment-554f4b674b-pwk8z  0/1     ContainerCreating   0
1m
mnist-deployment-554f4b674b-vn6sd  0/1     ContainerCreating   0
1m
mnist-deployment-554f4b674b-zt4xt  0/1     ContainerCreating   0
1m
nginx-ingress-controller-724n5     1/1     Running             2
9d

$ kubectl get services
NAME                   TYPE           CLUSTER-IP       EXTERNAL-IP
PORT(S)          AGE
default-http-backend   ClusterIP      10.152.183.223   <none>
80/TCP           9d
kubernetes             ClusterIP      10.152.183.1     <none>
443/TCP          9d
mnist-service          LoadBalancer   10.152.183.66    <pending>
8500:32414/TCP   1m

$ kubectl describe service mnist-service
Name:                     mnist-service
Namespace:                default
Labels:                   run=mnist-service
Annotations:              <none>
Selector:                 app=mnist-server
Type:                     LoadBalancer
IP:                       10.152.183.66
Port:                     <unset>  8500/TCP
TargetPort:               8500/TCP
NodePort:                 <unset>  32414/TCP
Endpoints:
10.1.43.122:8500,10.1.43.123:8500,10.1.43.124:8500
Session Affinity:         None
External Traffic Policy:  Cluster
Events:                   <none>
```

4. Wait until the status of all the pods is `Running`:

```
$ kubectl get pods
NAME                               READY     STATUS     RESTARTS
AGE
default-http-backend-bbchw         1/1       Running    3
9d
mnist-deployment-554f4b674b-pwk8z  1/1       Running    0
3m
mnist-deployment-554f4b674b-vn6sd  1/1       Running    0
3m
```

```
mnist-deployment-554f4b674b-zt4xt   1/1     Running   0
3m
nginx-ingress-controller-724n5      1/1     Running   2
9d
```

5. Check the logs of one of the pods, and you should see something like this :

```
$ kubectl logs mnist-deployment-59dfc5df64-g7prf
I tensorflow_serving/model_servers/main.cc:147] Building single
TensorFlow model file config: model_name: mnist model_base_path:
/tmp/mnist_model
I tensorflow_serving/model_servers/server_core.cc:441]
Adding/updating models.
I tensorflow_serving/model_servers/server_core.cc:492] (Re-)adding
model: mnist
I tensorflow_serving/core/basic_manager.cc:705] Successfully
reserved resources to load servable {name: mnist version: 1}
I tensorflow_serving/core/loader_harness.cc:66] Approving load for
servable version {name: mnist version: 1}
I tensorflow_serving/core/loader_harness.cc:74] Loading servable
version {name: mnist version: 1}
I
external/org_tensorflow/tensorflow/contrib/session_bundle/bundle_sh
im.cc:360] Attempting to load native SavedModelBundle in bundle-
shim from: /tmp/mnist_model/1
I external/org_tensorflow/tensorflow/cc/saved_model/loader.cc:236]
Loading SavedModel from: /tmp/mnist_model/1
I
external/org_tensorflow/tensorflow/core/platform/cpu_feature_guard.
cc:137] Your CPU supports instructions that this TensorFlow binary
was not compiled to use: AVX2 FMA
I external/org_tensorflow/tensorflow/cc/saved_model/loader.cc:155]
Restoring SavedModel bundle.
I external/org_tensorflow/tensorflow/cc/saved_model/loader.cc:190]
Running LegacyInitOp on SavedModel bundle.
I external/org_tensorflow/tensorflow/cc/saved_model/loader.cc:284]
Loading SavedModel: success. Took 45319 microseconds.
I tensorflow_serving/core/loader_harness.cc:86] Successfully loaded
servable version {name: mnist version: 1}
E1122 12:18:04.566415410 6 ev_epoll1_linux.c:1051] grpc epoll fd: 3
I tensorflow_serving/model_servers/main.cc:288] Running ModelServer
at 0.0.0.0:8500 ...
```

6. You can also look at the UI console with the following command:

```
$ kubectl proxy xdg-open http://localhost:8001/ui
```

The Kubernetes UI console looks like the following images:

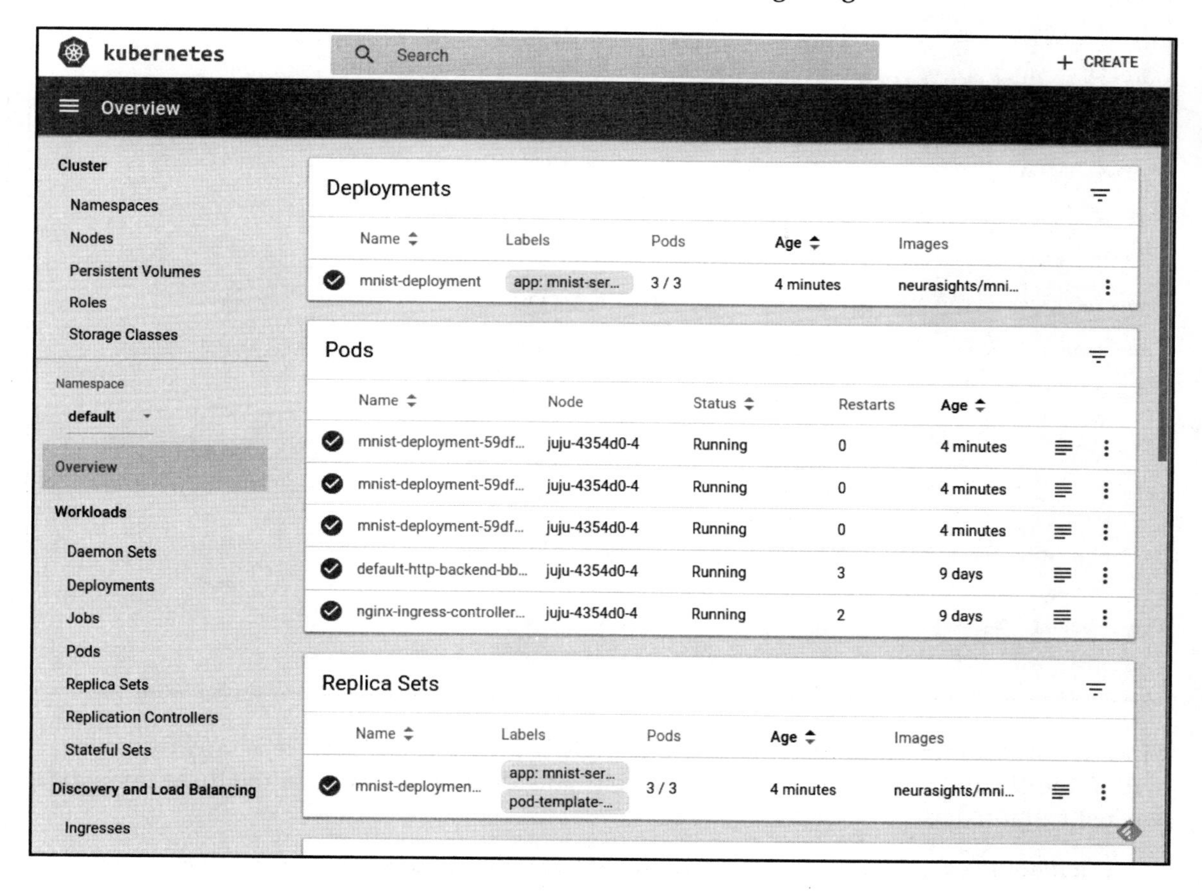

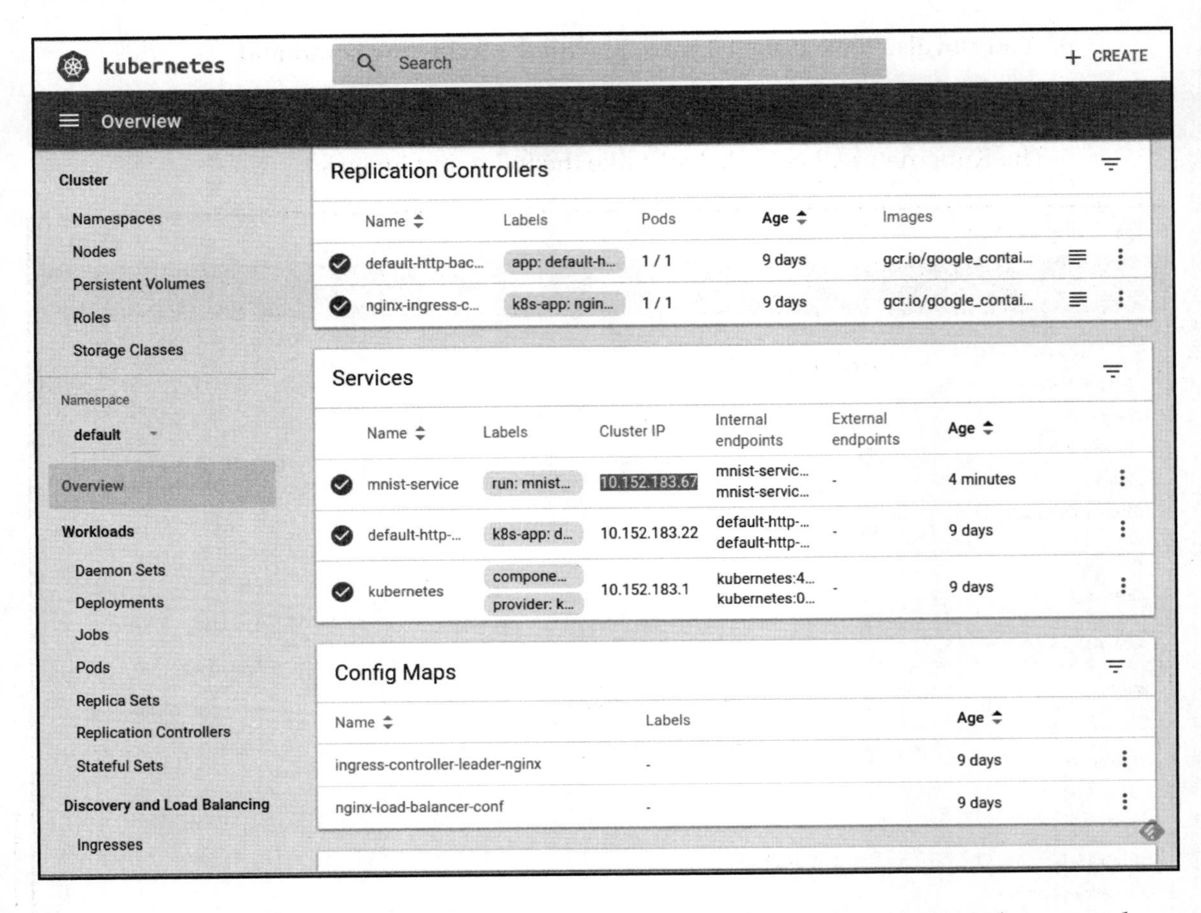

Since we are running the cluster locally on single Node, thus our service is only exposed within the cluster and cannot be accessed from outside. Log in to one of the three pods that we just instantiated:

```
$ kubectl exec -it mnist-deployment-59dfc5df64-bb24q -- /bin/bash
```

Change to the home directory and run the MNIST client to test the service:

```
$ kubectl exec -it mnist-deployment-59dfc5df64-bb24q -- /bin/bash
root@mnist-deployment-59dfc5df64-bb24q:/# cd
root@mnist-deployment-59dfc5df64-bb24q:~# python
serving/tensorflow_serving/example/mnist_client.py --num_tests=100 --
server=10.152.183.67:8500
Extracting /tmp/train-images-idx3-ubyte.gz
Extracting /tmp/train-labels-idx1-ubyte.gz
Extracting /tmp/t10k-images-idx3-ubyte.gz
```

```
Extracting /tmp/t10k-labels-idx1-ubyte.gz
......................................................................
........................
Inference error rate: 7.0%
root@mnist-deployment-59dfc5df64-bb24q:~#
```

We learned how to deploy TensorFlow serving on the Kubernetes cluster running on single node locally. You can use the same conceptual knowledge to deploy the serving across the public clouds or private clouds on your premises.

Summary

In this chapter, we learned how to leverage the TensorFlow Serving to serve the models in production environments. We also learned how to save and restore full models or selective models using both TensorFlow and Keras. We built a Docker container and served the sample MNIST example code in the Docker container from the official TensorFlow Serving repository. We also installed a local Kubernetes cluster and deployed the MNIST model to serve from TensorFlow Serving running in Kubernetes pods. We encourage the reader to build upon these examples and try out serving different models. TF Serving documentation describes various options and provides additional information enabling you to explore this topic further.

In the coming chapters, we will continue our journey with advanced models using transfer learning. The pre-trained models available in the TensorFlow repository are the best candidates to practice serving the TF models using TF serving. We installed the TF Serving using Ubuntu packages but you may want to try building from the source to optimize TF Serving installation for your production environment.

12
Transfer Learning and Pre-Trained Models

In simple words, **transfer learning** means that you take a pre-trained model trained to predict one kind of class, and then either use it directly or re-train only a small part of it, in order to predict another kind of class. For example, you can take a pre-trained model to identify types of cats, and then retrain only small parts of the model on the types of dogs and then use it to predict the types of dogs.

Without transfer learning, training a huge model on large datasets would take several days or even months. However, with transfer learning, by taking a pre-trained model, and only training the last couple of layers, we save a lot of time in training the model from scratch.

Transfer learning is also useful when you don't have a huge dataset. The models trained on small datasets may not be able to detect features that a model trained on large datasets can. Thus with transfer learning, you can get better models, even with smaller datasets.

In this chapter, we shall take pre-trained models and train them on new objects. We show examples of pre-trained models with images and apply them to image classification problems. You should try to find other pre-trained models and apply them to different problems such as object detection, text generation or machine translation. The following topics will be covered in this chapter:

- ImageNet dataset
- Retraining or fine-tuning the models
- COCO animals dataset and preprocessing
- Image classification using pre-trained VGG16 in TensorFlow

- Image preprocessing in TensorFlow for pre-trained VGG16
- Image classification using ***retrained*** VGG16 in TensorFlow
- Image classification using pre-trained VGG16 in Keras
- Image classification using ***retrained*** VGG16 in Keras
- Image classification using Inception v3 in TensorFlow
- Image classification using ***retrained*** Inception v3 in TensorFlow

ImageNet dataset

According to `http://image-net.org`:

> *ImageNet is an image dataset organized according to the WordNet hierarchy. Each meaningful concept in WordNet, possibly described by multiple words or word phrases, is called a synonym set or synset.*

The ImageNet has about 100 K synsets and on average, about 1,000 human-annotated images for each synset. ImageNet only stores the references to the images while the images are stored in their original locations on the internet. In deep learning papers, the ImageNet-1K refers to the dataset released as part of ImageNet's **Large Scale Visual Recognition Challenge** (**ILSVRC**) to classify the dataset into 1,000 categories:

The 1,000 categories for the challenges can be found at the following URLs:
`http://image-net.org/challenges/LSVRC/2017/browse-synsets`
`http://image-net.org/challenges/LSVRC/2016/browse-synsets`
`http://image-net.org/challenges/LSVRC/2015/browse-synsets`
`http://image-net.org/challenges/LSVRC/2014/browse-synsets`
`http://image-net.org/challenges/LSVRC/2013/browse-synsets`
`http://image-net.org/challenges/LSVRC/2012/browse-synsets`
`http://image-net.org/challenges/LSVRC/2011/browse-synsets`
`http://image-net.org/challenges/LSVRC/2010/browse-synsets`.

We have written a custom function to download the ImageNet labels from Google:

```
def build_id2label(self):
    base_url =
'https://raw.githubusercontent.com/tensorflow/models/master/research/incept
ion/inception/data/'
    synset_url = '{}/imagenet_lsvrc_2015_synsets.txt'.format(base_url)
    synset_to_human_url = '{}/imagenet_metadata.txt'.format(base_url)

    filename, _ = urllib.request.urlretrieve(synset_url)
    synset_list = [s.strip() for s in open(filename).readlines()]
```

```
    num_synsets_in_ilsvrc = len(synset_list)
    assert num_synsets_in_ilsvrc == 1000

    filename, _ = urllib.request.urlretrieve(synset_to_human_url)
    synset_to_human_list = open(filename).readlines()
    num_synsets_in_all_imagenet = len(synset_to_human_list)
    assert num_synsets_in_all_imagenet == 21842

    synset2name = {}
    for s in synset_to_human_list:
        parts = s.strip().split('\t')
        assert len(parts) == 2
        synset = parts[0]
        name = parts[1]
        synset2name[synset] = name

    if self.n_classes == 1001:
        id2label={0:'empty'}
        id=1
    else:
        id2label = {}
        id=0

    for synset in synset_list:
        label = synset2name[synset]
        id2label[id] = label
        id += 1

    return id2label
```

We load these labels into our Jupyter Notebook as follows:

```
### Load ImageNet dataset for labels
from datasetslib.imagenet import imageNet
inet = imageNet()
inet.load_data(n_classes=1000)
#n_classes is 1001 for Inception models and 1000 for VGG models
```

Popular pre-trained image classification models that have been trained on the ImageNet-1K dataset are listed in the following table:

Model Name	**Top-1 Accuracy**	**Top-5 Accuracy**	**Top-5 Error Rate**	**Link to Original Paper**
AlexNet			**15.3%**	`https://www.cs.toronto.edu/~fritz/absps/imagenet.pdf`
Inception also known as Inception V1	69.8	89.6	6.67%	`https://arxiv.org/abs/1409.4842`
BN-Inception-v2 also known as Inception V2	73.9	91.8	4.9%	`https://arxiv.org/abs/1502.03167`
Inception V3	78.0	93.9	3.46%	`https://arxiv.org/abs/1512.00567`
Inception V4	80.2	95.2		`http://arxiv.org/abs/1602.07261`
Inception-Resenet-V2	80.4	95.2		`http://arxiv.org/abs/1602.07261`
VGG16	71.5	89.8	7.4%	`https://arxiv.org/abs/1409.1556`
VGG19	71.1	89.8	7.3%	`https://arxiv.org/abs/1409.1556`
ResNet V1 50	75.2	92.2	7.24%	`https://arxiv.org/abs/1512.03385`
Resnet V1 101	76.4	92.9		`https://arxiv.org/abs/1512.03385`
Resnet V1 152	76.8	93.2		`https://arxiv.org/abs/1512.03385`
ResNet V2 50	75.6	92.8		`https://arxiv.org/abs/1603.05027`
ResNet V2 101	77.0	93.7		`https://arxiv.org/abs/1603.05027`
ResNet V2 152	77.8	94.1		`https://arxiv.org/abs/1603.05027`

Model Name	Top-1 Accuracy	Top-5 Accuracy	Top-5 Error Rate	Link to Original Paper
ResNet V2 200	79.9	95.2		`https://arxiv.org/abs/1603.05027`
Xception	79.0	94.5		`https://arxiv.org/abs/1610.02357`
MobileNet V1 versions	41.3 to 70.7	66.2 to 89.5		`https://arxiv.org/pdf/1704.04861.pdf`

In the preceding table, Top-1 and Top-5 metrics refer to the model's performance on the ImageNet validation dataset.

Google Research recently released a newer kind of model known as MobileNets. MobileNets have been developed with a mobile-first strategy that sacrifices accuracy for low-resource usage. MobileNets are designed to consume low power and provide low latency to provide a better experience on mobile and embedded devices. Google has provided 16 pre-trained checkpoint files for MobileNet models, with each model offering a different number of parameters and **Multiply-Accumulates** (**MAC**). The higher the MAC and parameters, the higher the resource usage and latency. So you can pick and choose between higher accuracy versus higher-resource usage/latency.

Model Checkpoint	Million MACs	Million Parameters	Top-1 Accuracy	Top-5 Accuracy
`MobileNet_v1_1.0_224`	**569**	**4.24**	**70.7**	**89.5**
`MobileNet_v1_1.0_192`	418	4.24	69.3	88.9
`MobileNet_v1_1.0_160`	291	4.24	67.2	87.5
`MobileNet_v1_1.0_128`	186	4.24	64.1	85.3
`MobileNet_v1_0.75_224`	317	2.59	68.4	88.2
`MobileNet_v1_0.75_192`	233	2.59	67.4	87.3
`MobileNet_v1_0.75_160`	162	2.59	65.2	86.1
`MobileNet_v1_0.75_128`	104	2.59	61.8	83.6
`MobileNet_v1_0.50_224`	150	1.34	64.0	85.4
`MobileNet_v1_0.50_192`	110	1.34	62.1	84.0

Model Checkpoint	Million MACs	Million Parameters	Top-1 Accuracy	Top-5 Accuracy
`MobileNet_v1_0.50_160`	77	1.34	59.9	82.5
`MobileNet_v1_0.50_128`	49	1.34	56.2	79.6
`MobileNet_v1_0.25_224`	41	0.47	50.6	75.0
`MobileNet_v1_0.25_192`	34	0.47	49.0	73.6
`MobileNet_v1_0.25_160`	21	0.47	46.0	70.7
`MobileNet_v1_0.25_128`	14	0.47	41.3	66.2

More information about MobileNets can be found in the following resources:

`https://research.googleblog.com/2017/06/mobilenets-open-source-models-for.html`
`https://github.com/tensorflow/models/blob/master/research/slim/nets/mobilenet_v1.md`
`https://arxiv.org/pdf/1704.04861.pdf.`

Retraining or fine-tuning models

Models trained on large and diverse datasets like ImageNet are able to detect and capture some of the universal features such as curves, edges, and shapes. Some of these features are easily applicable to other kinds of datasets. Thus, in transfer learning we take such universal models and use some of the following techniques to fine-tune or retrain them to our datasets:

- **Repeal and replace the last layer:** The most common practice is to remove the last layer and add the new classification layer that matches our dataset. For example, ImageNet models are trained with 1,000 categories, but our COCO animals dataset is only 8 classes, thus we remove the softmax layer that generates probabilities for 1,000 classes with a softmax layer that generates probabilities for 8 classes. Generally, this technique is used when the new dataset is almost similar to the one that the model was trained on, thus needing only the last layer to be retrained.

- **Freeze the first few layers**: Another common practice is to freeze the first few layers such that the weights of only the last unfrozen layers get updated with the new dataset. We shall see an example of this where we freeze the first 15 layers while retraining only the last 10 layers. Generally, this technique is used when the new dataset is not very similar to the one that the model was trained on, thus requiring more than just the last layers to be trained.
- **Tune the hyperparameters**: You can also tune the hyperparameters before retraining, such as changing the learning rate or trying a different loss function or different optimizer.

The pre-trained models are available in both TensorFlow and Keras.

We shall demonstrate our examples through TensorFlow Slim which has several pre-trained models available at the time of writing, in the folder `tensorflow/models/research/slim/nets`. We will use TensorFlow Slim to instantiate the pre-trained models and then load the weights from a downloaded checkpoint file. The loaded models will then be used for prediction with the new dataset. We shall then retrain the models to fine-tune the predictions.

We shall also demonstrate the transfer learning through Keras pre-trained models available in the `keras.applications` module. While TensorFlow has about 20+ pre-trained models, `keras.appplications` has only the following 7 pre-trained models :

- `Xception` - `https://keras.io/applications/#xception`
- `VGG16` - `https://keras.io/applications/#vgg16`
- `VGG19` - `https://keras.io/applications/#vgg19`
- `ResNet50` - `https://keras.io/applications/#resnet50`
- `InceptionV3` - `https://keras.io/applications/#inceptionv3`
- `InceptionResNetV2` - `https://keras.io/applications/#inceptionresnetv2`
- `MobileNet` - `https://keras.io/applications/#mobilenet`

COCO animals dataset and pre-processing images

For our examples, we shall use the COCO animals dataset, which is a smaller subset of the COCO dataset made available by the researchers at the Stanford University at the following link: `http://cs231n.stanford.edu/coco-animals.zip`. The COCO animals dataset has 800 training images and 200 test images of 8 classes of animals: bear, bird, cat, dog, giraffe, horse, sheep, and zebra. The images are downloaded and pre-processed for the VGG16 and Inception models.

For the VGG model, the image size is 224 x 224 and the preprocessing steps are as follows:

1. Images are resized to 224 x 224 with a function similar to the `tf.image.resize_image_with_crop_or_pad` function from TensorFlow. We implemented this function as follows:

```
def resize_image(self,in_image:PIL.Image, new_width,
    new_height, crop_or_pad=True):
    img = in_image
    if crop_or_pad:
        half_width = img.size[0] // 2
        half_height = img.size[1] // 2
        half_new_width = new_width // 2
        half_new_height = new_height // 2
        img = img.crop((half_width-half_new_width,
                        half_height-half_new_height,
                        half_width+half_new_width,
                        half_height+half_new_height
                        ))
    img = img.resize(size=(new_width, new_height))

    return img
```

2. After resizing, convert the image from PIL.Image to NumPy Array and check if the image has depth channels, as some of the images in the dataset are only grayscale.

```
img = self.pil_to_nparray(img)
if len(img.shape)==2:
    # greyscale or no channels then add three channels
    h=img.shape[0]
    w=img.shape[1]
    img = np.dstack([img]*3)
```

3. Then we subtract the VGG dataset means from the image to center the data. The reason we center the data for our new training images is so that the features have a similar range as the initial data that was used to rain the model. By making features in the similar range, we make sure that the gradients during retraining do not become too high or too low. Also by centering the data, the learning process becomes faster because gradients become uniform for each channel centered around the zero mean.

```
means = np.array([[[123.68, 116.78, 103.94]]]) #shape=[1, 1, 3]
img = img - means
```

The complete preprocess function is as follows:

```
def preprocess_for_vgg(self,incoming, height, width):
    if isinstance(incoming, six.string_types):
        img = self.load_image(incoming)
    else:
        img=incoming
    img_size = vgg.vgg_16.default_image_size
    height = img_size
    width = img_size
    img = self.resize_image(img,height,width)
    img = self.pil_to_nparray(img)
    if len(img.shape)==2:
        # greyscale or no channels then add three channels
        h=img.shape[0]
        w=img.shape[1]
        img = np.dstack([img]*3)

    means = np.array([[[123.68, 116.78, 103.94]]]) #shape=[1, 1, 3]
    try:
        img = img - means
    except Exception as ex:
        print('Error preprocessing ',incoming)
        print(ex)

    return img
```

For the Inception model, the image size is 299 x 299 and the preprocessing steps are as follows:

1. Images are resized to 299 x 299 with a function similar to `tf.image.resize_image_with_crop_or_pad` function from TensorFlow. We implemented this function as defined earlier in the VGG preprocessing steps.
2. The image is then scaled to the range (-1,+1) with the following code:

```
img = ((img/255.0) - 0.5) * 2.0
```

The complete preprocess function is as follows:

```
def preprocess_for_inception(self,incoming):
    img_size = inception.inception_v3.default_image_size
    height = img_size
    width = img_size
    if isinstance(incoming, six.string_types):
        img = self.load_image(incoming)
    else:
        img=incoming
    img = self.resize_image(img,height,width)
    img = self.pil_to_nparray(img)
    if len(img.shape)==2:
        # greyscale or no channels then add three channels
        h=img.shape[0]
        w=img.shape[1]
        img = np.dstack([img]*3)
    img = ((img/255.0) - 0.5) * 2.0

    return img
```

Let us load the COCO animals dataset:

```
from datasetslib.coco import coco_animals
coco = coco_animals()
x_train_files, y_train, x_val_files, x_val = coco.load_data()
```

We take one image from each class in the validation set, to make the list, `x_test` and preprocess the images to make the list `images_test`:

```
x_test = [x_val_files[25*x] for x in range(8)]
images_test=np.array([coco.preprocess_for_vgg(x) for x in x_test])
```

We use this helper function to display the images and probabilities of the top five classes associated with the image:

```
# helper function
def disp(images,id2label=None,probs=None,n_top=5,scale=False):
    if scale:
        imgs = np.abs(images + np.array([[[[123.68,
                116.78, 103.94]]]]))/255.0
    else:
        imgs = images
    ids={}
    for j in range(len(images)):
        if scale:
            plt.figure(figsize=(5,5))
            plt.imshow(imgs[j])
        else:
            plt.imshow(imgs[j].astype(np.uint8) )
        plt.show()
        if probs is not None:
            ids[j] = [i[0] for i in sorted(enumerate(-probs[j]),
                        key=lambda x:x[1])]
            for k in range(n_top):
                id = ids[j][k]
                print('Probability {0:1.2f}% of[{1:}]'
                    .format(100*probs[j,id],id2label[id]))
```

The following code in the preceding function reverts back to the effect of preprocessing, so as to display the original image instead of the preprocessed image:

```
imgs = np.abs(images + np.array([[[[123.68, 116.78, 103.94]]]]))/255.0
```

In the case of the Inception model, the code for reversing the preprocessing is as follows:

```
imgs = (images / 2.0) + 0.5
```

You can look at the test images with the following code:

```
images=np.array([mpimg.imread(x) for x in x_test])
disp(images)
```

Follow the code in the Jupyter notebook to see the images. They all appear to have different dimensions, so let us print their original dimensions:

```
print([x.shape for x in images])
```

And the dimensions are:

```
[(640, 425, 3), (373, 500, 3), (367, 640, 3), (427, 640, 3), (428, 640, 3),
(426, 640, 3), (480, 640, 3), (612, 612, 3)]
```

Let us preprocess the test images and see the dimensions:

```
images_test=np.array([coco.preprocess_for_vgg(x) for x in x_test])
print(images_test.shape)
```

And the dimensions are:

```
(8, 224, 224, 3)
```

In the case of Inception, the dimensions are:

```
(8, 299, 299, 3)
```

The preprocessed images for Inception are not visible but let us print the preprocessed images for VGG to get an idea of what they look like:

```
disp(images_test)
```

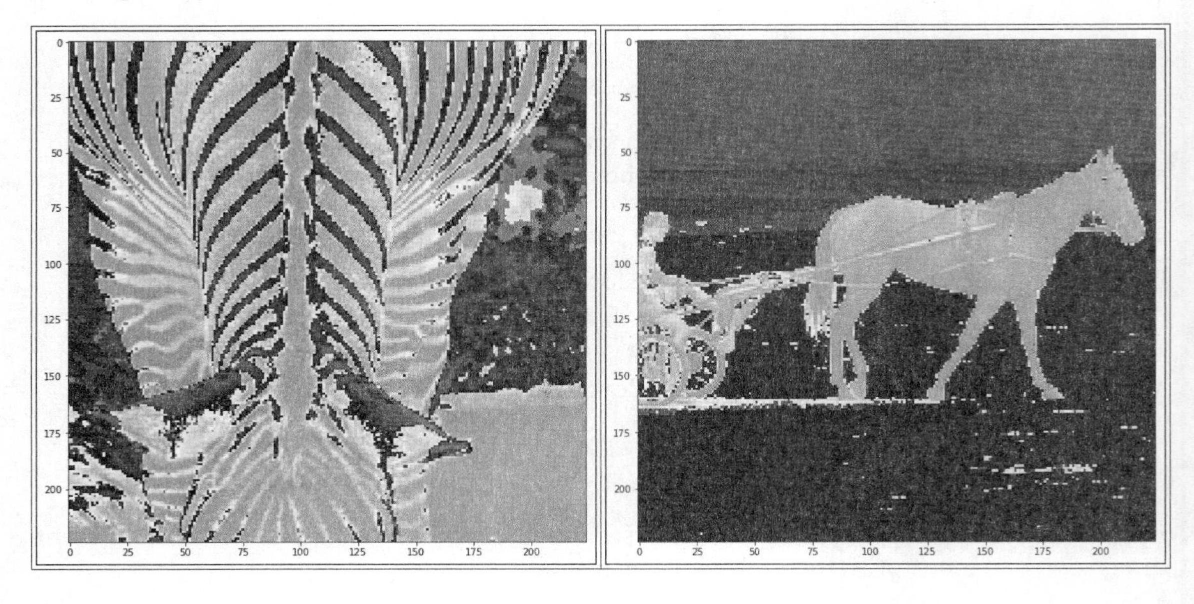

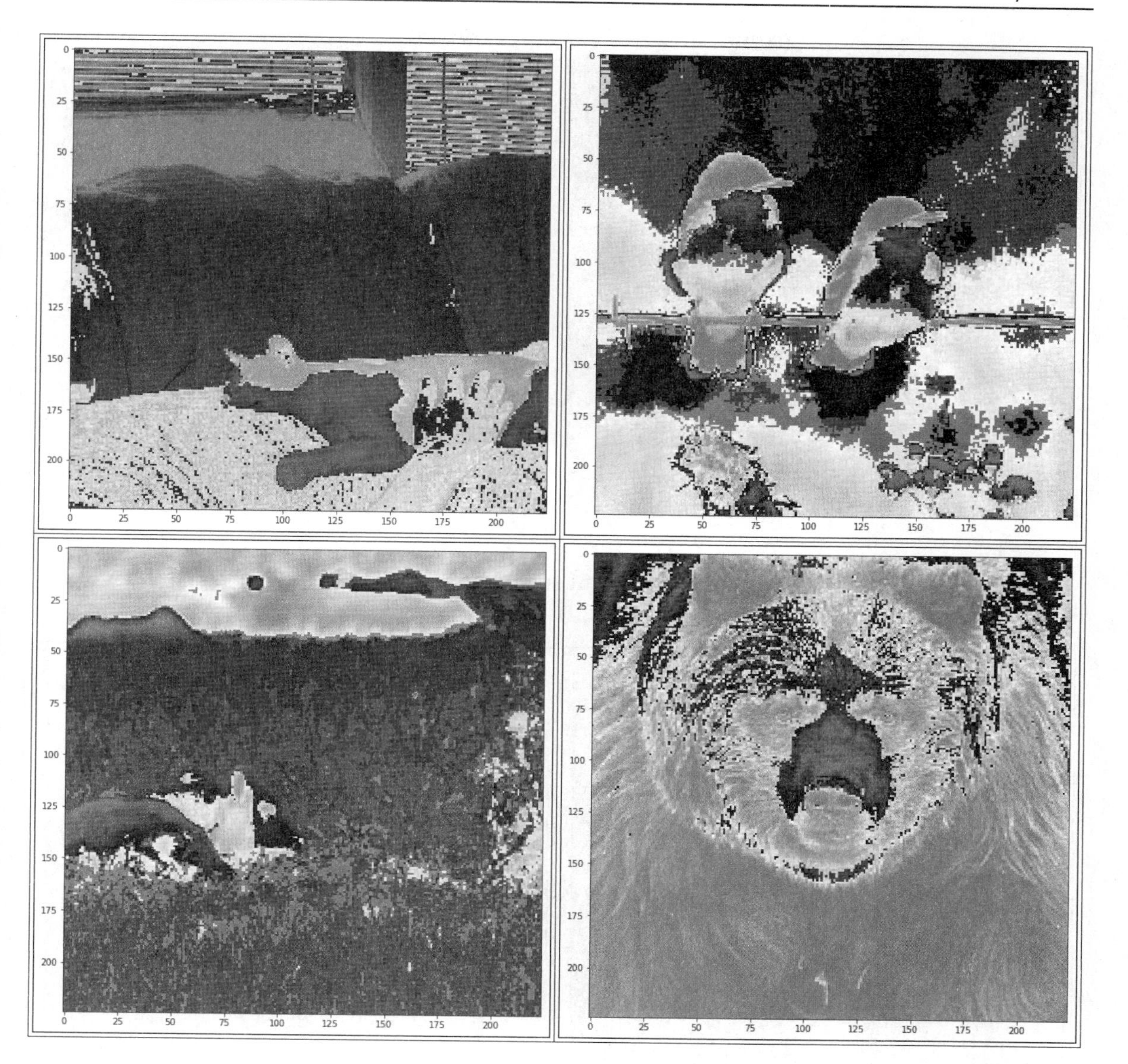
0
25
50
75
100
125
150
175
200

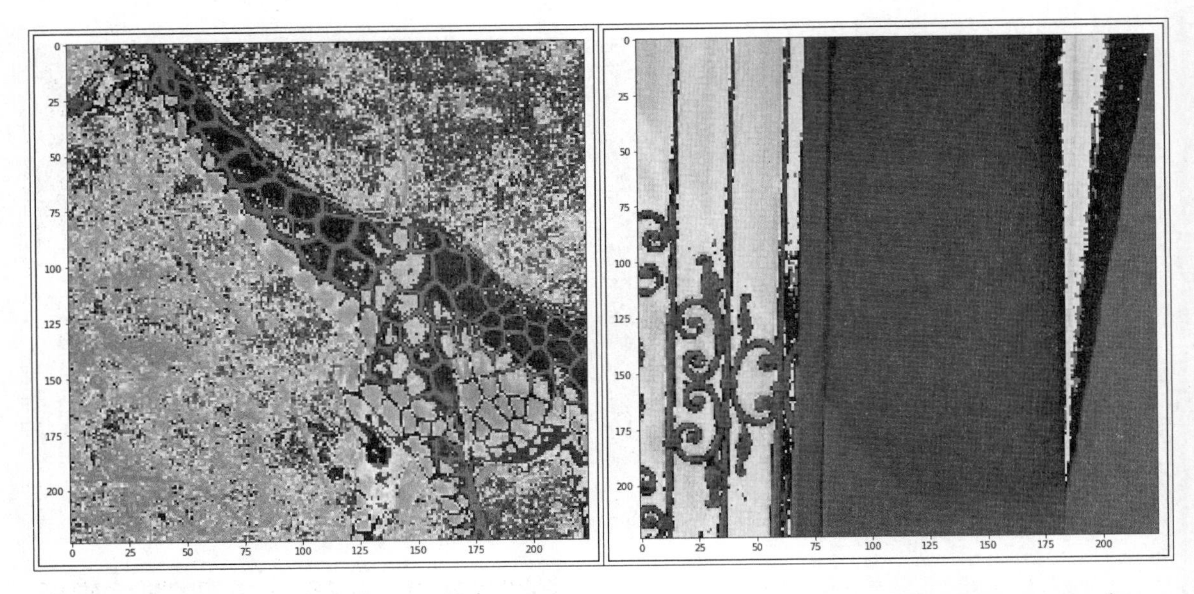

The images actually get cropped and we can see what they look like when we reverse the preprocessing while keeping the cropping:

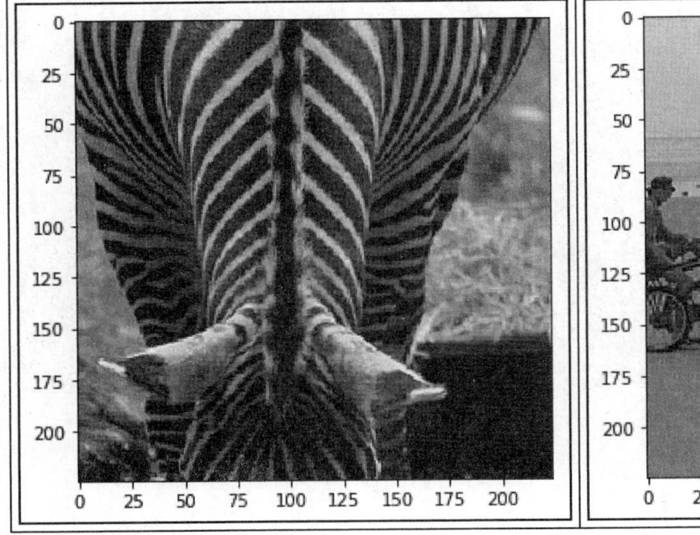

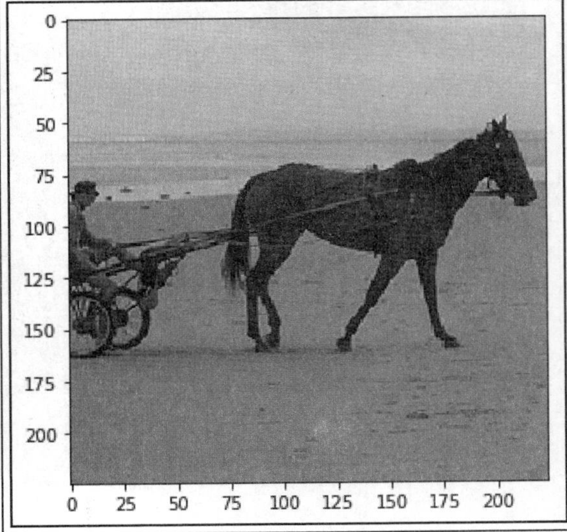

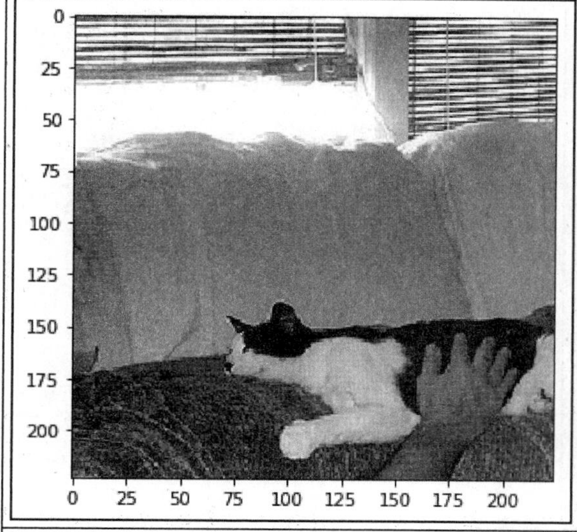
0
25
50
75
100
125
150
175
200
0 25 50 75 100 125 150 175 200

0
25
50
75
100
125
150
175
200
0 25 50 75 100 125 150 175 200

0
25
50
75
100
125
150
175
200
0 25 50 75 100 125 150 175 200

0
25
50
75
100
125
150
175
200
0 25 50 75 100 125 150 175 200

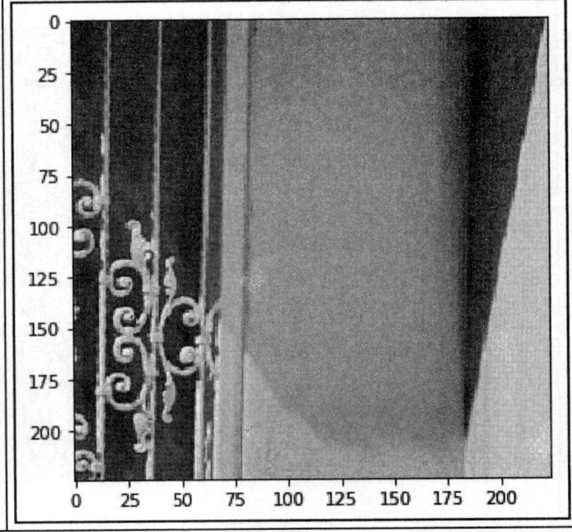

Now that we have labels from ImageNet and both images and labels from the COCO images dataset loaded, let us try the transfer learning examples.

VGG16 in TensorFlow

You can follow along with the code in the Jupyter notebook `ch-12a_VGG16_TensorFlow`.

For all examples of VGG16 in TensorFlow, we first download the checkpoint file from `http://download.tensorflow.org/models/vgg_16_2016_08_28.tar.gz` and initialize variables with the following code:

```
model_name='vgg_16'
model_url='http://download.tensorflow.org/models/'
model_files=['vgg_16_2016_08_28.tar.gz']
model_home=os.path.join(models_root,model_name)

dsu.download_dataset(source_url=model_url,
    source_files=model_files,
    dest_dir = model_home,
    force=False,
```

```
        extract=True)
```

We also define some common imports and variables:

```
from tensorflow.contrib import slim
from tensorflow.contrib.slim.nets import vgg
image_height=vgg.vgg_16.default_image_size
image_width=vgg.vgg_16.default_image_size
```

Image classification using pre-trained VGG16 in TensorFlow

Now let us first try to predict the classes of our test images, without retraining. First, we clear the default graph and define a placeholder for images:

```
tf.reset_default_graph()
x_p = tf.placeholder(shape=(None,image_height, image_width,3),
                     dtype=tf.float32,name='x_p')
```

The shape of placeholder `x_p` is `(?, 224, 224, 3)`. Next, load the `vgg16` model:

```
with slim.arg_scope(vgg.vgg_arg_scope()):
    logits,_ = vgg.vgg_16(x_p,num_classes=inet.n_classes,
                          is_training=False)
```

Add the softmax layer for producing probabilities over the classes:

```
probabilities = tf.nn.softmax(logits)
```

Define the initialization function to restore the variables, such as weights and biases from the checkpoint file.

```
init = slim.assign_from_checkpoint_fn(
    os.path.join(model_home, '{}.ckpt'.format(model_name)),
    slim.get_variables_to_restore())
```

Within a TensorFlow session, initialize the variables and run the probabilities tensor to get the probabilities of each image:

```
with tf.Session() as tfs:
    init(tfs)
    probs = tfs.run([probabilities],feed_dict={x_p:images_test})
    probs=probs[0]
```

Let us see what classes we get:

```
disp(images_test,id2label=inet.id2label,probs=probs,scale=True)
```

Input image	**Output probabilities**

	`Probability 99.15% of [zebra]`
	`Probability 0.37% of [tiger cat]`
	`Probability 0.33% of [tiger, Panthera tigris]`
	`Probability 0.04% of [goose]`
	`Probability 0.02% of [tabby, tabby cat]`

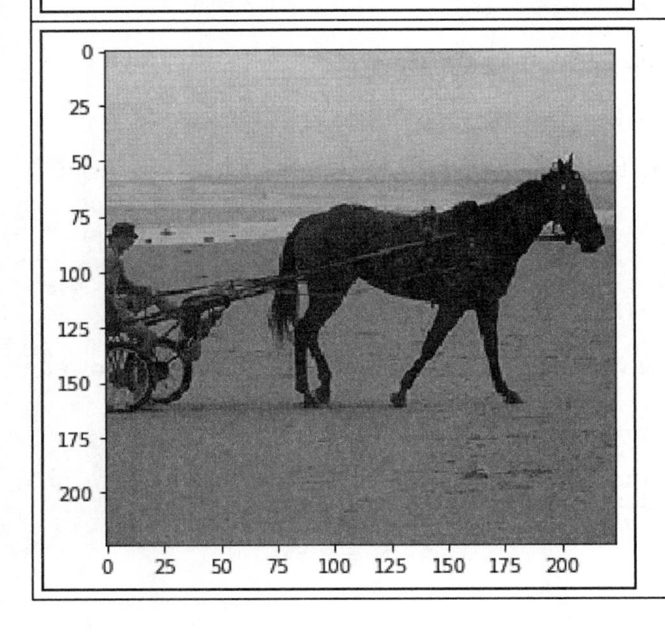
 	
	`Probability 99.50% of [horse cart, horse-cart]`
	`Probability 0.37% of [plow, plough]`
	`Probability 0.06% of [Arabian camel, dromedary, Camelus dromedarius]`
	`Probability 0.05% of [sorrel]`
	`Probability 0.01% of [barrel, cask]`

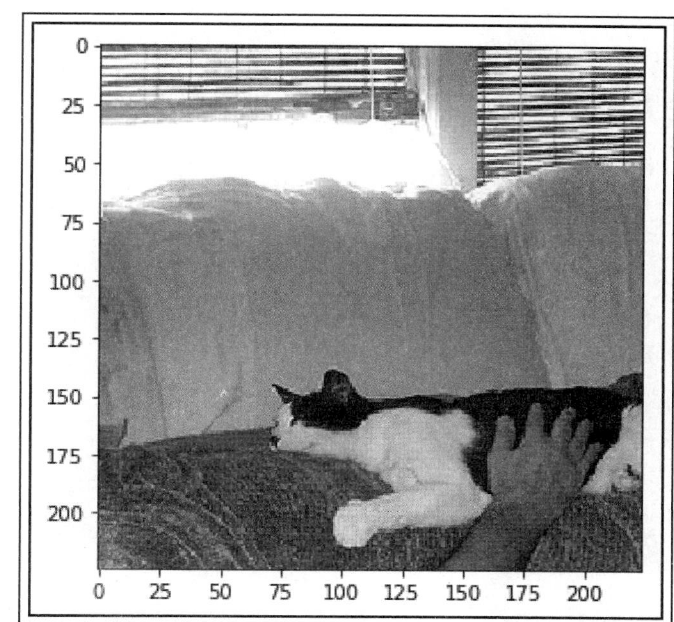


```
Probability 19.32% of [Cardigan,
Cardigan Welsh corgi]
Probability 11.78% of [papillon]
Probability 9.01% of [Shetland
sheepdog, Shetland sheep dog,
Shetland] Probability 7.09% of
[Siamese cat, Siamese]
Probability 6.27% of [Pembroke,
Pembroke Welsh corgi]
```

```
Probability 97.09% of
[chickadee]
Probability 2.52% of [water
ouzel, dipper]
Probability 0.23% of [junco,
snowbird]
Probability 0.09% of
[hummingbird]
Probability 0.04% of [bulbul]
```

	Probability 24.98% of [whippet] Probability 16.48% of [lion, king of beasts, Panthera leo] Probability 5.54% of [Saluki, gazelle hound] Probability 4.99% of [brown bear, bruin, Ursus arctos] Probability 4.11% of [wire-haired fox terrier]
	Probability 98.56% of [brown bear, bruin, Ursus arctos] Probability 1.40% of [American black bear, black bear, Ursus americanus, Euarctos americanus] Probability 0.03% of [sloth bear, Melursus ursinus, Ursus ursinus] Probability 0.00% of [wombat] Probability 0.00% of [beaver]

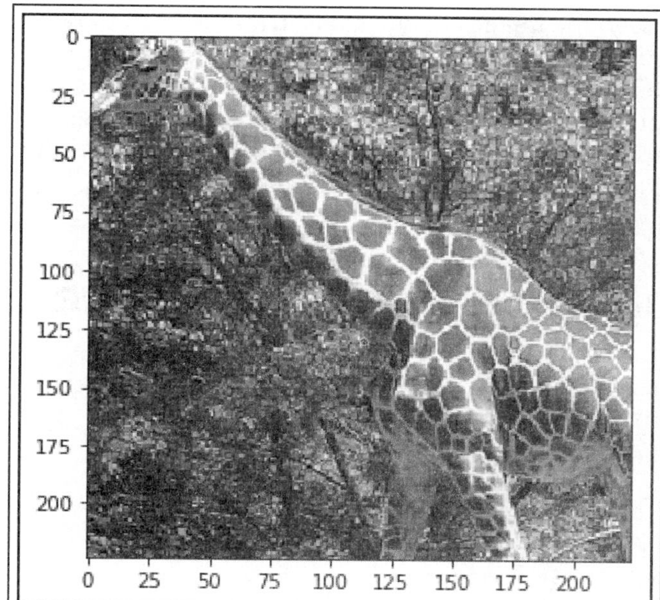	Probability 20.84% of [leopard, Panthera pardus] Probability 12.81% of [cheetah, chetah, Acinonyx jubatus] Probability 12.26% of [banded gecko] Probability 10.28% of [jaguar, panther, Panthera onca, Felis onca] Probability 5.30% of [gazelle]
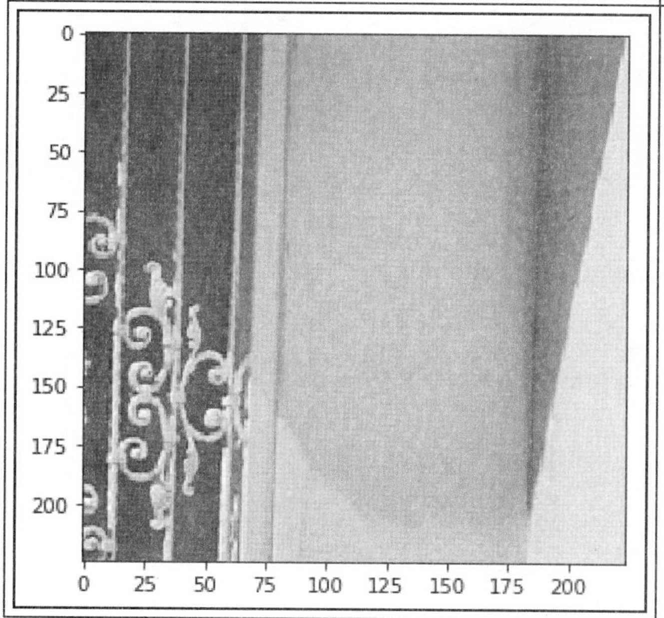	Probability 8.09% of [shower curtain] Probability 3.59% of [binder, ring-binder] Probability 3.32% of [accordion, piano accordion, squeeze box] Probability 3.12% of [radiator] Probability 1.81% of [abaya]

The pre-trained model which has never seen the images in our dataset, and does not know anything about the classes in our dataset, has correctly identified the zebra, horse-cart, bird, and bear. It failed to recognize the giraffe because it has never seen a giraffe before. We shall retrain this model on our dataset with much less effort and with the smaller dataset size of 800 images. But before we do that, let us have a look at doing the same image preprocessing in TensorFlow.

Image preprocessing in TensorFlow for pre-trained VGG16

We define a function for the preprocessing steps in TensorFlow as follows:

```
def tf_preprocess(filelist):
    images=[]
    for filename in filelist:
        image_string = tf.read_file(filename)
        image_decoded = tf.image.decode_jpeg(image_string, channels=3)
        image_float = tf.cast(image_decoded, tf.float32)
        resize_fn = tf.image.resize_image_with_crop_or_pad
        image_resized = resize_fn(image_float, image_height, image_width)
        means = tf.reshape(tf.constant([123.68, 116.78, 103.94]),
                                       [1, 1, 3])
        image = image_resized - means
        images.append(image)
    images = tf.stack(images)
    return images
```

Here, we create the images variable instead of a placeholder:

```
images=tf_preprocess([x for x in x_test])
```

We follow the same process as before to define the VGG16 model, restore the variables and then run the predictions:

```
with slim.arg_scope(vgg.vgg_arg_scope()):
    logits,_ = vgg.vgg_16(images,
                          num_classes=inet.n_classes,
                          is_training=False
                         )
probabilities = tf.nn.softmax(logits)

init = slim.assign_from_checkpoint_fn(
        os.path.join(model_home, '{}.ckpt'.format(model_name)),
```

```
        slim.get_variables_to_restore())
```

We get the same class probabilities as before. We just wanted to demonstrate that preprocessing can also be done in TensorFlow. However, preprocessing in TensorFlow is limited to the functions provided by TensorFlow and ties you deeply to the framework.

We recommend that you keep the preprocessing pipelines separate from the TensorFlow Model Training and Predictions code. Keeping it separate makes it modular and has other advantages, such as you can save the data for reuse in multiple models.

Image classification using retrained VGG16 in TensorFlow

Now we are going to do the magic of retraining the VGG16 model for the COCO animals dataset. Let us start by defining three placeholders:

- `is_training` placeholder specifies if we are using the model for training or for predictions
- `x_p` is a placeholder of shape (None,image_height, image_width,3) for inputs
- `y_p` is a placeholder of shape (None,1) for outputs

```
is_training = tf.placeholder(tf.bool,name='is_training')
x_p = tf.placeholder(shape=(None,image_height, image_width,3),
                     dtype=tf.float32,name='x_p')
y_p = tf.placeholder(shape=(None,1),dtype=tf.int32,name='y_p')
```

As we explained in the strategies section, we shall restore the layers from the checkpoint file except for the last layer, which is called `vgg/fc8` layer:

```
with slim.arg_scope(vgg.vgg_arg_scope()):
    logits, _ = vgg.vgg_16(x_p,num_classes=coco.n_classes,
                           is_training=is_training)

probabilities = tf.nn.softmax(logits)
# restore except last last layer fc8
fc7_variables=tf.contrib.framework.get_variables_to_restore(exclude=['vgg_1
6/fc8'])
fc7_init = tf.contrib.framework.assign_from_checkpoint_fn(
    os.path.join(model_home, '{}.ckpt'.format(model_name)),
    fc7_variables)
```

Next, define the last layer's variables to be initialized, not restored:

```
# fc8 layer
fc8_variables = tf.contrib.framework.get_variables('vgg_16/fc8')
fc8_init = tf.variables_initializer(fc8_variables)
```

As we have learned in previous chapters, define the loss function with `tf.losses.sparse_softmax_cross_entropy()`.

```
tf.losses.sparse_softmax_cross_entropy(labels=y_p, logits=logits)
loss = tf.losses.get_total_loss()
```

Train the last layer for a few epochs and then train the full network for a few layers. Thus, define two separate optimizer and training operations.

```
learning_rate = 0.001
fc8_optimizer = tf.train.GradientDescentOptimizer(learning_rate)
fc8_train_op = fc8_optimizer.minimize(loss, var_list=fc8_variables)

full_optimizer = tf.train.GradientDescentOptimizer(learning_rate)
full_train_op = full_optimizer.minimize(loss)
```

We decided to use the same learning rate for both optimizer functions, but you can define a separate learning rate if you decide to tune the hyper-parameters further.

Define the accuracy function as usual:

```
y_pred = tf.to_int32(tf.argmax(logits, 1))
n_correct_pred = tf.equal(y_pred, y_p)
accuracy = tf.reduce_mean(tf.cast(n_correct_pred, tf.float32))
```

Finally, we run the training for the last layer for 10 epochs, and then for the full network for 10 epochs, using a batch size of 32. We also use the same session to predict the classes:

```
fc8_epochs = 10
full_epochs = 10
coco.y_onehot = False
coco.batch_size = 32
coco.batch_shuffle = True

total_images = len(x_train_files)
n_batches = total_images // coco.batch_size

with tf.Session() as tfs:
        fc7_init(tfs)
        tfs.run(fc8_init)
        for epoch in range(fc8_epochs):
```

```
        print('Starting fc8 epoch ',epoch)
        coco.reset_index()
        epoch_accuracy=0
        for batch in range(n_batches):
            x_batch, y_batch = coco.next_batch()
            images=np.array([coco.preprocess_for_vgg(x) \
                        for x in x_batch])
            feed_dict={x_p:images,y_p:y_batch,is_training:True}
            tfs.run(fc8_train_op, feed_dict = feed_dict)
            feed_dict={x_p:images,y_p:y_batch,is_training:False}
            batch_accuracy = tfs.run(accuracy,feed_dict=feed_dict)
            epoch_accuracy += batch_accuracy
            except Exception as ex:
        epoch_accuracy /= n_batches
        print('Train accuracy in epoch {}:{}'
            .format(epoch,epoch_accuracy))
    for epoch in range(full_epochs):
        print('Starting full epoch ',epoch)
        coco.reset_index()
        epoch_accuracy=0
        for batch in range(n_batches):
            x_batch, y_batch = coco.next_batch()
            images=np.array([coco.preprocess_for_vgg(x) \
                    for x in x_batch])
            feed_dict={x_p:images,y_p:y_batch,is_training:True}
            tfs.run(full_train_op, feed_dict = feed_dict )
            feed_dict={x_p:images,y_p:y_batch,is_training:False}
            batch_accuracy = tfs.run(accuracy,feed_dict=feed_dict)
            epoch_accuracy += batch_accuracy
        epoch_accuracy /= n_batches
        print('Train accuracy in epoch {}:{}'
                .format(epoch,epoch_accuracy))
    # now run the predictions
    feed_dict={x_p:images_test,is_training: False}
    probs = tfs.run([probabilities],feed_dict=feed_dict)
    probs=probs[0]
```

Let us see print the results of our predictions:

```
disp(images_test,id2label=coco.id2label,probs=probs,scale=True)
```

Input image	Output probabilities
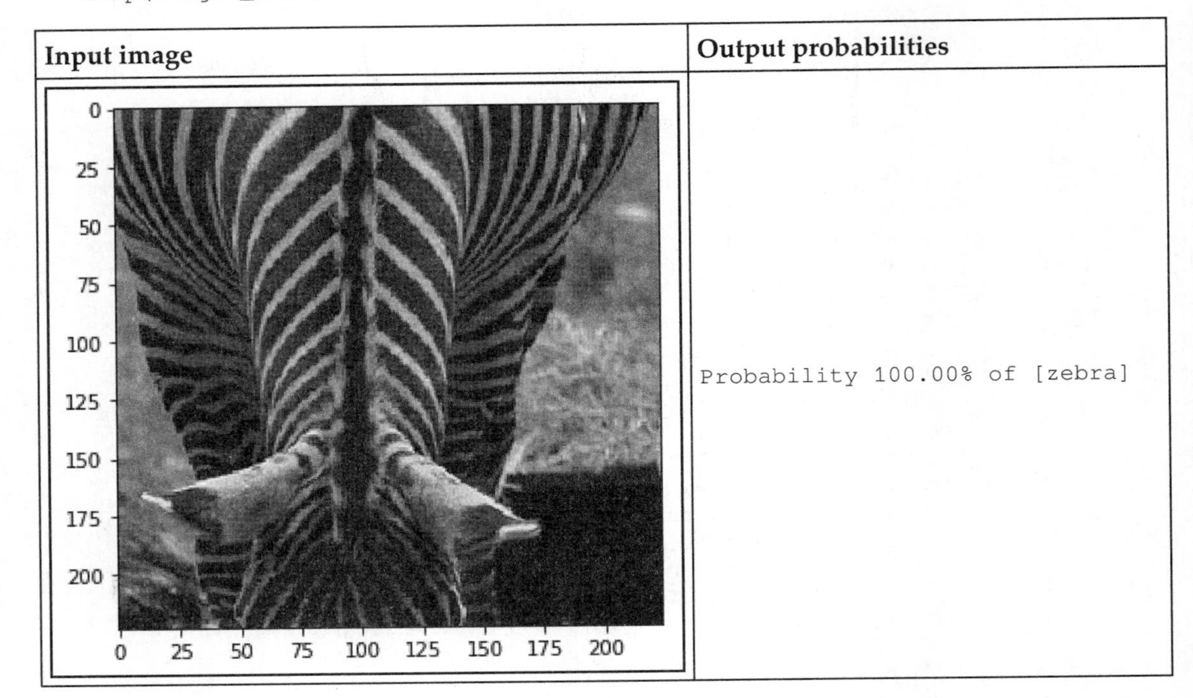	`Probability 100.00% of [zebra]`

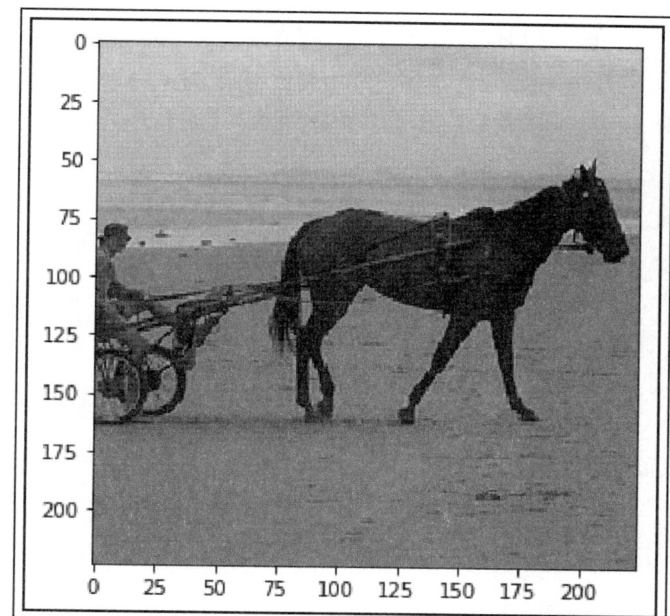


```
Probability 100.00% of [horse]
```

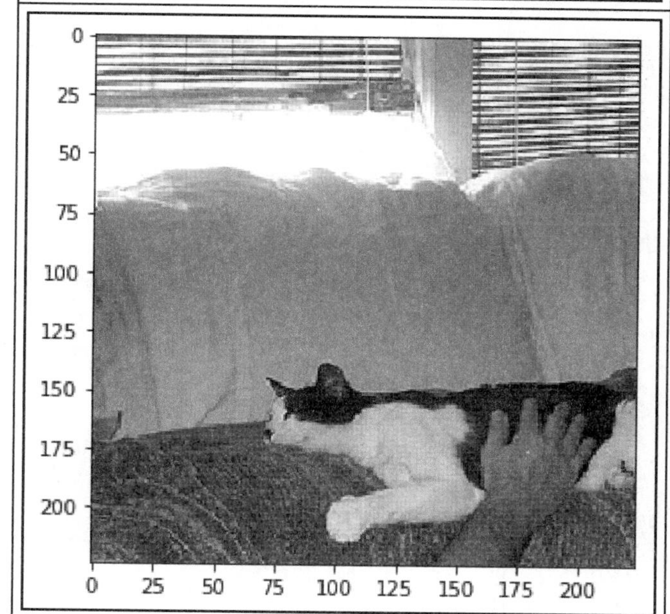


```
Probability 98.88% of [cat]
```

```
Probability 100.00% of [bird]
```

```
Probability 68.88% of [bear]
Probability 31.06% of [sheep]
Probability 0.02% of [dog]
Probability 0.02% of [bird]
Probability 0.01% of [horse]
```

	Probability 100.00% of [bear] Probability 0.00% of [dog] Probability 0.00% of [bird] Probability 0.00% of [sheep] Probability 0.00% of [cat]
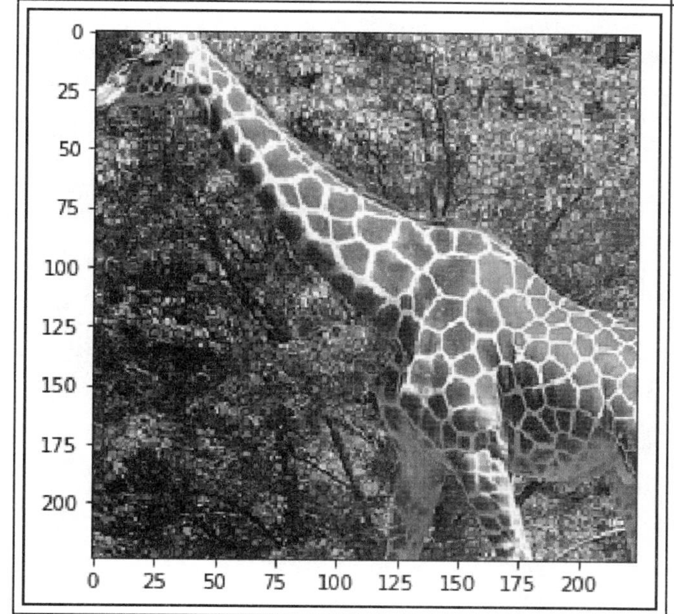	Probability 100.00% of [giraffe]

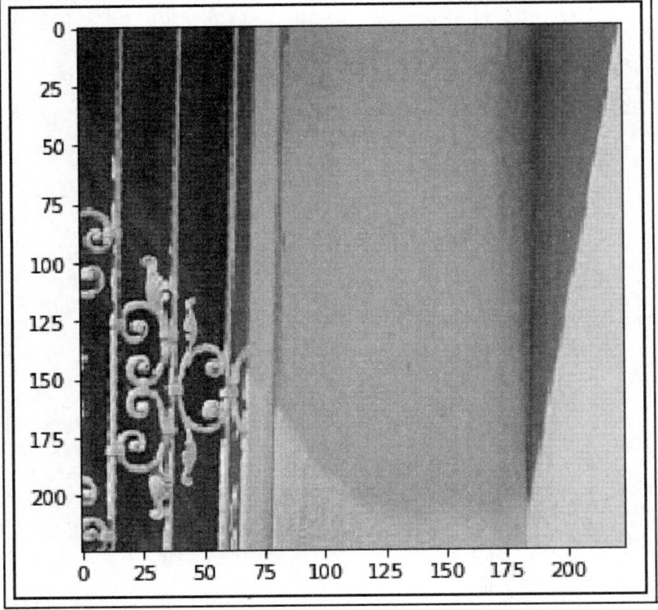


```
Probability 61.36% of [cat]
Probability 16.70% of [dog]
Probability 7.46% of [bird]
Probability 5.34% of [bear]
Probability 3.65% of [giraffe]
```

It correctly identified the cat and the giraffe and improved the other probabilities to 100%. It still made some mistakes, as the last picture was classified as the cat, which was actually a noise picture after cropping. We leave it up to you to improve upon these results.

VGG16 in Keras

You can follow along with the code in the Jupyter notebook `ch-12a_VGG16_Keras`.

Now let us do the same classification and retraining with Keras. You will see how easily we can use the VGG16 pre-trained model in Keras with the lesser amount of code.

Image classification using pre-trained VGG16 in Keras

Loading the model is a one-line operation:

```
from keras.applications import VGG16
model=VGG16(weights='imagenet')
```

We can use this model to predict the probabilities of classes:

```
probs = model.predict(images_test)
```

Here are the results of this classification:

Input image	Output probabilities
0 25 50 75 100 125 150 175 200 / 0 25 50 75 100 125 150 175 200	Probability 99.41% of [zebra] Probability 0.19% of [tiger cat] Probability 0.13% of [goose] Probability 0.09% of [tiger, Panthera tigris] Probability 0.02% of [mushroom]

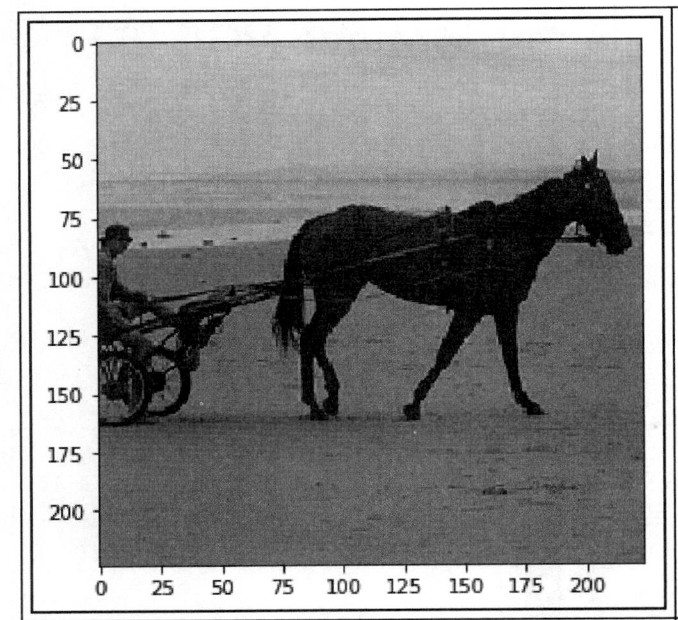	Probability 87.50% of [horse cart, horse-cart] Probability 5.58% of [Arabian camel, dromedary, Camelus dromedarius] Probability 4.72% of [plow, plough] Probability 1.03% of [dogsled, dog sled, dog sleigh] Probability 0.31% of [wreck]
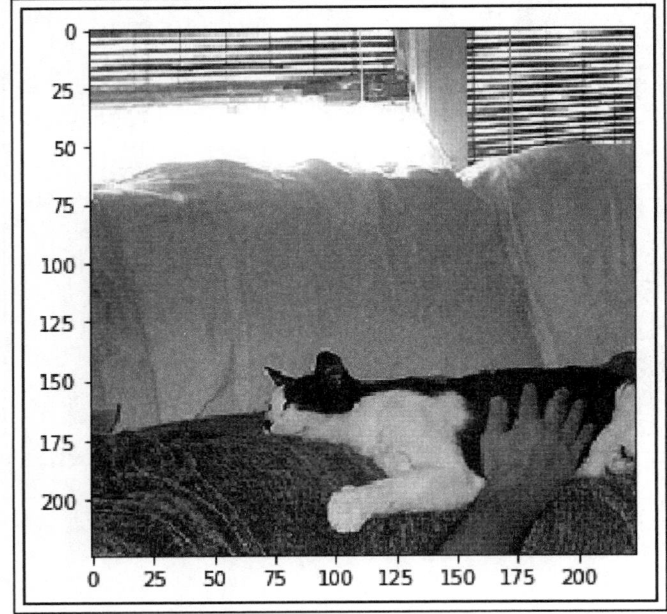	Probability 34.96% of [Siamese cat, Siamese] Probability 12.71% of [toy terrier] Probability 10.15% of [Boston bull, Boston terrier] Probability 6.53% of [Italian greyhound] Probability 6.01% of [Cardigan, Cardigan Welsh corgi]

	Probability 56.41% of [junco, snowbird] Probability 38.08% of [chickadee] Probability 1.93% of [bulbul] Probability 1.35% of [hummingbird] Probability 1.09% of [house finch, linnet, Carpodacus mexicanus]
	Probability 54.19% of [brown bear, bruin, Ursus arctos] Probability 28.07% of [lion, king of beasts, Panthera leo] Probability 0.87% of [Norwich terrier] Probability 0.82% of [Lakeland terrier] Probability 0.73% of [wild boar, boar, Sus scrofa]

<table>
<tr>
<td></td>
<td>Probability 88.64% of [brown bear, bruin, Ursus arctos]
Probability 7.22% of [American black bear, black bear, Ursus americanus, Euarctos americanus]
Probability 4.13% of [sloth bear, Melursus ursinus, Ursus ursinus]
Probability 0.00% of [badger]
Probability 0.00% of [wombat]</td>
</tr>
<tr>
<td></td>
<td>Probability 38.70% of [jaguar, panther, Panthera onca, Felis onca]
Probability 33.78% of [leopard, Panthera pardus]
Probability 14.22% of [cheetah, chetah, Acinonyx jubatus]
Probability 6.15% of [banded gecko]
Probability 1.53% of [snow leopard, ounce, Panthera uncia]</td>
</tr>
</table>

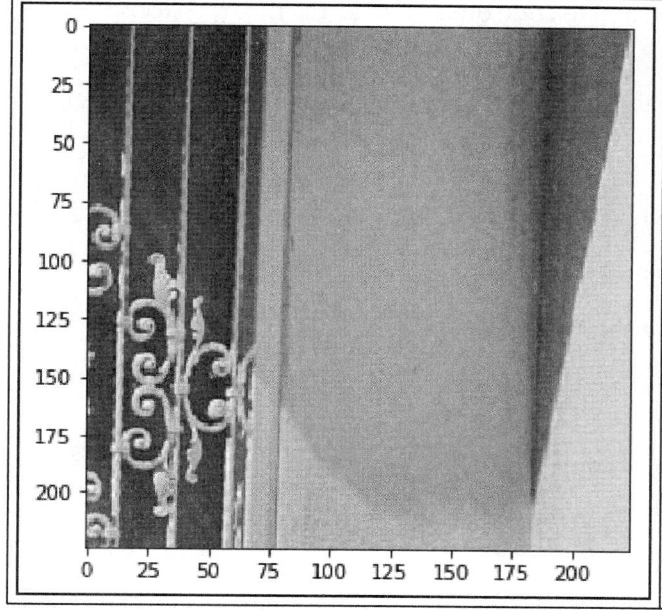


```
Probability 12.54% of [shower
curtain]
Probability 2.82% of [binder,
ring-binder]
Probability 2.28% of [toilet
tissue,
    toilet paper, bathroom
tissue]
Probability 2.12% of [accordion,
piano accordion, squeeze box]
Probability 2.05% of [bath
towel]
```

It failed to identify the sheep, giraffe and the last noise image from which the dog's image was cropped out. Now, let us retrain the model in Keras with our dataset.

Image classification using retrained VGG16 in Keras

Let us use the COCO images dataset to retrain the model to fine-tune the classification task. We shall remove the last layer in the Keras model and add our own fully connected layer with `softmax` activation for 8 classes. We shall also demonstrate freezing the first few layers by setting the `trainable` attribute of the first 15 layers to `False`.

1. Start by importing the VGG16 model without the variables for the top layer, by setting the `include_top` to `False`:

```
# load the vgg model
from keras.applications import VGG16
base_model=VGG16(weights='imagenet',include_top=False,
input_shape=(224,224,3))
```

We also specify the `input_shape` in the above code otherwise Keras throws exceptions later.

2. Now we build the classifier model to put on top of the imported VGG model:

```
top_model = Sequential()
top_model.add(Flatten(input_shape=base_model.output_shape[1:]))
top_model.add(Dense(256, activation='relu'))
top_model.add(Dropout(0.5))
top_model.add(Dense(coco.n_classes, activation='softmax'))
```

3. Next, add the model on top of the VGG base:

```
model=Model(inputs=base_model.input,
outputs=top_model(base_model.output))
```

4. Freeze the first 15 layers:

```
for layer in model.layers[:15]:
    layer.trainable = False
```

5. We randomly picked 15 layers to freeze, you may want to play around with this number. Let us compile the model and print the model summary:

```
model.compile(loss='categorical_crossentropy',
      optimizer=optimizers.SGD(lr=1e-4, momentum=0.9),
      metrics=['accuracy'])
model.summary()
```

```
_________________________________________________________________
Layer (type)                 Output Shape              Param #
=================================================================
input_1 (InputLayer)         (None, 224, 224, 3)       0
_________________________________________________________________
block1_conv1 (Conv2D)        (None, 224, 224, 64)      1792
_________________________________________________________________
block1_conv2 (Conv2D)        (None, 224, 224, 64)      36928
_________________________________________________________________
block1_pool (MaxPooling2D)   (None, 112, 112, 64)      0
_________________________________________________________________
block2_conv1 (Conv2D)        (None, 112, 112, 128)     73856
_________________________________________________________________
block2_conv2 (Conv2D)        (None, 112, 112, 128)     147584
_________________________________________________________________
block2_pool (MaxPooling2D)   (None, 56, 56, 128)       0
_________________________________________________________________
block3_conv1 (Conv2D)        (None, 56, 56, 256)       295168
```

```
_________________________________________________________________
block3_conv2 (Conv2D)        (None, 56, 56, 256)       590080
_________________________________________________________________
block3_conv3 (Conv2D)        (None, 56, 56, 256)       590080
_________________________________________________________________
block3_pool (MaxPooling2D)   (None, 28, 28, 256)       0
_________________________________________________________________
block4_conv1 (Conv2D)        (None, 28, 28, 512)       1180160
_________________________________________________________________
block4_conv2 (Conv2D)        (None, 28, 28, 512)       2359808
_________________________________________________________________
block4_conv3 (Conv2D)        (None, 28, 28, 512)       2359808
_________________________________________________________________
block4_pool (MaxPooling2D)   (None, 14, 14, 512)       0
_________________________________________________________________
block5_conv1 (Conv2D)        (None, 14, 14, 512)       2359808
_________________________________________________________________
block5_conv2 (Conv2D)        (None, 14, 14, 512)       2359808
_________________________________________________________________
block5_conv3 (Conv2D)        (None, 14, 14, 512)       2359808
_________________________________________________________________
block5_pool (MaxPooling2D)   (None, 7, 7, 512)         0
_________________________________________________________________
sequential_1 (Sequential)    (None, 8)                 6424840
=================================================================
Total params: 21,139,528
Trainable params: 13,504,264
Non-trainable params: 7,635,264
```

We see that almost 40% of parameters are frozen and non-trainable.

6. Next, train the Keras model for 20 epochs with a batch size of 32:

```
from keras.utils import np_utils

batch_size=32
n_epochs=20

total_images = len(x_train_files)
n_batches = total_images // batch_size
for epoch in range(n_epochs):
 print('Starting epoch ',epoch)
 coco.reset_index_in_epoch()
 for batch in range(n_batches):
 try:
     x_batch, y_batch = coco.next_batch(batch_size=batch_size)
     images=np.array([coco.preprocess_image(x) for x in x_batch])
```

```
        y_onehot = np_utils.to_categorical(y_batch,
                       num_classes=coco.n_classes)
        model.fit(x=images,y=y_onehot,verbose=0)
    except Exception as ex:
        print('error in epoch {} batch {}'.format(epoch,batch))
        print(ex)
```

7. Let us classify the images using the newly retrained model:

```
probs = model.predict(images_test)
```

Following are the classification results:

Input image	**Output probabilities**

	`Probability 100.00% of [zebra]`
	`Probability 0.00% of [dog]`
	`Probability 0.00% of [horse]`
	`Probability 0.00% of [giraffe]`
	`Probability 0.00% of [bear]`

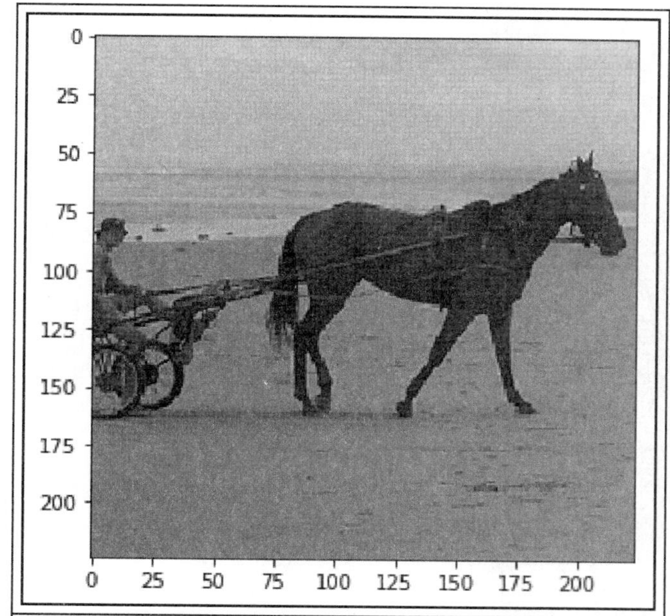	Probability 96.11% of [horse] Probability 1.85% of [cat] Probability 0.77% of [bird] Probability 0.43% of [giraffe] Probability 0.40% of [sheep]
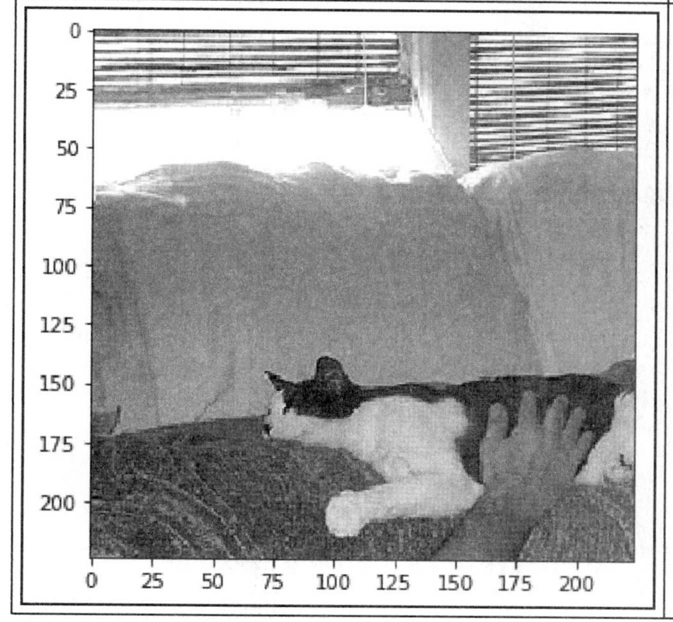	Probability 99.75% of [dog] Probability 0.22% of [cat] Probability 0.03% of [horse] Probability 0.00% of [bear] Probability 0.00% of [zebra]

```
Probability 99.88% of [bird]
Probability 0.11% of [horse]
Probability 0.00% of [giraffe]
Probability 0.00% of [bear]
Probability 0.00% of [cat]
```

```
Probability 65.28% of [bear]
Probability 27.09% of [sheep]
Probability 4.34% of [bird]
Probability 1.71% of [giraffe]
Probability 0.63% of [dog]
```

<table>
<tr>
<td>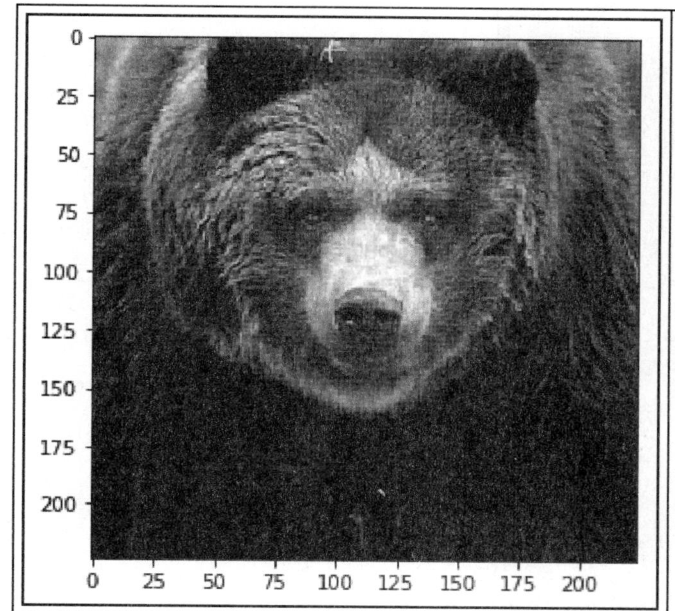
</td>
<td>

```
Probability 100.00% of [bear]
Probability 0.00% of [sheep]
Probability 0.00% of [dog]
Probability 0.00% of [cat]
Probability 0.00% of [giraffe]
```

</td>
</tr>
<tr>
<td>
</td>
<td>

```
Probability 100.00% of [giraffe]
Probability 0.00% of [bird]
Probability 0.00% of [bear]
Probability 0.00% of [sheep]
Probability 0.00% of [zebra]
```

</td>
</tr>
</table>

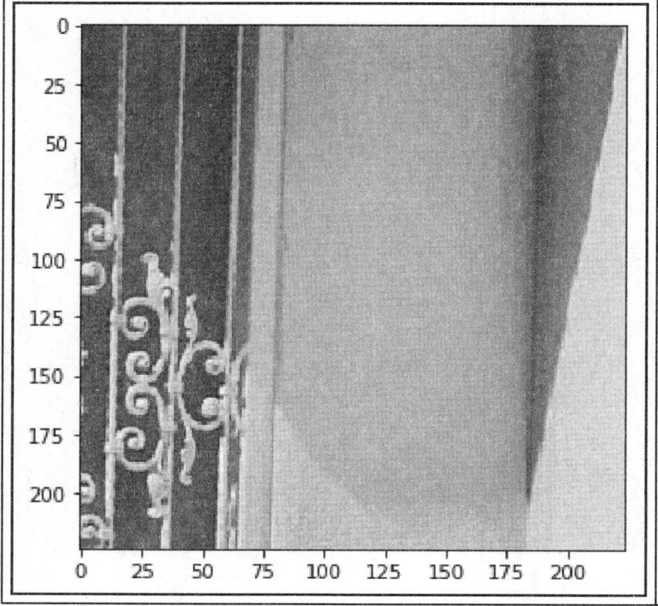


```
Probability 81.05% of [cat]
Probability 15.68% of [dog]
Probability 1.64% of [bird]
Probability 0.90% of [horse]
Probability 0.43% of [bear]
```

All the classes have been identified correctly except the last noisy image. With proper hyperparameter tuning, that can also be improved.

By now you have seen examples of classifying, using the pre-trained model and fine-tuning the pre-trained model. Next we shall show the classification example using the Inception v3 model.

Inception v3 in TensorFlow

You can follow along with the code in the Jupyter notebook `ch-12c_InceptionV3_TensorFlow`.

TensorFlow's Inception v3 is trained on 1,001 labels instead of 1,000. Also, the images used for training are pre-processed differently. We showed the preprocessing code in previous sections. Let us dive directly into restoring the Inception v3 model using TensorFlow.

Let us download the checkpoint file for the Inception v3:

```
# load the inception V3 model
model_name='inception_v3'
model_url='http://download.tensorflow.org/models/'
model_files=['inception_v3_2016_08_28.tar.gz']
model_home=os.path.join(models_root,model_name)

dsu.download_dataset(source_url=model_url,
    source_files=model_files,
    dest_dir = model_home,
    force=False,
    extract=True)
```

Define the common imports for inception module and variables:

```
### define common imports and variables
from tensorflow.contrib.slim.nets import inception
image_height=inception.inception_v3.default_image_size
image_width=inception.inception_v3.default_image_size
```

Image classification using Inception v3 in TensorFlow

The image classification is the same as explained in the previous section using the VGG 16 model. The full code for the Inception v3 model is as follows:

```
x_p = tf.placeholder(shape=(None,
                            image_height,
                            image_width,
                            3
                           ),
                     dtype=tf.float32,
                     name='x_p')
with slim.arg_scope(inception.inception_v3_arg_scope()):
    logits,_ = inception.inception_v3(x_p,
                                      num_classes=inet.n_classes,
                                      is_training=False
                                     )
probabilities = tf.nn.softmax(logits)

init = slim.assign_from_checkpoint_fn(
        os.path.join(model_home, '{}.ckpt'.format(model_name)),
        slim.get_variables_to_restore())
```

```
with tf.Session() as tfs:
    init(tfs)
    probs = tfs.run([probabilities],feed_dict={x_p:images_test})
    probs=probs[0]
```

Let us see how our model did with the test images:

Input image	**Output probabilities**
 0 25 50 75 100 125 150 175 200	Probability 95.15% of [zebra] Probability 0.07% of [ostrich, Struthio camelus] Probability 0.07% of [hartebeest] Probability 0.03% of [sock] Probability 0.03% of [warthog]

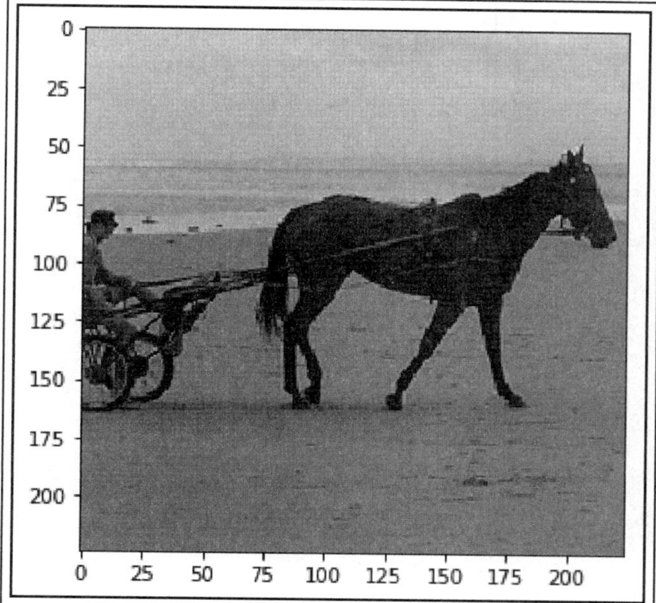


```
Probability 93.09% of [horse cart, horse-cart]
Probability 0.47% of [plow, plough]
Probability 0.07% of [oxcart]
Probability 0.07% of [seashore, coast, seacoast, sea-coast]
Probability 0.06% of [military uniform]
```

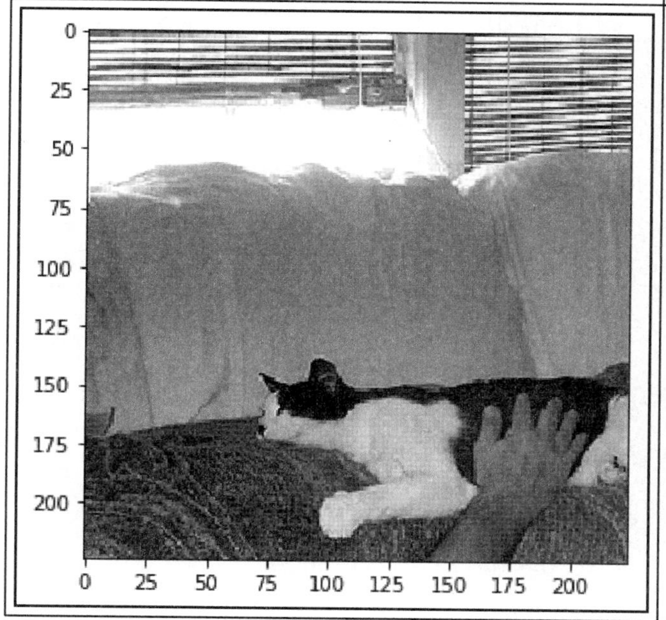


```
Probability 18.94% of [Cardigan, Cardigan Welsh corgi]
Probability 8.19% of [Pembroke, Pembroke Welsh corgi]
Probability 7.86% of [studio couch, day bed]
Probability 5.36% of [English springer, English springer spaniel]
Probability 4.16% of [Border collie]
```

	Probability 27.18% of [water ouzel, dipper] Probability 24.38% of [junco, snowbird] Probability 6.91% of [chickadee] Probability 0.99% of [magpie] Probability 0.73% of [brambling, Fringilla montifringilla]
	Probability 93.00% of [hog, pig, grunter, squealer, Sus scrofa] Probability 2.23% of [wild boar, boar, Sus scrofa] Probability 0.65% of [ram, tup] Probability 0.43% of [ox] Probability 0.23% of [marmot]

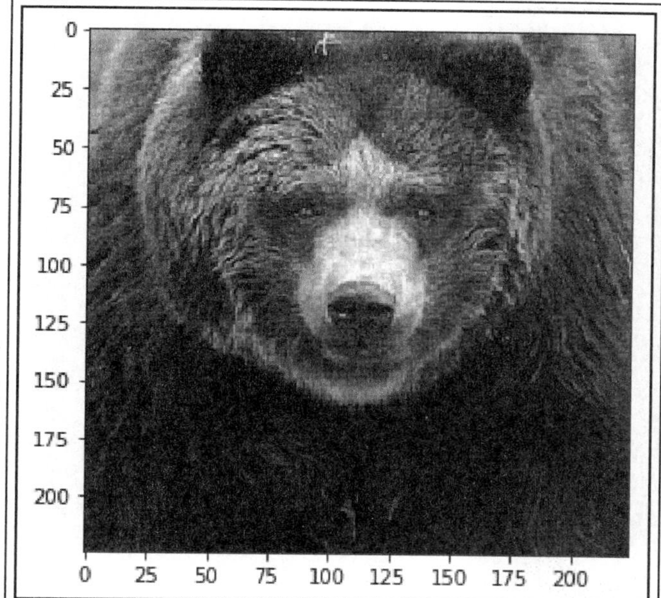	Probability 84.27% of [brown bear, bruin, Ursus arctos] Probability 1.57% of [American black bear, black bear, Ursus americanus, Euarctos americanus] Probability 1.34% of [sloth bear, Melursus ursinus, Ursus ursinus] Probability 0.13% of [lesser panda, red panda, panda, bear cat, cat bear, Ailurus fulgens] Probability 0.12% of [ice bear, polar bear, Ursus Maritimus, Thalarctos maritimus]
	Probability 20.20% of [honeycomb] Probability 6.52% of [gazelle] Probability 5.14% of [sorrel] Probability 3.72% of [impala, Aepyceros melampus] Probability 2.44% of [Saluki, gazelle hound]

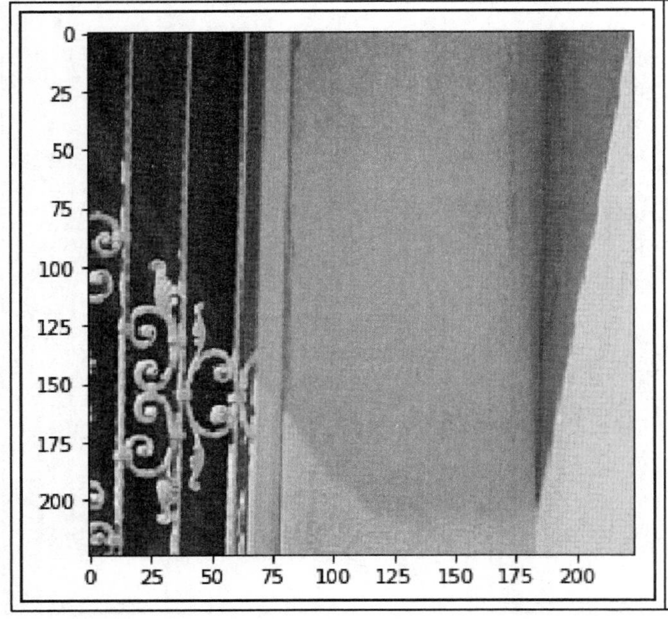


```
Probability 41.17% of [harp]
Probability 13.64% of
[accordion, piano accordion,
squeeze box]
Probability 2.97% of [window
shade]
Probability 1.59% of [chain]
Probability 1.55% of [pay-phone,
pay-station]
```

Not too bad, although it failed at almost the same places as the VGG model. Now let us retrain this model with the COCO animals images and labels.

Image classification using retrained Inception v3 in TensorFlow

The retraining of Inception v3 differs from the VGG16, as we use the softmax activation layer for the output with `tf.losses.softmax_cross_entropy()` as the loss function.

1. First define the placeholders:

```
is_training = tf.placeholder(tf.bool,name='is_training')
x_p = tf.placeholder(shape=(None,
                            image_height,
                            image_width,
                            3
                           ),
                     dtype=tf.float32,
                     name='x_p')
y_p = tf.placeholder(shape=(None,coco.n_classes),
                     dtype=tf.int32,
```

```
                            name='y_p')
```

2. Next, load the model:

```
with slim.arg_scope(inception.inception_v3_arg_scope()):
    logits,_ = inception.inception_v3(x_p,
                                  num_classes=coco.n_classes,
                                  is_training=True
                                 )
probabilities = tf.nn.softmax(logits)
```

3. Next, define functions to restore the variables except for the last layer:

```
with slim.arg_scope(inception.inception_v3_arg_scope()):
    logits,_ = inception.inception_v3(x_p,
                                  num_classes=coco.n_classes,
                                  is_training=True
                                 )
probabilities = tf.nn.softmax(logits)
# restore except last layer
checkpoint_exclude_scopes=["InceptionV3/Logits",
                           "InceptionV3/AuxLogits"]
exclusions = [scope.strip() for scope in checkpoint_exclude_scopes]

variables_to_restore = []
for var in slim.get_model_variables():
    excluded = False
    for exclusion in exclusions:
        if var.op.name.startswith(exclusion):
            excluded = True
            break
    if not excluded:
        variables_to_restore.append(var)

init_fn = slim.assign_from_checkpoint_fn(
    os.path.join(model_home, '{}.ckpt'.format(model_name)),
    variables_to_restore)
```

4. Define the loss, optimizer, and training operation:

```
tf.losses.softmax_cross_entropy(onehot_labels=y_p, logits=logits)
loss = tf.losses.get_total_loss()
learning_rate = 0.001
optimizer = tf.train.GradientDescentOptimizer(learning_rate)
train_op = optimizer.minimize(loss)
```

5. Train the model and run the prediction once the training is done in the same session:

```
n_epochs=10
coco.y_onehot = True
coco.batch_size = 32
coco.batch_shuffle = True
total_images = len(x_train_files)
n_batches = total_images // coco.batch_size

with tf.Session() as tfs:
    tfs.run(tf.global_variables_initializer())
    init_fn(tfs)
    for epoch in range(n_epochs):
        print('Starting epoch ',epoch)
        coco.reset_index()
        epoch_accuracy=0
        epoch_loss=0
        for batch in range(n_batches):
            x_batch, y_batch = coco.next_batch()
            images=np.array([coco.preprocess_for_inception(x) \
                    for x in x_batch])
            feed_dict={x_p:images,y_p:y_batch,is_training:True}
            batch_loss,_ = tfs.run([loss,train_op],
                                feed_dict = feed_dict)
            epoch_loss += batch_loss
      epoch_loss /= n_batches
      print('Train loss in epoch {}:{}'
            .format(epoch,epoch_loss))
   # now run the predictions
   feed_dict={x_p:images_test,is_training: False}
   probs = tfs.run([probabilities],feed_dict=feed_dict)
   probs=probs[0]
```

We see the loss reduces with each epoch:

```
INFO:tensorflow:Restoring parameters from
/home/armando/models/inception_v3/inception_v3.ckpt
Starting epoch  0
Train loss in epoch 0:2.7896385192871094
Starting epoch  1
Train loss in epoch 1:1.6651896286010741
Starting epoch  2
Train loss in epoch 2:1.2332031989097596
Starting epoch  3
Train loss in epoch 3:0.9912329530715942
Starting epoch  4
```

```
Train loss in epoch 4:0.8110128355026245
Starting epoch  5
Train loss in epoch 5:0.7177265572547913
Starting epoch  6
Train loss in epoch 6:0.6175705575942994
Starting epoch  7
Train loss in epoch 7:0.5542363750934601
Starting epoch  8
Train loss in epoch 8:0.523461252450943
Starting epoch  9
Train loss in epoch 9:0.4923107647895813
```

This time the results identified the sheep correctly but incorrectly identified the cat picture as a dog:

Input image	**Output probabilities**

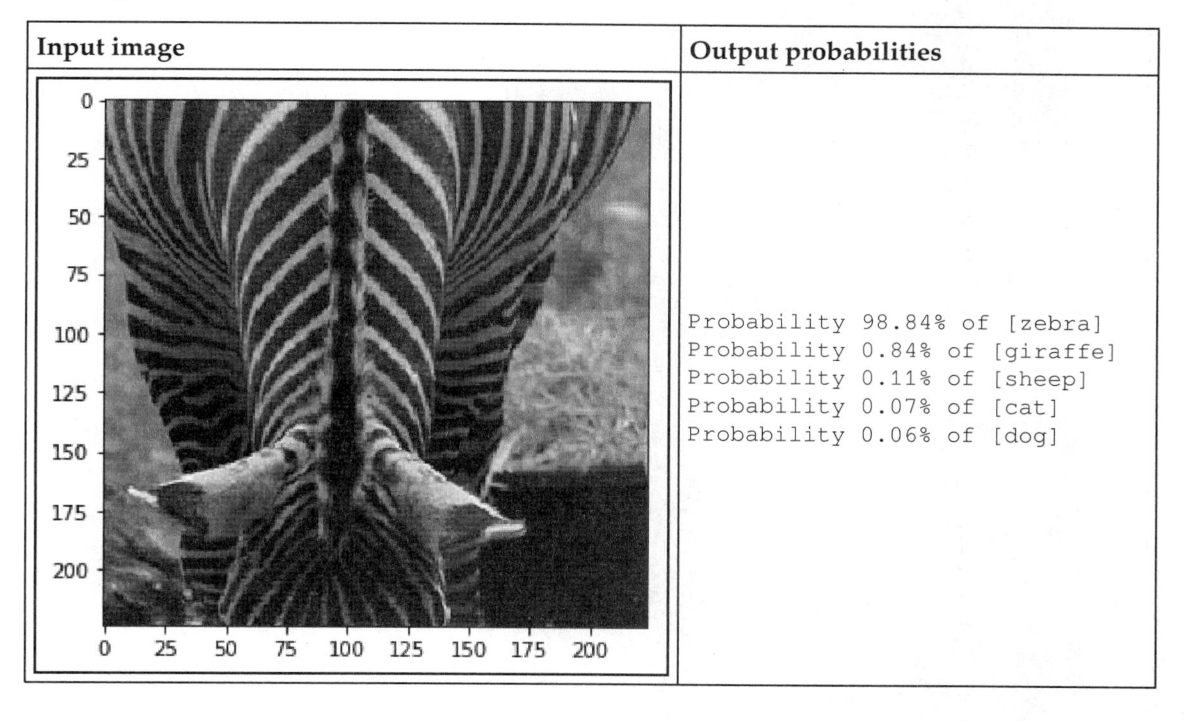	`Probability 98.84% of [zebra]`
	`Probability 0.84% of [giraffe]`
	`Probability 0.11% of [sheep]`
	`Probability 0.07% of [cat]`
	`Probability 0.06% of [dog]`

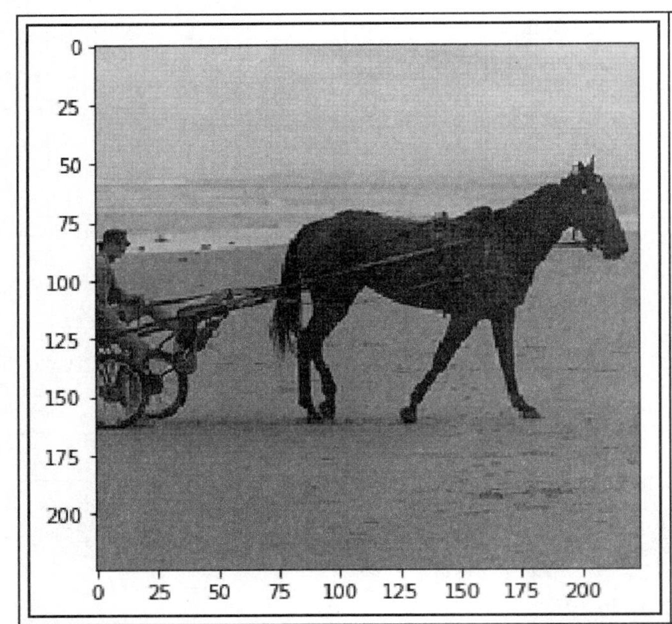


```
Probability 95.77% of [horse]
Probability 1.34% of [dog]
Probability 0.89% of [zebra]
Probability 0.68% of [bird]
Probability 0.61% of [sheep]
```

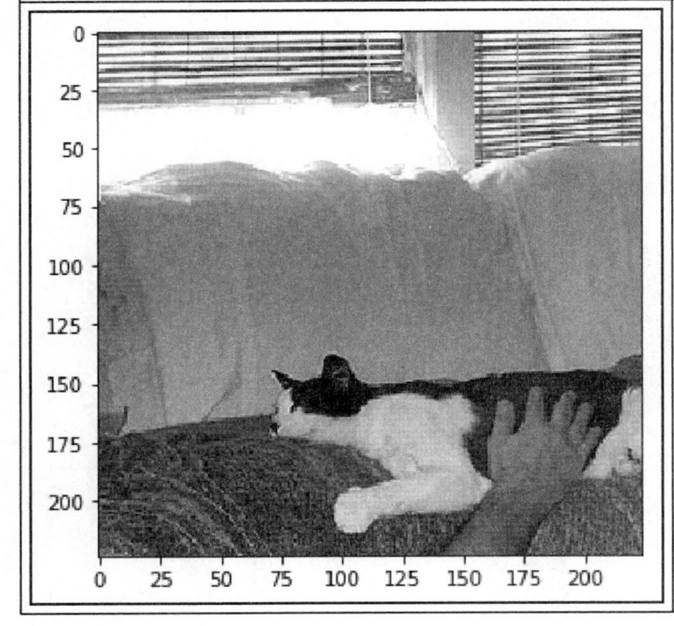


```
Probability 94.83% of [dog]
Probability 4.53% of [cat]
Probability 0.56% of [sheep]
Probability 0.04% of [bear]
Probability 0.02% of [zebra]
```

```
Probability 42.80% of [bird]
Probability 25.64% of [cat]
Probability 15.56% of [bear]
Probability 8.77% of [giraffe]
Probability 3.39% of [sheep]
```

```
Probability 72.58% of [sheep]
Probability 8.40% of [bear]
Probability 7.64% of [giraffe]
Probability 4.02% of [horse]
Probability 3.65% of [bird]
```

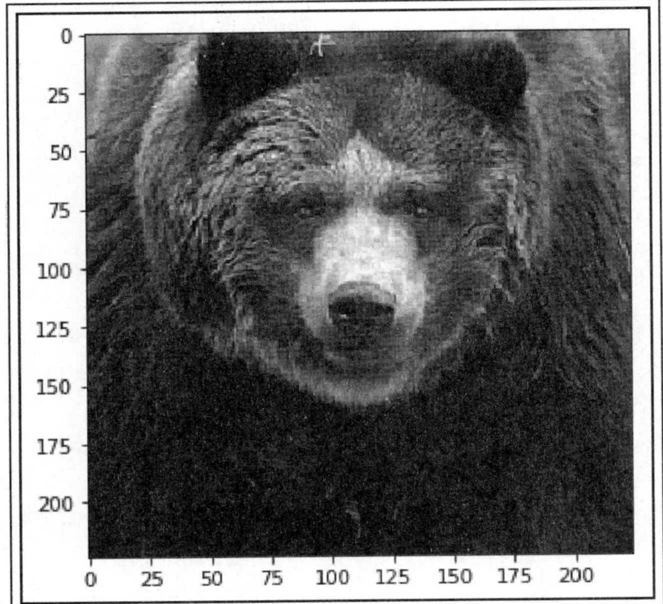	Probability 98.03% of [bear] Probability 0.74% of [cat] Probability 0.54% of [sheep] Probability 0.28% of [bird] Probability 0.17% of [horse]
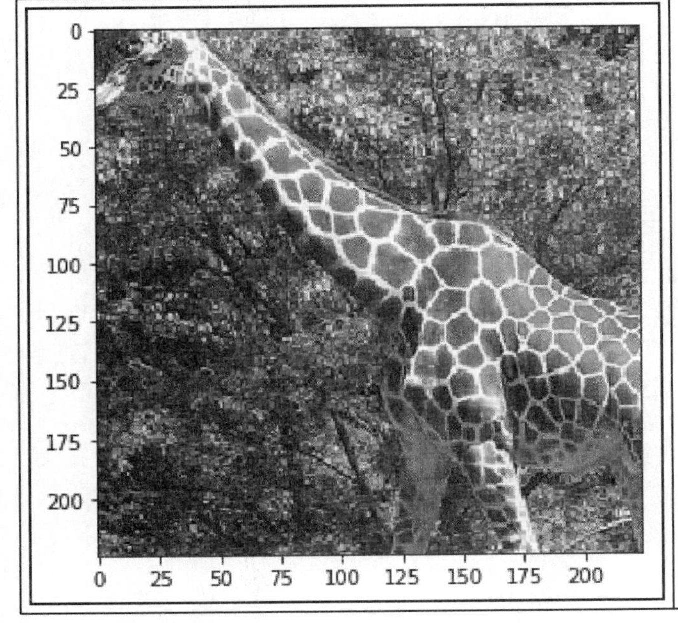	Probability 96.43% of [giraffe] Probability 1.78% of [bird] Probability 1.10% of [sheep] Probability 0.32% of [zebra] Probability 0.14% of [bear]

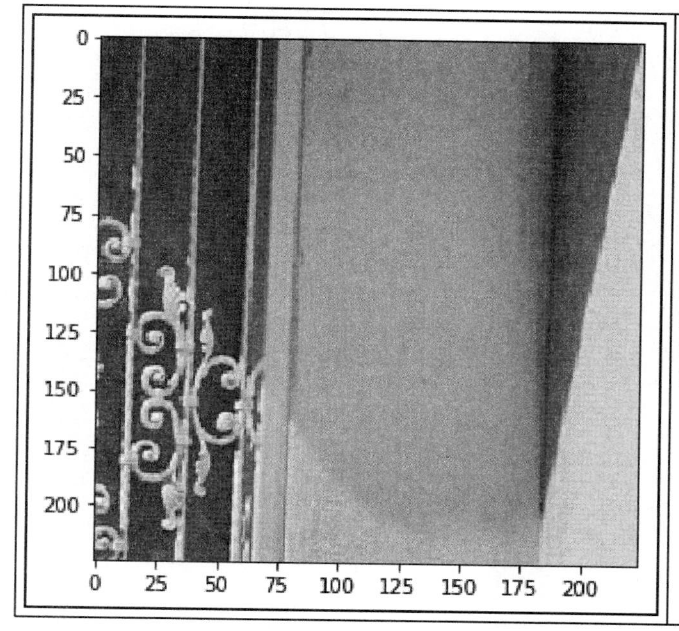


```
Probability 34.43% of [horse]
Probability 23.53% of [dog]
Probability 16.03% of [zebra]
Probability 9.76% of [cat]
Probability 9.02% of [giraffe]
```

Summary

Transfer learning is a great discovery that allows us to save time by applying the models trained in bigger datasets to different datasets. Transfer learning also helps warm start the training process when the dataset is small. In this chapter, we learned how to use pre-trained models, such as VGG16 and Inception v3 to classify the images in a different dataset to the dataset they were trained on. We also learned how to retrain the pre-trained models with examples in both TensorFlow and Keras, and how to preprocess the images for feeding into both the models.

We also learned that there are several models that are trained on the ImageNet dataset. Try to find some other models that are trained on different datasets, such as video datasets, speech datasets or text/NLP datasets. Try using these models to retrain and use for your own deep learning problems on your own datasets.

13
Deep Reinforcement Learning

Reinforcement learning is a form of learning in which a software agent observes the environment and takes actions so as to maximize its rewards from the environment, as depicted in the following diagram:

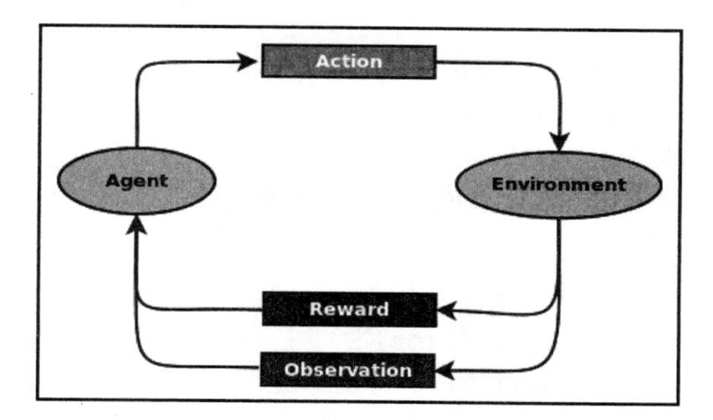

This metaphor can be used to represent real-life situations such as the following:

- A stock trading agent observes the trade information, news, analysis, and other form information, and takes actions to buy or sell trades so as to maximize the reward in the form of short-term profit or long-term profit.
- An insurance agent observes the information about the customer and then takes action to define the amount of insurance premium, so as to maximize the profit and minimize the risk.
- A humanoid robot observes the environment and then takes action, such as walking, running, or picking up objects, so as to maximize the reward in terms of the goal achieved.

Reinforcement learning has been successfully applied to many applications such as advertising optimization, stock market trading, self-driving vehicles, robotics, and games, to name a few.

Reinforcement learning is different from supervised learning in the sense that there are no labels in advance to tune the parameters of the model. The model learns from the rewards received from the runs. Although the short-term rewards are available instantly, long-term rewards are only available after a couple of steps. This phenomenon is also known as **delayed feedback**.

Reinforcement learning is also different from unsupervised learning because in unsupervised learning there are no labels available, whereas in reinforcement learning the feedback is available in terms of the rewards.

In this chapter, we shall learn about reinforcement learning and its implementation in TensorFlow and Keras by covering the following topics:

- OpenAI Gym 101
- Applying simple policies to a cartpole game
- Reinforcement Learning 101
    - Q function
    - Exploration and exploitation
    - V function
    - RL techniques
- Simple neural network policy for RL
- Implementing Q-Learning
    - Initializing and discretizing for Q-Learning
    - Q-Learning with Q-Table
    - Deep Q networks: Q-Learning with Q-Network

We shall demonstrate our examples in OpenAI Gym, so let us first learn about OpenAI Gym.

OpenAI Gym 101

OpenAI Gym is a Python-based toolkit for the research and development of reinforcement learning algorithms. OpenAI Gym provides more than 700 opensource contributed environments at the time of writing. With OpenAI, you can also create your own environments. The biggest advantage is that OpenAI provides a unified interface for working with these environments, and takes care of running the simulation while you focus on the reinforcement learning algorithms.

The research paper describing OpenAI Gym is available at this link: `http://arxiv.org/abs/1606.01540`.

You can install OpenAI Gym using the following command:

```
pip3 install gym
```

If the above command does not work, then you can find further help with installation at the following link: `https://github.com/openai/gym#installation`.

1. Let us print the number of available environments in OpenAI Gym:

You can follow along with the code in the Jupyter notebook `ch-13a_Reinforcement_Learning_NN` in the code bundle of this book.

```
all_env = list(gym.envs.registry.all())
print('Total Environments in Gym version {} : {}'
    .format(gym.__version__,len(all_env)))

Total Environments in Gym version 0.9.4 : 777
```

2. Let us print the list of all environments:

```
for e in list(all_env):
    print(e)
```

The partial list from the output is as follows:

```
EnvSpec(Carnival-ramNoFrameskip-v0)
EnvSpec(EnduroDeterministic-v0)
EnvSpec(FrostbiteNoFrameskip-v4)
EnvSpec(Taxi-v2)
EnvSpec(Pooyan-ram-v0)
EnvSpec(Solaris-ram-v4)
EnvSpec(Breakout-ramDeterministic-v0)
EnvSpec(Kangaroo-ram-v4)
EnvSpec(StarGunner-ram-v4)
EnvSpec(Enduro-ramNoFrameskip-v4)
EnvSpec(DemonAttack-ramDeterministic-v0)
EnvSpec(TimePilot-ramNoFrameskip-v0)
EnvSpec(Amidar-v4)
```

Each environment, represented by the `env` object, has a standardized interface, for example:

- An `env` object can be created with the `env.make(<game-id-string>)` function by passing the id string.
- Each `env` object contains the following main functions:
    - The `step()` function takes an action object as an argument and returns four objects:
        - *observation*: An object implemented by the environment, representing the observation of the environment.
        - *reward*: A signed float value indicating the gain (or loss) from the previous action.
        - *done*: A Boolean value representing if the scenario is finished.
        - *info*: A Python dictionary object representing the diagnostic information.
    - The `render()` function creates a visual representation of the environment.
    - The `reset()` function resets the environment to the original state.
- Each `env` object comes with well-defined actions and observations, represented by `action_space` and `observation_space`.

One of the most popular games in the gym to learn reinforcement learning is CartPole. In this game, a pole attached to a cart has to be balanced so that it doesn't fall. The game ends if either the pole tilts by more than 15 degrees or the cart moves by more than 2.4 units from the center. The home page of `OpenAI.com` emphasizes the game in these words:

> *The small size and simplicity of this environment make it possible to run very quick experiments, which is essential when learning the basics.*

The game has only four observations and two actions. The actions are to move a cart by applying a force of +1 or -1. The observations are the position of the cart, the velocity of the cart, the angle of the pole, and the rotation rate of the pole. However, knowledge of the semantics of observation is not necessary to learn to maximize the rewards of the game.

Now let us load a popular game environment, CartPole-v0, and play it with stochastic control:

1. Create the `env` object with the standard `make` function:

```
env = gym.make('CartPole-v0')
```

2. The number of episodes is the number of game plays. We shall set it to one, for now, indicating that we just want to play the game once. Since every episode is stochastic, in actual production runs you will run over several episodes and calculate the average values of the rewards. Additionally, we can initialize an array to store the visualization of the environment at every timestep:

```
n_episodes = 1
env_vis = []
```

3. Run two nested loops—an external loop for the number of episodes and an internal loop for the number of timesteps you would like to simulate for. You can either keep running the internal loop until the scenario is done or set the number of steps to a higher value.
    - At the beginning of every episode, reset the environment using `env.reset()`.
    - At the beginning of every timestep, capture the visualization using `env.render()`.

```
for i_episode in range(n_episodes):
    observation = env.reset()
    for t in range(100):
        env_vis.append(env.render(mode = 'rgb_array'))
        print(observation)
```

```
                action = env.action_space.sample()
                observation, reward, done, info = env.step(action)
                if done:
                    print("Episode finished at t{}".format(t+1))
                    break
```

4. Render the environment using the helper function:

```
env_render(env_vis)
```

5. The code for the helper function is as follows:

```
def env_render(env_vis):
    plt.figure()
    plot = plt.imshow(env_vis[0])
    plt.axis('off')
    def animate(i):
        plot.set_data(env_vis[i])

    anim = anm.FuncAnimation(plt.gcf(),
                             animate,
                             frames=len(env_vis),
                             interval=20,
                             repeat=True,
                             repeat_delay=20)
    display(display_animation(anim, default_mode='loop'))
```

We get the following output when we run this example:

```
[-0.00666995 -0.03699492 -0.00972623  0.00287713]
[-0.00740985  0.15826516 -0.00966868 -0.29285861]
[-0.00424454 -0.03671761 -0.01552586 -0.00324067]
[-0.0049789  -0.2316135  -0.01559067  0.28450351]
[-0.00961117 -0.42650966 -0.0099006   0.57222875]
[-0.01814136 -0.23125029  0.00154398  0.27644332]
[-0.02276636 -0.0361504   0.00707284 -0.01575223]
[-0.02348937  0.1588694   0.0067578  -0.30619523]
[-0.02031198 -0.03634819  0.00063389 -0.01138875]
[-0.02103895  0.15876466  0.00040612 -0.3038716 ]
[-0.01786366  0.35388083 -0.00567131 -0.59642642]
[-0.01078604  0.54908168 -0.01759984 -0.89089036]
[  1.95594914e-04   7.44437934e-01  -3.54176495e-02  -1.18905344e+00]
[ 0.01508435  0.54979251 -0.05919872 -0.90767902]
[ 0.0260802   0.35551978 -0.0773523  -0.63417465]
[ 0.0331906   0.55163065 -0.09003579 -0.95018025]
[ 0.04422321  0.74784161 -0.1090394  -1.26973934]
[ 0.05918004  0.55426764 -0.13443418 -1.01309691]
[ 0.0702654   0.36117014 -0.15469612 -0.76546874]
```

```
[ 0.0774888   0.16847818 -0.1700055  -0.52518186]
[ 0.08085836  0.3655333  -0.18050913 -0.86624457]
[ 0.08816903  0.56259197 -0.19783403 -1.20981195]
Episode finished at t22
```

It took 22 time-steps for the pole to become unbalanced. At every run, we get a different time-step value because we picked the action scholastically by using `env.action_space.sample()`.

Since the game results in a loss so quickly, randomly picking an action and applying it is probably not the best strategy. There are many algorithms for finding solutions to keeping the pole straight for a longer number of time-steps that you can use, such as Hill Climbing, Random Search, and Policy Gradient.

Some of the algorithms for solving the Cartpole game are available at the following links:
`https://openai.com/requests-for-research/#cartpole`
`http://kvfrans.com/simple-algoritms-for-solving-cartpole/`
`https://github.com/kvfrans/openai-cartpole`

Applying simple policies to a cartpole game

So far, we have randomly picked an action and applied it. Now let us apply some logic to picking the action instead of random chance. The third observation refers to the angle. If the angle is greater than zero, that means the pole is tilting right, thus we move the cart to the right (1). Otherwise, we move the cart to the left (0). Let us look at an example:

1. We define two policy functions as follows:

```
def policy_logic(env,obs):
    return 1 if obs[2] > 0 else 0
def policy_random(env,obs):
    return env.action_space.sample()
```

2. Next, we define an experiment function that will run for a specific number of episodes; each episode runs until the game is lost, namely when `done` is `True`. We use `rewards_max` to indicate when to break out of the loop as we do not wish to run the experiment forever:

```
def experiment(policy, n_episodes, rewards_max):
    rewards=np.empty(shape=(n_episodes))
    env = gym.make('CartPole-v0')
    for i in range(n_episodes):
```

```
        obs = env.reset()
        done = False
        episode_reward = 0
        while not done:
            action = policy(env,obs)
            obs, reward, done, info = env.step(action)
            episode_reward += reward
            if episode_reward > rewards_max:
                break
        rewards[i]=episode_reward
    print('Policy:{}, Min reward:{}, Max reward:{}'
          .format(policy.__name__,
                  min(rewards),
                  max(rewards)))
```

3. We run the experiment 100 times, or until the rewards are less than or equal to `rewards_max`, that is set to 10,000:

```
n_episodes = 100
rewards_max = 10000
experiment(policy_random, n_episodes, rewards_max)
experiment(policy_logic, n_episodes, rewards_max)
```

We can see that the logically selected actions do better than the randomly selected ones, but not that much better:

```
Policy:policy_random, Min reward:9.0, Max reward:63.0, Average reward:20.26
Policy:policy_logic, Min reward:24.0, Max reward:66.0, Average reward:42.81
```

Now let us modify the process of selecting the action further—to be based on parameters. The parameters will be multiplied by the observations and the action will be chosen based on whether the multiplication result is zero or one. Let us modify the random search method in which we initialize the parameters randomly. The code looks as follows:

```
def policy_logic(theta,obs):
    # just ignore theta
    return 1 if obs[2] > 0 else 0

def policy_random(theta,obs):
    return 0 if np.matmul(theta,obs) < 0 else 1

def episode(env, policy, rewards_max):
    obs = env.reset()
    done = False
    episode_reward = 0
    if policy.__name__ in ['policy_random']:
        theta = np.random.rand(4) * 2 - 1
```

```
        else:
            theta = None
        while not done:
            action = policy(theta,obs)
            obs, reward, done, info = env.step(action)
            episode_reward += reward
            if episode_reward > rewards_max:
                break
        return episode_reward
    def experiment(policy, n_episodes, rewards_max):
        rewards=np.empty(shape=(n_episodes))
        env = gym.make('CartPole-v0')
        for i in range(n_episodes):
            rewards[i]=episode(env,policy,rewards_max)
            #print("Episode finished at t{}".format(reward))
        print('Policy:{}, Min reward:{}, Max reward:{}, Average reward:{}'
              .format(policy.__name__,
                      np.min(rewards),
                      np.max(rewards),
                      np.mean(rewards)))

n_episodes = 100
rewards_max = 10000
experiment(policy_random, n_episodes, rewards_max)
experiment(policy_logic, n_episodes, rewards_max)
```

We can see that random search does improve the results:

```
Policy:policy_random, Min reward:8.0, Max reward:200.0, Average
reward:40.04
Policy:policy_logic, Min reward:25.0, Max reward:62.0, Average reward:43.03
```

With the random search, we have improved our results to get the max rewards of 200. On average, the rewards for random search are lower because random search tries various bad parameters that bring the overall results down. However, we can select the best parameters from all the runs and then, in production, use the best parameters. Let us modify the code to train the parameters first:

```
def policy_logic(theta,obs):
    # just ignore theta
    return 1 if obs[2] > 0 else 0

def policy_random(theta,obs):
    return 0 if np.matmul(theta,obs) < 0 else 1

def episode(env,policy, rewards_max,theta):
    obs = env.reset()
```

```
        done = False
        episode_reward = 0

        while not done:
            action = policy(theta,obs)
            obs, reward, done, info = env.step(action)
            episode_reward += reward
            if episode_reward > rewards_max:
                break
        return episode_reward

    def train(policy, n_episodes, rewards_max):

        env = gym.make('CartPole-v0')
        theta_best = np.empty(shape=[4])
        reward_best = 0

        for i in range(n_episodes):
            if policy.__name__ in ['policy_random']:
                theta = np.random.rand(4) * 2 - 1
            else:
                theta = None
            reward_episode=episode(env,policy,rewards_max, theta)
            if reward_episode > reward_best:
                reward_best = reward_episode
                theta_best = theta.copy()
        return reward_best,theta_best
    def experiment(policy, n_episodes, rewards_max, theta=None):
        rewards=np.empty(shape=[n_episodes])
        env = gym.make('CartPole-v0')
        for i in range(n_episodes):
            rewards[i]=episode(env,policy,rewards_max,theta)
            #print("Episode finished at t{}".format(reward))
        print('Policy:{}, Min reward:{}, Max reward:{}, Average reward:{}'
              .format(policy.__name__,
                      np.min(rewards),
                      np.max(rewards),
                      np.mean(rewards)))

    n_episodes = 100
    rewards_max = 10000

    reward,theta = train(policy_random, n_episodes, rewards_max)
    print('trained theta: {}, rewards: {}'.format(theta,reward))
    experiment(policy_random, n_episodes, rewards_max, theta)
    experiment(policy_logic, n_episodes, rewards_max)
```

We train for 100 episodes and then use the best parameters to run the experiment for the random search policy:

```
n_episodes = 100
rewards_max = 10000

reward,theta = train(policy_random, n_episodes, rewards_max)
print('trained theta: {}, rewards: {}'.format(theta,reward))
experiment(policy_random, n_episodes, rewards_max, theta)
experiment(policy_logic, n_episodes, rewards_max)
```

We find the that the training parameters gives us the best results of 200:

```
trained theta: [-0.14779543  0.93269603  0.70896423  0.84632461], rewards: 
200.0
Policy:policy_random, Min reward:200.0, Max reward:200.0, Average 
reward:200.0
Policy:policy_logic, Min reward:24.0, Max reward:63.0, Average reward:41.94
```

We may optimize the training code to continue training until we reach a maximum reward. The code for this optimization is provided in the notebook `ch-13a_Reinforcement_Learning_NN`.

Now that we have learned the basics of OpenAI Gym, let us learn about reinforcement learning.

Reinforcement learning 101

Reinforcement learning is described by an agent getting inputs of the *observation* and *reward* from the previous time-step and producing output as an *action* with the goal of maximizing cumulative rewards.

The agent has a policy, value function, and model:

- The algorithm used by the agent to pick the next action is known as the **policy**. In the previous section, we wrote a policy that would take a set of parameters theta and would return the next action based on the multiplication between the observation and the parameters. The policy is represented by the following equation:
  $\pi(s) : S \rightarrow A$,
  S is set of states and A is set of actions.

- A policy is deterministic or stochastic.
    - A deterministic policy returns the same action for the same state in each run:
    $\pi(s) = a$
    - A stochastic policy returns the different probabilities for the same action for the same state in each run:
    $\pi(a|s) = P(A = a|S = s)$
- The **value function** predicts the amount of long-term reward based on the selected action in the current state. Thus, the value function is specific to the policy used by the agent. The reward indicates the immediate gain from the action while the value function indicates the cumulative or long-term future gain from the action. The reward is returned by the environment and the value function is estimated by the agent at every time-step.
- The **model** is a representation of the environment kept internally by the agent. The model could be an imperfect representation of the environment. The agent uses the model to estimate the reward and the next state from the selected action.

The goal of an agent can also be to find the optimal policy for the Markovian Decision Process (MDP). MDP is a mathematical representation of the observations, actions, rewards, and transitions from one state to another. We will omit the discussion of MDP for the sake of brevity and advise the curious reader to search for resources on the internet for diving deeper into MDP.

Q function (learning to optimize when the model is not available)

If the model is not available then the agent learns the model and optimal policy by trial and error. When the model is not available, the agent uses a Q function, which is defined as follows:

$$Q : S \times A \rightarrow \mathbb{R}$$

The Q function basically maps the pairs of states and actions to a real number that denotes the expected total reward if the agent at state s selects an action a.

Exploration and exploitation in the RL algorithms

In the absence of a model, at every step the agent either explores or exploits. **Exploration** means that the agent selects an unknown action to find out the reward and the model. **Exploitation** means that the agent selects the best-known action to get the maximum reward. If the agent always decides to exploit then it might get stuck in a local optimal value. Hence, sometimes the agent takes a detour from learned policy to explore unknown actions. Similarly, if an agent always decides to explore then it may fail to find an optimal policy. Thus, it is important to have a balance of exploration and exploitation. In our code, we implement this by using a probability p to select a random action and probability $1\text{-}p$ to select the optimal action.

V function (learning to optimize when the model is available)

If the model is known beforehand then the agent can perform a **policy search** to find the optimal policy that maximizes the value function. When the model is available, the agent uses a value function that can be defined naively as a sum of the rewards of the future states:

$$V^{\pi}(s) = \sum_{i} R_i \quad \forall s \in S$$

Thus, the value at time-step t for selecting actions using the policy p would be:

$$V_t^{\pi} = R_t + R_{t+1} + \ldots + R_{t+n}$$

V is the value and R is the reward, and the value function is estimated only up to n time-steps in the future.

When the agent estimates the reward with this approach, it treats that reward as a result of all actions equally. In the pole cart example, if the poll falls at step 50, it will treat all the steps up to the 50th step as being equally responsible for the fall. Hence, instead of adding the future rewards, the weighted sum of future rewards is estimated. Usually, the weights are a discount rate raised to the power of the time-step. If the discount rate is zero then the value function becomes the naive function discussed above, and if the value of the discount rate is close to one, such as 0.9 or 0.92, then the future rewards have less effect when compared to the current rewards.

Thus, now the value at time-step t for action a would be:

$$V_t^{\pi} = R_t + r \times R_{t+1} = R_t + rR_{t+1} + r^2 R_{t+2} + \ldots + r^n R_{t+n}$$

V is the value, R is the rewards, and r is the discount rate.

The relationship between the V function and the Q function:

V*(s) is the optimal value function at state s that gives the maximum reward, and Q*(s,a) is the optimal Q function at state s that gives the maximum expected reward by selecting action a. Thus, V*(s) is the maximum of all optimal Q functions Q*(s,a) over all possible actions:

$$V^*(s) = \max_a Q^*(s, a) \quad \forall s \in S$$

Reinforcement learning techniques

Reinforcement learning techniques can be categorized on the basis of the availability of the model as follows:

- **Model is available**: If the model is available then the agent can plan offline by iterating over policies or the value function to find the optimal policy that gives the maximum reward.
    - **Value-iteration learning**: In the value-iteration learning approach, the agent starts by initializing the *V(s)* to a random value and then repeatedly updates the *V(s)* until a maximum reward is found.
    - **Policy-iterative learning**: In the policy-iteration learning approach, the agent starts by initializing a random policy *p*, and then repeatedly updates the policy until a maximum reward is found.

- **Model is not available**: If the model is not available, then an agent can only learn by observing the results of its actions. Thus, from the history of observations, actions, and rewards, an agent either tries to estimate the model or tries to directly derive the optimal policy:
    - **Model-based learning**: In model-based learning, the agent first estimates the model from the history, and then uses a policy or value-based approach to find the optimal policy.
    - **Model-free learning**: In model-free learning, the agent does not estimate the model, but rather estimates the optimal policy directly from the history. Q-Learning is an example of model-free learning.

As an example, the algorithm for the value-iteration learning is as follows:

```
initialize V(s) to random values for all states
Repeat
    for s in states
        for a in actions
            compute Q[s,a]
        V(s) = max(Q[s])    # maximum of Q for all actions for that state
Until optimal value of V(s) is found for all states
```

The algorithm for the policy-iteration learning is as follows:

```
initialize a policy P_new to random sequence of actions for all states
Repeat
    P = P_new
    for s in states
        compute V(s) with P[s]
        P_new[s] = policy of optimal V(s)
Until P == P_new
```

Naive Neural Network policy for Reinforcement Learning

We proceed with the policy as follows:

1. Let us implement a naive neural network-based policy. Define a new policy to use the neural network based predictions to return the actions:

    ```
    def policy_naive_nn(nn,obs):
        return np.argmax(nn.predict(np.array([obs])))
    ```

2. Define `nn` as a simple one layer MLP network that takes the observations having four dimensions as input, and produces the probabilities of the two actions:

    ```
    from keras.models import Sequential
    from keras.layers import Dense
    model = Sequential()
    model.add(Dense(8,input_dim=4, activation='relu'))
    model.add(Dense(2, activation='softmax'))
    model.compile(loss='categorical_crossentropy',optimizer='adam')
    model.summary()
    ```

This is what the model looks like:

```
Layer (type)                 Output Shape              Param #
=================================================================
dense_16 (Dense)             (None, 8)                 40
_________________________________________________________________
dense_17 (Dense)             (None, 2)                 18
=================================================================
Total params: 58
Trainable params: 58
Non-trainable params: 0
```

3. This model needs to be trained. Run the simulation for 100 episodes and collect the training data only for those episodes where the score is more than 100. If the score is less then 100, then those states and actions are not worth recording since they are not examples of good play:

```
# create training data
env = gym.make('CartPole-v0')
n_obs = 4
n_actions = 2
theta = np.random.rand(4) * 2 - 1
n_episodes = 100
r_max = 0
t_max = 0

x_train, y_train = experiment(env,
                              policy_random,
                              n_episodes,
                              theta,r_max,t_max,
                              return_hist_reward=100 )
y_train = np.eye(n_actions)[y_train]
print(x_train.shape,y_train.shape)
```

We are able to collect 5732 samples for training:

```
(5732, 4) (5732, 2)
```

4. Next, train the model:

```
model.fit(x_train, y_train, epochs=50, batch_size=10)
```

5. The trained model can be used to play the game. However, the model will not learn from the further plays of the game until we incorporate a loop updating the training data:

```
n_episodes = 200
```

```
r_max = 0
t_max = 0

_ = experiment(env,
               policy_naive_nn,
               n_episodes,
               theta=model,
               r_max=r_max,
               t_max=t_max,
               return_hist_reward=0 )

_ = experiment(env,
               policy_random,
               n_episodes,
               theta,r_max,t_max,
               return_hist_reward=0 )
```

We can see that this naive policy performs almost in the same way, albeit a little better than the random policy:

```
Policy:policy_naive_nn, Min reward:37.0, Max reward:200.0, Average
reward:71.05
Policy:policy_random, Min reward:36.0, Max reward:200.0, Average
reward:68.755
```

We can improve the results further with network tuning and hyper-parameters tuning, or by learning from more gameplay. However, there are better algorithms, such as Q-Learning.

In the rest of this chapter, we shall focus on the Q-Learning algorithm since most real-life problems involve model-free learning.

Implementing Q-Learning

Q-Learning is a model-free method of finding the optimal policy that can maximize the reward of an agent. During initial gameplay, the agent learns a Q value for each pair of (state, action), also known as the exploration strategy, as explained in previous sections. Once the Q values are learned, then the optimal policy will be to select an action with the largest Q-value in every state, also known as the exploitation strategy. The learning algorithm may end in locally optimal solutions, hence we keep using the exploration policy by setting an `exploration_rate` parameter.

The Q-Learning algorithm is as follows:

```
initialize Q(shape=[#s,#a]) to random values or zeroes
Repeat (for each episode)
    observe current state s
    Repeat
        select an action a (apply explore or exploit strategy)
        observe state s_next as a result of action a
        update the Q-Table using bellman's equation
        set current state s = s_next
    until the episode ends or a max reward / max steps condition is reached
Until a number of episodes or a condition is reached
        (such as max consecutive wins)
```

The *Q(s, a)* in the preceding algorithm represents the Q function that we described in the previous sections. The values of this functions are used for selecting the action instead of the rewards, thus this function represents the reward or discounted rewards. The values for the Q-function are updated using the values of the Q function in the future state. The well-known *bellman equation* captures this update:

$$Q(s_t, a_t) = r_t + \gamma \max_a Q(s_{t+1}, a)$$

This basically means that at time step t, in state s, for action a, the maximum future reward (Q) is equal to the reward from the current state plus the max future reward from the next state.

Q(s,a) can be implemented as a Q-Table or as a neural network known as a Q-Network. In both cases, the task of the Q-Table or the Q-Network is to provide the best possible action based on the Q value of the given input. The Q-Table-based approach generally becomes intractable as the Q-Table becomes large, thus making neural networks the best candidate for approximating the Q-function through Q-Network. Let us look at both of these approaches in action.

You can follow along with the code in the Jupyter notebook `ch-13b_Reinforcement_Learning_DQN` in the code bundle of this book.

Initializing and discretizing for Q-Learning

The observations returned by the pole-cart environment involves the state of the environment. The state of pole-cart is represented by continuous values that we need to discretize.

If we discretize these values into small state-space, then the agent gets trained faster, but with the caveat of risking the convergence to the optimal policy.

We use the following helper function to discretize the state-space of the pole-cart environment:

```
# discretize the value to a state space
def discretize(val,bounds,n_states):
    discrete_val = 0
    if val <= bounds[0]:
        discrete_val = 0
    elif val >= bounds[1]:
        discrete_val = n_states-1
    else:
        discrete_val = int(round( (n_states-1) *
                                  ((val-bounds[0])/
                                   (bounds[1]-bounds[0]))
                                ))
    return discrete_val

def discretize_state(vals,s_bounds,n_s):
    discrete_vals = []
    for i in range(len(n_s)):
        discrete_vals.append(discretize(vals[i],s_bounds[i],n_s[i]))
    return np.array(discrete_vals,dtype=np.int)
```

We discretize the space into 10 units for each of the observation dimensions. You may want to try out different discretization spaces. After the discretization, we find the upper and lower bounds of the observations, and change the bounds of velocity and angular velocity to be between -1 and +1, instead of -Inf and +Inf. The code is as follows:

```
env = gym.make('CartPole-v0')
n_a = env.action_space.n
# number of discrete states for each observation dimension
n_s = np.array([10,10,10,10])   # position, velocity, angle, angular
velocity
s_bounds = np.array(list(zip(env.observation_space.low,
env.observation_space.high)))
# the velocity and angular velocity bounds are
# too high so we bound between -1, +1
```

```
s_bounds[1] = (-1.0,1.0)
s_bounds[3] = (-1.0,1.0)
```

Q-Learning with Q-Table

You may follow the code for this section in `ch-13b.ipynb`. Since our discretised space is of the dimensions [10,10,10,10], our Q-Table is of [10,10,10,10,2] dimensions:

```
# create a Q-Table of shape (10,10,10,10, 2) representing S X A -> R
q_table = np.zeros(shape = np.append(n_s,n_a))
```

We define a Q-Table policy that exploits or explores based on the `exploration_rate`:

```
def policy_q_table(state, env):
    # Exploration strategy - Select a random action
    if np.random.random() < explore_rate:
        action = env.action_space.sample()
    # Exploitation strategy - Select the action with the highest q
    else:
        action = np.argmax(q_table[tuple(state)])
    return action
```

Define the `episode()` function that runs a single episode as follows:

1. Start with initializing the variables and the first state:

```
obs = env.reset()
state_prev = discretize_state(obs,s_bounds,n_s)

episode_reward = 0
done = False
t = 0
```

2. Select the action and observe the next state:

```
action = policy(state_prev, env)
obs, reward, done, info = env.step(action)
state_new = discretize_state(obs,s_bounds,n_s)
```

3. Update the Q-Table:

```
best_q = np.amax(q_table[tuple(state_new)])
bellman_q = reward + discount_rate * best_q
indices = tuple(np.append(state_prev,action))
q_table[indices] += learning_rate*( bellman_q - q_table[indices])
```

4. Set the next state as the previous state and add the rewards to the episode's rewards:

```
state_prev = state_new
episode_reward += reward
```

The `experiment()` function calls the episode function and accumulates the rewards for reporting. You may want to modify the function to check for consecutive wins and other logic specific to your play or games:

```
# collect observations and rewards for each episode
def experiment(env, policy, n_episodes,r_max=0, t_max=0):
    rewards=np.empty(shape=[n_episodes])
    for i in range(n_episodes):
        val = episode(env, policy, r_max, t_max)
        rewards[i]=val
    print('Policy:{}, Min reward:{}, Max reward:{}, Average reward:{}'
      .format(policy.__name__,
              np.min(rewards),
              np.max(rewards),
              np.mean(rewards)))
```

Now, all we have to do is define the parameters, such as `learning_rate`, `discount_rate`, and `explore_rate`, and run the `experiment()` function as follows:

```
learning_rate = 0.8
discount_rate = 0.9
explore_rate = 0.2
n_episodes = 1000
experiment(env, policy_q_table, n_episodes)
```

For 1000 episodes, the Q-Table-based policy's maximum reward is 180 based on our simple implementation:

```
Policy:policy_q_table, Min reward:8.0, Max reward:180.0, Average
reward:17.592
```

Our implementation of the algorithm is very simple to explain. However, you can modify the code to set the explore rate high initially and then decay as the time-steps pass. Similarly, you can also implement the decay logic for the learning and discount rates. Let us see if we can get a higher reward with fewer episodes as our Q function learns faster.

Q-Learning with Q-Network or Deep Q Network (DQN)

In the DQN, we replace the Q-Table with a neural network (Q-Network) that will learn to respond with the optimal action as we train it continuously with the explored states and their Q-Values. Thus, for training the network we need a place to store the game memory:

1. Implement the game memory using a deque of size 1000:

```
memory = deque(maxlen=1000)
```

2. Next, build a simple hidden layer neural network model, `q_nn`:

```
from keras.models import Sequential
from keras.layers import Dense
model = Sequential()
model.add(Dense(8,input_dim=4, activation='relu'))
model.add(Dense(2, activation='linear'))
model.compile(loss='mse',optimizer='adam')
model.summary()
q_nn = model
```

The Q-Network looks like this:

```
_________________________________________________________________
Layer (type)                 Output Shape              Param #
=================================================================
dense_1 (Dense)              (None, 8)                 40
_________________________________________________________________
dense_2 (Dense)              (None, 2)                 18
=================================================================
Total params: 58
Trainable params: 58
Non-trainable params: 0
_________________________________________________________________
```

The `episode()` function that executes one episode of the game, incorporates the following changes for the Q-Network-based algorithm:

1. After generating the next state, add the states, action, and rewards to the game memory:

```
action = policy(state_prev, env)
obs, reward, done, info = env.step(action)
state_next = discretize_state(obs,s_bounds,n_s)
```

```
# add the state_prev, action, reward, state_new, done to memory
memory.append([state_prev,action,reward,state_next,done])
```

2. Generate and update the `q_values` with the maximum future rewards using the bellman function:

```
states = np.array([x[0] for x in memory])
states_next = np.array([np.zeros(4) if x[4] else x[3] for x in
memory])
q_values = q_nn.predict(states)
q_values_next = q_nn.predict(states_next)

for i in range(len(memory)):
    state_prev,action,reward,state_next,done = memory[i]
    if done:
        q_values[i,action] = reward
    else:
        best_q = np.amax(q_values_next[i])
        bellman_q = reward + discount_rate * best_q
        q_values[i,action] = bellman_q
```

3. Train the `q_nn` with the states and the `q_values` we received from memory:

```
q_nn.fit(states,q_values,epochs=1,batch_size=50,verbose=0)
```

The process of saving gameplay in memory and using it to train the model is also known as **memory replay** in deep reinforcement learning literature. Let us run our DQN-based gameplay as follows:

```
learning_rate = 0.8
discount_rate = 0.9
explore_rate = 0.2
n_episodes = 100
experiment(env, policy_q_nn, n_episodes)
```

We get a max reward of 150 that you can improve upon with hyper-parameter tuning, network tuning, and by using rate decay for the discount rate and explore rate:

```
Policy:policy_q_nn, Min reward:8.0, Max reward:150.0, Average reward:41.27
```

We calculated and trained the model in every step; you may want to explore changing it to training after the episode. Also, you can change the code to discard the memory replay and retraining the model for the episodes that return smaller rewards. However, implement this option with caution as it may slow down your learning as initial gameplay would generate smaller rewards more often.

Summary

In this chapter, we learned how to implement reinforcement learning algorithms in Keras. For the sake of keeping the examples simple, we used Keras; you can implement the same networks and models with TensorFlow as well. We only used a one-layer MLP, as our example game was very simple, but for complex examples, you may end up using complex CNN, RNN, or Sequence to Sequence models.

We also learned about OpenAI Gym, a framework that provides an environment to simulate many popular games in order to implement and practice the reinforcement learning algorithms. We touched on deep reinforcement learning concepts, and we encourage you to explore books specifically written about reinforcement learning to learn deeply about the theories and concepts.

Reinforcement Learning is an advanced technique that you will find is often used for solving complex problems. In the next chapter, we shall learn another family of advanced deep learning techniques: Generative Adversarial Networks.

14
Generative Adversarial Networks

Generative models are trained to generate more data similar to the one they are trained on, and adversarial models are trained to distinguish the real versus fake data by providing adversarial examples.

The **Generative Adversarial Networks** (**GAN**) combine the features of both the models. The GANs have two components:

- A generative model that learns how to generate similar data
- A discriminative model that learns how to distinguish between the real and generated data (from the generative model)

GANs have been successfully applied to various complex problems such as:

- Generating photo-realistic resolution images from low-resolution images
- Synthesizing images from the text
- Style transfer
- Completing the incomplete images and videos

In this chapter, we shall study the following topics for learning how to implement GANs in TensorFlow and Keras:

- Generative Adversarial Networks
- Simple GAN in TensorFlow
- Simple GAN in Keras
- Deep Convolutional GAN with TensorFlow and Keras

Generative Adversarial Networks 101

As shown in the following diagram, the Generative Adversarial Networks, popularly known as GANs, have two models working in sync to learn and train on complex data such as images, videos or audio files:

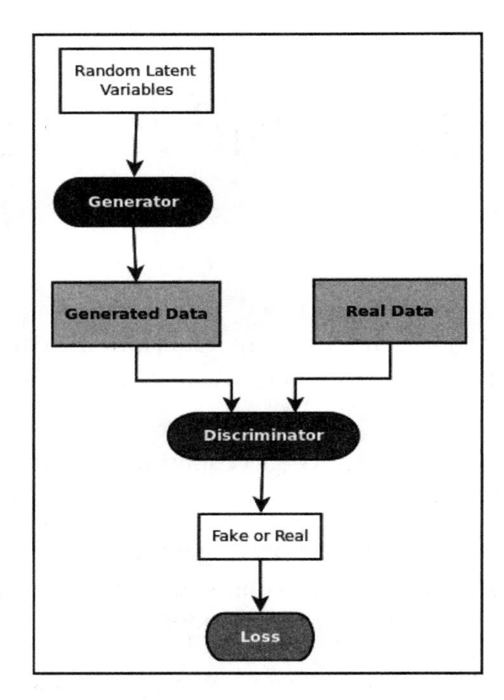

Intuitively, the generator model generates data starting from random noise but slowly learns how to generate more realistic data. The generator output and the real data is fed into the discriminator that learns how to differentiate fake data from real data.

Thus, both generator and discriminator play an adversarial game where the generator tries to fool the discriminator by generating as real data as possible, and the discriminator tries not to be fooled by identifying fake data from real data, thus the discriminator tries to minimize the classification loss. Both the models are trained in a lockstep fashion.

Mathematically, the generative model $G(z)$ learns the probability distribution $p(z)$ such that the discriminator $D(G(z), x)$ is unable to identify between the probability distributions, $p(z)$ and $p(x)$. The objective function of the GAN can be described by the following equation describing the value function V, (from `https://papers.nips.cc/paper/5423-generative-adversarial-nets.pdf`):

$$\min_G \max_D V(D, G) = E_{x \sim p_{data}(x)}[logD(x)] + E_{z \sim p_z(z)}[log(1 - D(G(z)))]$$

The seminal tutorial at NIPS 2016 on GANs by Ian Goodfellow can be found at the following link: `https://arxiv.org/pdf/1701.00160.pdf`.

This description represents a simple GAN (also known as a vanilla GAN in literature), first introduced by Goodfellow in the seminal paper available at this link: `https://arxiv.org/abs/1406.2661`. Since then, there has been tremendous research in deriving different architectures based on GANs and applying them to different application areas.

For example, in conditional GANs the generator and the discriminator networks are provided with the labels such that the objective function of the conditional GAN can be described by the following equation describing the value function V:

$$\min_G \max_D V(D, G) = E_{x \sim p_{data}(x)}[logD(x)] + E_{z \sim p_z(z)}[log(1 - D(G(z, y), y))]$$

The original paper describing the conditional GANs is located at the following link: `https://arxiv.org/abs/1411.1784`.

Several other derivatives and their originating papers used in applications, such as Text to Image, Image Synthesis, Image Tagging, Style Transfer, and Image Transfer and so on are listed in the following table:

GAN Derivative	**Originating Paper**	**Demonstrated Application**
StackGAN	`https://arxiv.org/abs/1710.10916`	**Text to Image**
StackGAN++	`https://arxiv.org/abs/1612.03242`	Photo-realistic Image Synthesis
DCGAN	`https://arxiv.org/abs/1511.06434`	Image Synthesis
HR-DCGAN	`https://arxiv.org/abs/1711.06491`	High-Resolution Image Synthesis
Conditonal GAN	`https://arxiv.org/abs/1411.1784`	Image Tagging

GAN Derivative	Originating Paper	Demonstrated Application
InfoGAN	`https://arxiv.org/abs/1606.03657`	Style Identification
Wasserstein GAN	`https://arxiv.org/abs/1701.07875` `https://arxiv.org/abs/1704.00028`	Image Generation
Coupled GAN	`https://arxiv.org/abs/1606.07536`	Image Transformation, Domain Adaptation
BE GAN	`https://arxiv.org/abs/1703.10717`	Image Generation
DiscoGAN	`https://arxiv.org/abs/1703.05192`	Style Transfer
CycleGAN	`https://arxiv.org/abs/1703.10593`	Style Transfer

Let us practice creating a simple GAN using the MNIST dataset. For this exercise, we shall normalize the MNIST dataset to lie between [-1,+1], using the following function:

```
def norm(x):
    return (x-0.5)/0.5
```

We also define the random noise with 256 dimensions that would be used to test the generator models:

```
n_z = 256
z_test = np.random.uniform(-1.0,1.0,size=[8,n_z])
```

The function to display the generated images that would be used in all the examples in this chapter:

```
def display_images(images):
    for i in range(images.shape[0]):
        plt.subplot(1, 8, i + 1)
        plt.imshow(images[i])
        plt.axis('off')
    plt.tight_layout()
    plt.show()
```

Best practices for building and training GANs

For the dataset we selected for this demonstration, the discriminator was becoming very good at classifying the real and fake images, and therefore not providing much of the feedback in terms of gradients to the generator. Hence we had to make the discriminator weak with the following best practices:

- The learning rate of the discriminator is kept much higher than the learning rate of the generator.
- The optimizer for the discriminator is `GradientDescent` and the optimizer for the generator is `Adam`.
- The discriminator has dropout regularization while the generator does not.
- The discriminator has fewer layers and fewer neurons as compared to the generator.
- The output of the generator is `tanh` while the output of the discriminator is sigmoid.
- In the Keras model, we use a value of 0.9 instead of 1.0 for labels of real data and we use 0.1 instead of 0.0 for labels of fake data, in order to introduce a little bit of noise in the labels

You are welcome to explore and try other best practices.

Simple GAN with TensorFlow

You can follow along with the code in the Jupyter notebook `ch-14a_SimpleGAN`.

For building the GAN with TensorFlow, we build three networks, two discriminator models, and one generator model with the following steps:

1. Start by adding the hyper-parameters for defining the network:

```
# graph hyperparameters
g_learning_rate = 0.00001
d_learning_rate = 0.01
```

```
n_x = 784  # number of pixels in the MNIST image

# number of hidden layers for generator and discriminator
g_n_layers = 3
d_n_layers = 1
# neurons in each hidden layer
g_n_neurons = [256, 512, 1024]
d_n_neurons = [256]

# define parameter ditionary
d_params = {}
g_params = {}

activation = tf.nn.leaky_relu
w_initializer = tf.glorot_uniform_initializer
b_initializer = tf.zeros_initializer
```

2. Next, define the generator network:

```
z_p = tf.placeholder(dtype=tf.float32, name='z_p',
        shape=[None, n_z])
layer = z_p

# add generator network weights, biases and layers
with tf.variable_scope('g'):
    for i in range(0, g_n_layers):
        w_name = 'w_{0:04d}'.format(i)
        g_params[w_name] = tf.get_variable(
            name=w_name,
            shape=[n_z if i == 0 else g_n_neurons[i - 1],
                    g_n_neurons[i]],
            initializer=w_initializer())
        b_name = 'b_{0:04d}'.format(i)
        g_params[b_name] = tf.get_variable(
            name=b_name, shape=[g_n_neurons[i]],
            initializer=b_initializer())
        layer = activation(
            tf.matmul(layer, g_params[w_name]) + g_params[b_name])
    # output (logit) layer
    i = g_n_layers
    w_name = 'w_{0:04d}'.format(i)
    g_params[w_name] = tf.get_variable(
        name=w_name,
        shape=[g_n_neurons[i - 1], n_x],
        initializer=w_initializer())
    b_name = 'b_{0:04d}'.format(i)
    g_params[b_name] = tf.get_variable(
```

```
            name=b_name, shape=[n_x], initializer=b_initializer())
    g_logit = tf.matmul(layer, g_params[w_name]) + g_params[b_name]
    g_model = tf.nn.tanh(g_logit)
```

3. Next, define the weights and biases for the two discriminator networks that we shall build:

```
with tf.variable_scope('d'):
    for i in range(0, d_n_layers):
        w_name = 'w_{0:04d}'.format(i)
        d_params[w_name] = tf.get_variable(
            name=w_name,
            shape=[n_x if i == 0 else d_n_neurons[i - 1],
                   d_n_neurons[i]],
            initializer=w_initializer())

        b_name = 'b_{0:04d}'.format(i)
        d_params[b_name] = tf.get_variable(
            name=b_name, shape=[d_n_neurons[i]],
            initializer=b_initializer())

    #output (logit) layer
    i = d_n_layers
    w_name = 'w_{0:04d}'.format(i)
    d_params[w_name] = tf.get_variable(
        name=w_name, shape=[d_n_neurons[i - 1], 1],
        initializer=w_initializer())

    b_name = 'b_{0:04d}'.format(i)
    d_params[b_name] = tf.get_variable(
        name=b_name, shape=[1], initializer=b_initializer())
```

4. Now using these parameters, build the discriminator that takes the real images as input and outputs the classification:

```
# define discriminator_real

# input real images
x_p = tf.placeholder(dtype=tf.float32, name='x_p',
        shape=[None, n_x])

layer = x_p

with tf.variable_scope('d'):
    for i in range(0, d_n_layers):
        w_name = 'w_{0:04d}'.format(i)
        b_name = 'b_{0:04d}'.format(i)
```

```
            layer = activation(
                tf.matmul(layer, d_params[w_name]) + d_params[b_name])
            layer = tf.nn.dropout(layer,0.7)
        #output (logit) layer
        i = d_n_layers
        w_name = 'w_{0:04d}'.format(i)
        b_name = 'b_{0:04d}'.format(i)
        d_logit_real = tf.matmul(layer,
            d_params[w_name]) + d_params[b_name]
        d_model_real = tf.nn.sigmoid(d_logit_real)
```

5. Next, build another discriminator network, with the same parameters, but providing the output of generator as input:

```
    # define discriminator_fake

    # input generated fake images
    z = g_model
    layer = z

    with tf.variable_scope('d'):
        for i in range(0, d_n_layers):
            w_name = 'w_{0:04d}'.format(i)
            b_name = 'b_{0:04d}'.format(i)
            layer = activation(
                tf.matmul(layer, d_params[w_name]) + d_params[b_name])
            layer = tf.nn.dropout(layer,0.7)
        #output (logit) layer
        i = d_n_layers
        w_name = 'w_{0:04d}'.format(i)
        b_name = 'b_{0:04d}'.format(i)
        d_logit_fake = tf.matmul(layer,
            d_params[w_name]) + d_params[b_name]
        d_model_fake = tf.nn.sigmoid(d_logit_fake)
```

6. Now that we have the three networks built, the connection between them is made using the loss, optimizer and training functions. While training the generator, we only train the generator's parameters and while training the discriminator, we only train the discriminator's parameters. We specify this using the `var_list` parameter to the optimizer's `minimize()` function. Here is the complete code for defining the loss, optimizer and training function for both kinds of network:

```
    g_loss = -tf.reduce_mean(tf.log(d_model_fake))
    d_loss = -tf.reduce_mean(tf.log(d_model_real) + tf.log(1 -
    d_model_fake))
```

```
g_optimizer = tf.train.AdamOptimizer(g_learning_rate)
d_optimizer = tf.train.GradientDescentOptimizer(d_learning_rate)

g_train_op = g_optimizer.minimize(g_loss,
                var_list=list(g_params.values()))
d_train_op = d_optimizer.minimize(d_loss,
                var_list=list(d_params.values()))
```

7. Now that we have defined the models, we have to train the models. The training is done as per the following algorithm:

```
For each epoch:
  For each batch:
    get real images x_batch
    generate noise z_batch
    train discriminator using z_batch and x_batch
    generate noise z_batch
    train generator using z_batch
```

The complete code for training from the notebook is as follows:

```
n_epochs = 400
batch_size = 100
n_batches = int(mnist.train.num_examples / batch_size)
n_epochs_print = 50

with tf.Session() as tfs:
    tfs.run(tf.global_variables_initializer())
    for epoch in range(n_epochs):
        epoch_d_loss = 0.0
        epoch_g_loss = 0.0
        for batch in range(n_batches):
            x_batch, _ = mnist.train.next_batch(batch_size)
            x_batch = norm(x_batch)
            z_batch = np.random.uniform(-1.0,1.0,size=[batch_size,n_z])
            feed_dict = {x_p: x_batch,z_p: z_batch}
            _,batch_d_loss = tfs.run([d_train_op,d_loss],
                                    feed_dict=feed_dict)
            z_batch = np.random.uniform(-1.0,1.0,size=[batch_size,n_z])
            feed_dict={z_p: z_batch}
            _,batch_g_loss = tfs.run([g_train_op,g_loss],
                                    feed_dict=feed_dict)
            epoch_d_loss += batch_d_loss
            epoch_g_loss += batch_g_loss
        if epoch%n_epochs_print == 0:
            average_d_loss = epoch_d_loss / n_batches
            average_g_loss = epoch_g_loss / n_batches
```

```
    print('epoch: {0:04d}   d_loss = {1:0.6f}  g_loss = {2:0.6f}'
          .format(epoch,average_d_loss,average_g_loss))
    # predict images using generator model trained
    x_pred = tfs.run(g_model,feed_dict={z_p:z_test})
    display_images(x_pred.reshape(-1,pixel_size,pixel_size))
```

We printed the generated images every 50 epochs:

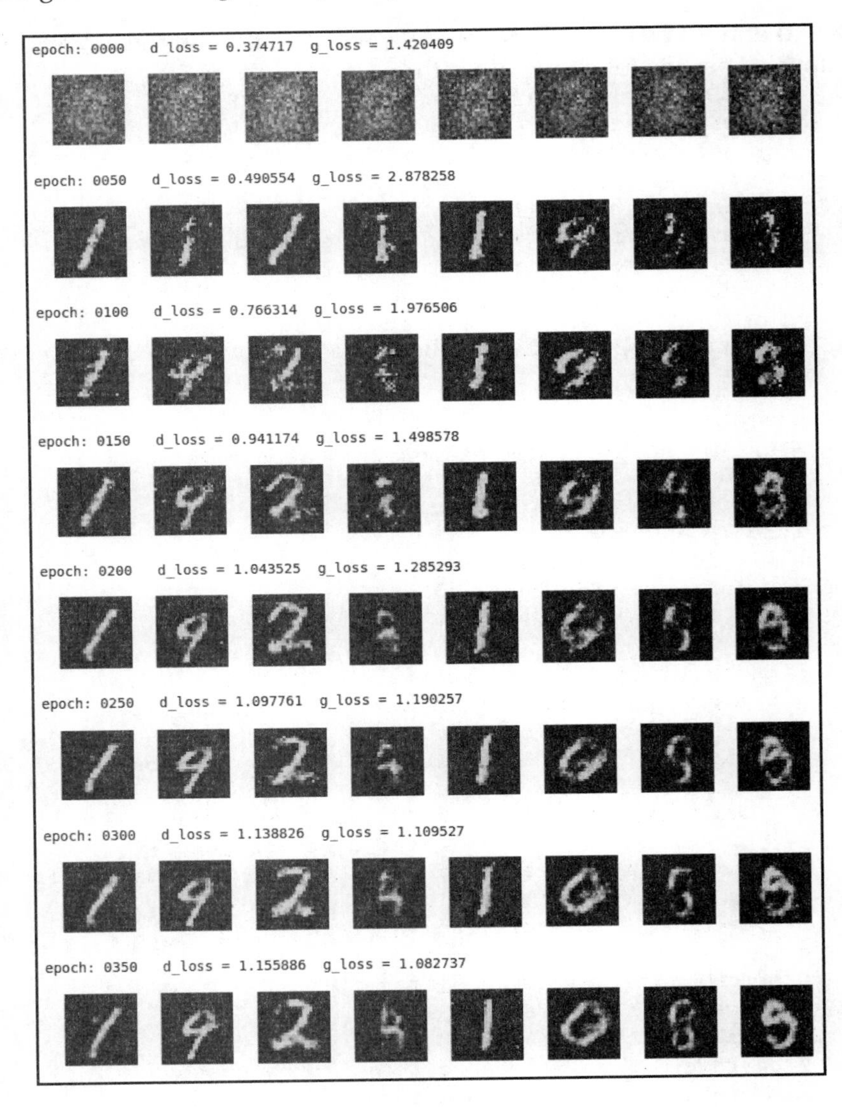

As we can see the generator was producing just noise in epoch 0, but by epoch 350, it got trained to produce much better shapes of handwritten digits. You can try experimenting with epochs, regularization, network architecture and other hyper-parameters to see if you can produce even faster and better results.

Simple GAN with Keras

You can follow along with the code in the Jupyter notebook `ch-14a_SimpleGAN`.

Now let us implement the same model in Keras:

1. The hyper-parameter definitions remain the same as the last section:

```
# graph hyperparameters
g_learning_rate = 0.00001
d_learning_rate = 0.01
n_x = 784  # number of pixels in the MNIST image
# number of hidden layers for generator and discriminator
g_n_layers = 3
d_n_layers = 1
# neurons in each hidden layer
g_n_neurons = [256, 512, 1024]
d_n_neurons = [256]
```

2. Next, define the generator network:

```
# define generator

g_model = Sequential()
g_model.add(Dense(units=g_n_neurons[0],
                  input_shape=(n_z,),
                  name='g_0'))
g_model.add(LeakyReLU())
for i in range(1,g_n_layers):
    g_model.add(Dense(units=g_n_neurons[i],
                      name='g_{}'.format(i)
                     ))
    g_model.add(LeakyReLU())
g_model.add(Dense(units=n_x, activation='tanh',name='g_out'))
print('Generator:')
g_model.summary()
```

```
g_model.compile(loss='binary_crossentropy',
                optimizer=keras.optimizers.Adam(lr=g_learning_rate)
               )
```

This is what the generator model looks like:

```
Generator:
_________________________________________________________________
Layer (type)                 Output Shape              Param #
=================================================================
g_0 (Dense)                  (None, 256)               65792
_________________________________________________________________
leaky_re_lu_1 (LeakyReLU)    (None, 256)               0
_________________________________________________________________
g_1 (Dense)                  (None, 512)               131584
_________________________________________________________________
leaky_re_lu_2 (LeakyReLU)    (None, 512)               0
_________________________________________________________________
g_2 (Dense)                  (None, 1024)              525312
_________________________________________________________________
leaky_re_lu_3 (LeakyReLU)    (None, 1024)              0
_________________________________________________________________
g_out (Dense)                (None, 784)               803600
=================================================================
Total params: 1,526,288
Trainable params: 1,526,288
Non-trainable params: 0
_________________________________________________________________
```

3. In the Keras example, we do not define two discriminator networks as we defined in the TensorFlow example. Instead, we define one discriminator network and then stitch the generator and discriminator network into the GAN network. The GAN network is then used to train the generator parameters only, and the discriminator network is used to train the discriminator parameters:

```
# define discriminator

d_model = Sequential()
d_model.add(Dense(units=d_n_neurons[0],
                  input_shape=(n_x,),
                  name='d_0'
                 ))
d_model.add(LeakyReLU())
d_model.add(Dropout(0.3))
for i in range(1,d_n_layers):
    d_model.add(Dense(units=d_n_neurons[i],
                      name='d_{}'.format(i)
```

```
                              ))
    d_model.add(LeakyReLU())
    d_model.add(Dropout(0.3))
d_model.add(Dense(units=1, activation='sigmoid',name='d_out'))
print('Discriminator:')
d_model.summary()
d_model.compile(loss='binary_crossentropy',
                optimizer=keras.optimizers.SGD(lr=d_learning_rate)
               )
```

This is what the discriminator models look:

```
Discriminator:
_________________________________________________________________
Layer (type)                 Output Shape              Param #
=================================================================
d_0 (Dense)                  (None, 256)               200960
_________________________________________________________________
leaky_re_lu_4 (LeakyReLU)    (None, 256)               0
_________________________________________________________________
dropout_1 (Dropout)          (None, 256)               0
_________________________________________________________________
d_out (Dense)                (None, 1)                 257
=================================================================
Total params: 201,217
Trainable params: 201,217
Non-trainable params: 0
_________________________________________________________________
```

4. Next, define the GAN Network, and turn the trainable property of the discriminator model to `false`, since GAN would only be used to train the generator:

```
# define GAN network
d_model.trainable=False
z_in = Input(shape=(n_z,),name='z_in')
x_in = g_model(z_in)
gan_out = d_model(x_in)

gan_model = Model(inputs=z_in,outputs=gan_out,name='gan')
print('GAN:')
gan_model.summary()
```

```
gan_model.compile(loss='binary_crossentropy',
                  optimizer=keras.optimizers.Adam(lr=g_learning_rate)
                 )
```

This is what the GAN model looks:

```
GAN:
_________________________________________________________________
Layer (type)                 Output Shape              Param #
=================================================================
z_in (InputLayer)            (None, 256)               0
_________________________________________________________________
sequential_1 (Sequential)    (None, 784)               1526288
_________________________________________________________________
sequential_2 (Sequential)    (None, 1)                 201217
=================================================================
Total params: 1,727,505
Trainable params: 1,526,288
Non-trainable params: 201,217
_________________________________________________________________
```

5. Great, now that we have defined the three models, we have to train the models. The training is as per the following algorithm:

```
For each epoch:
  For each batch:
    get real images x_batch
    generate noise z_batch
    generate images g_batch using generator model
    combine g_batch and x_batch into x_in and create labels y_out

    set discriminator model as trainable
    train discriminator using x_in and y_out
    generate noise z_batch
    set x_in = z_batch and labels y_out = 1
    set discriminator model as non-trainable
    train gan model using x_in and y_out,
        (effectively training generator model)
```

For setting the labels, we apply the labels as 0.9 and 0.1 for real and fake images respectively. Generally, it is suggested that you use label smoothing by picking a random value from 0.0 to 0.3 for fake data and 0.8 to 1.0 for real data.

Here is the complete code for training from the notebook:

```
n_epochs = 400
batch_size = 100
n_batches = int(mnist.train.num_examples / batch_size)
n_epochs_print = 50

for epoch in range(n_epochs+1):
    epoch_d_loss = 0.0
    epoch_g_loss = 0.0
    for batch in range(n_batches):
        x_batch, _ = mnist.train.next_batch(batch_size)
        x_batch = norm(x_batch)
        z_batch = np.random.uniform(-1.0,1.0,size=[batch_size,n_z])
        g_batch = g_model.predict(z_batch)
        x_in = np.concatenate([x_batch,g_batch])
        y_out = np.ones(batch_size*2)
        y_out[:batch_size]=0.9
        y_out[batch_size:]=0.1
        d_model.trainable=True
        batch_d_loss = d_model.train_on_batch(x_in,y_out)

        z_batch = np.random.uniform(-1.0,1.0,size=[batch_size,n_z])
        x_in=z_batch
        y_out = np.ones(batch_size)
        d_model.trainable=False
        batch_g_loss = gan_model.train_on_batch(x_in,y_out)
        epoch_d_loss += batch_d_loss
        epoch_g_loss += batch_g_loss
    if epoch%n_epochs_print == 0:
        average_d_loss = epoch_d_loss / n_batches
        average_g_loss = epoch_g_loss / n_batches
        print('epoch: {0:04d}   d_loss = {1:0.6f}  g_loss = {2:0.6f}'
              .format(epoch,average_d_loss,average_g_loss))
        # predict images using generator model trained
        x_pred = g_model.predict(z_test)
        display_images(x_pred.reshape(-1,pixel_size,pixel_size))
```

We printed the results every 50 epochs, up to 350 epochs:

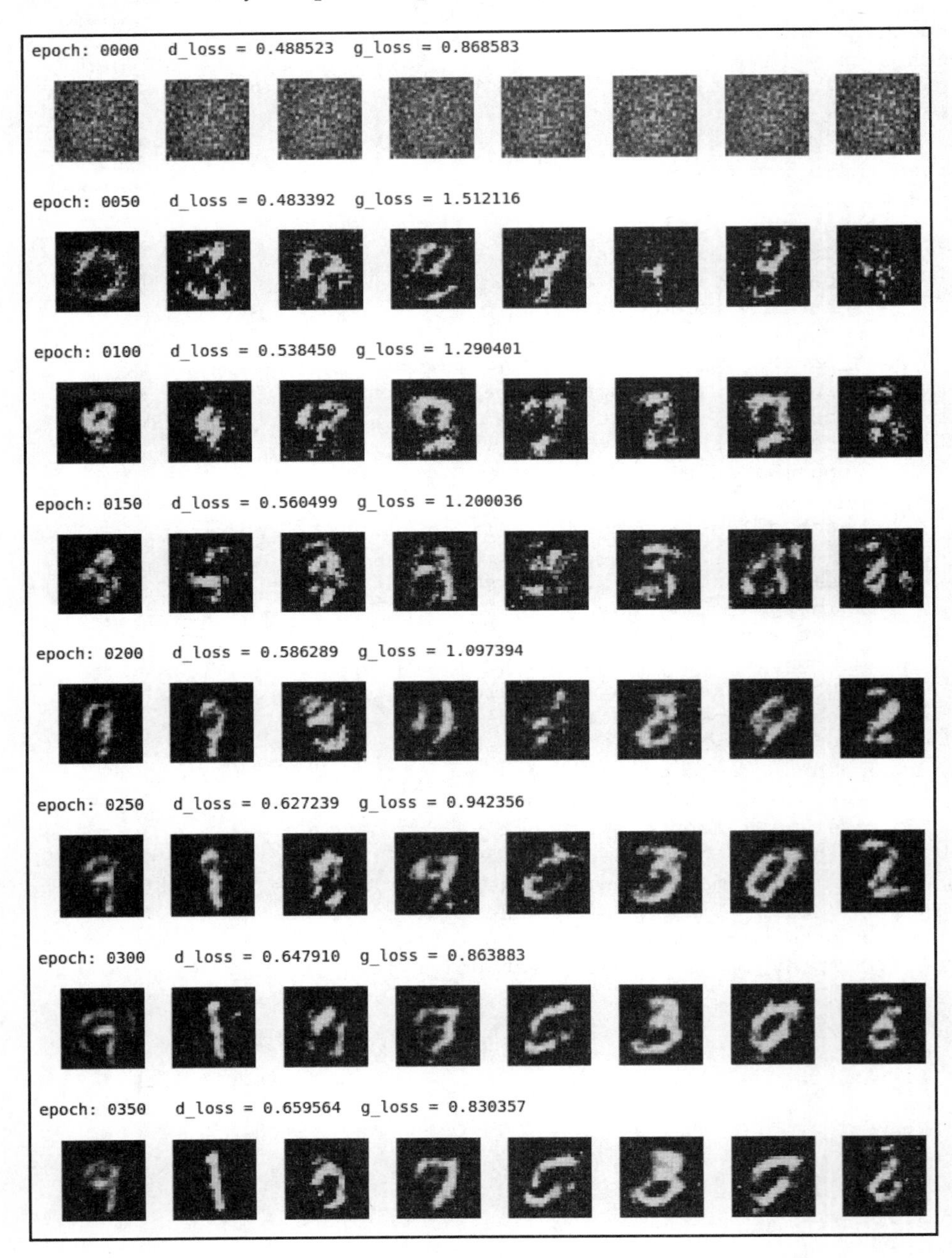

The model slowly learns to generate good quality images of handwritten digits from the random noise.

There are so many variations of the GANs that it will take another book to cover all the different kinds of GANs. However, the implementation techniques are almost similar to what we have shown here.

Deep Convolutional GAN with TensorFlow and Keras

You can follow along with the code in the Jupyter notebook `ch-14b_DCGAN`.

In DCGAN, both the discriminator and generator are implemented using a Deep Convolutional Network:

1. In this example, we decided to implement the generator as the following network:

```
Generator:
_________________________________________________________________
Layer (type)                 Output Shape              Param #
=================================================================
g_in (Dense)                 (None, 3200)              822400
_________________________________________________________________
g_in_act (Activation)        (None, 3200)              0
_________________________________________________________________
g_in_reshape (Reshape)       (None, 5, 5, 128)         0
_________________________________________________________________
g_0_up2d (UpSampling2D)      (None, 10, 10, 128)       0
_________________________________________________________________
g_0_conv2d (Conv2D)          (None, 10, 10, 64)        204864
_________________________________________________________________
g_0_act (Activation)         (None, 10, 10, 64)        0
_________________________________________________________________
g_1_up2d (UpSampling2D)      (None, 20, 20, 64)        0
_________________________________________________________________
g_1_conv2d (Conv2D)          (None, 20, 20, 32)        51232
_________________________________________________________________
g_1_act (Activation)         (None, 20, 20, 32)        0
```

```
_________________________________________________________________
g_2_up2d (UpSampling2D)      (None, 40, 40, 32)        0
_________________________________________________________________
g_2_conv2d (Conv2D)          (None, 40, 40, 16)        12816
_________________________________________________________________
g_2_act (Activation)         (None, 40, 40, 16)        0
_________________________________________________________________
g_out_flatten (Flatten)      (None, 25600)             0
_________________________________________________________________
g_out (Dense)                (None, 784)               20071184
=================================================================
Total params: 21,162,496
Trainable params: 21,162,496
Non-trainable params: 0
```

2. The generator is a stronger network having three convolutional layers followed by tanh activation. We define the discriminator network as follows:

```
Discriminator:
_________________________________________________________________
Layer (type)                 Output Shape              Param #
=================================================================
d_0_reshape (Reshape)        (None, 28, 28, 1)         0
_________________________________________________________________
d_0_conv2d (Conv2D)          (None, 28, 28, 64)        1664
_________________________________________________________________
d_0_act (Activation)         (None, 28, 28, 64)        0
_________________________________________________________________
d_0_maxpool (MaxPooling2D)   (None, 14, 14, 64)        0
_________________________________________________________________
d_out_flatten (Flatten)      (None, 12544)             0
_________________________________________________________________
d_out (Dense)                (None, 1)                 12545
=================================================================
Total params: 14,209
Trainable params: 14,209
Non-trainable params: 0
_________________________________________________________________
```

3. The GAN network is composed of the discriminator and generator as demonstrated previously:

```
GAN:
_________________________________________________________________
Layer (type)                 Output Shape              Param #
=================================================================
z_in (InputLayer)            (None, 256)               0
```

```
_________________________________________________________________
g (Sequential)               (None, 784)              21162496
_________________________________________________________________
d (Sequential)               (None, 1)                14209
=================================================================
Total params: 21,176,705
Trainable params: 21,162,496
Non-trainable params: 14,209
_________________________________________________________________
```

When we run this model for 400 epochs, we get the following output:

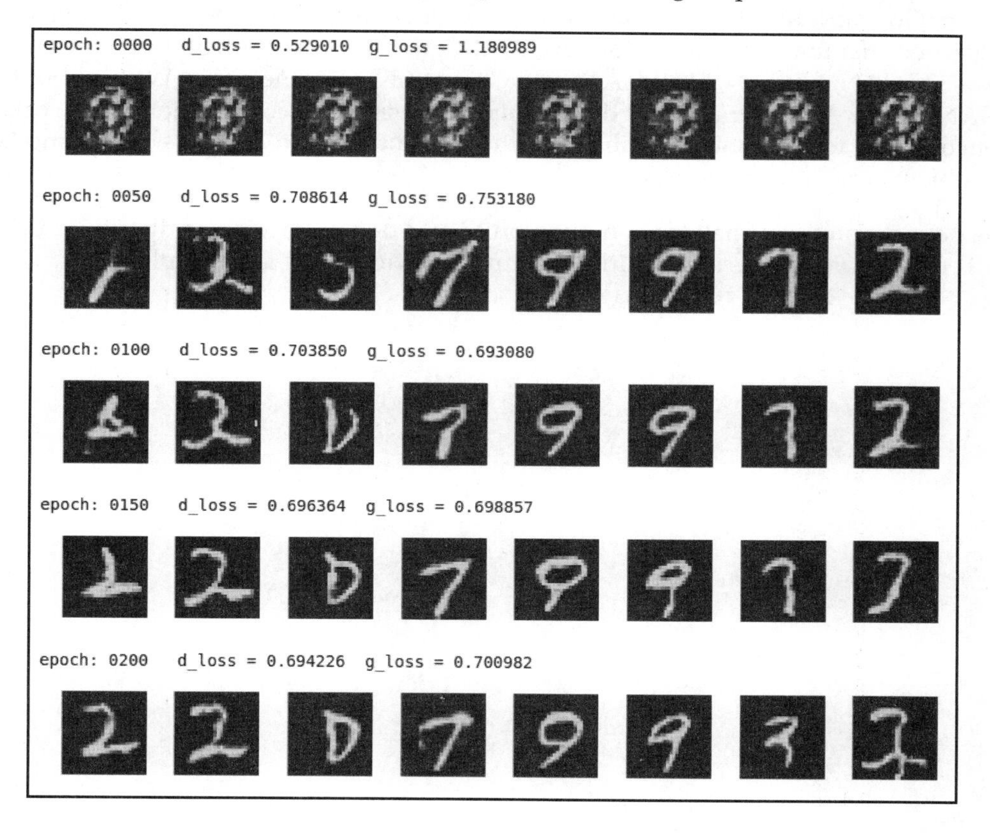

As you can see, the DCGAN is able to generate high-quality digits starting from epoch 100 itself. The DCGAN has been used for style transfer, generation of images and titles and for image algebra, namely taking parts of one image and adding that to parts of another image. The complete code for MNIST DCGAN is provided in the notebook `ch-14b_DCGAN`.

Summary

In this chapter, we learned about Generative Adversarial Networks. We built a simple GAN in TensorFlow and Keras and applied it to generate images from the MNIST dataset. We also learned that many different derivatives of GANs are being introduced continuously, such as DCGAN, SRGAN, StackGAN, and CycleGAN, to name a few. We also built a DCGAN where the generator and discriminator consisted of convolutional networks. You are encouraged to read and experiment with different derivatives to see which models fit the problems they are trying to solve.

In the next chapter, we shall learn how to build and deploy models in distributed clusters with TensorFlow clusters and multiple compute devices such as multiple GPUs.

15
Distributed Models with TensorFlow Clusters

Previously we learned how to run TensorFlow models at scale in production using Kubernetes, Docker and TensorFlow serving. TensorFlow serving is not the only way to run TensorFlow models at scale. TensorFlow provides another mechanism to not only run but also train the models on different nodes and different devices on multiple nodes or the same node. In `chapter 1`, *TensorFlow 101*, we also learned how to place variables and operations on different devices. In this chapter, we shall learn how to distribute the TensorFlow models to run on multiple devices across multiple nodes.

In this chapter, we shall cover the following topics:

- Strategies for distributed execution
- TensorFlow clusters
- Data parallel models
- Asynchronous and synchronous updates to distributed models

Strategies for distributed execution

For distributing the training of the single model across multiple devices or nodes, there are the following strategies:

- **Model Parallel:** Divide the model into multiple subgraphs and place the separate graphs on different nodes or devices. The subgraphs perform their computation and exchange the variables as required.

- **Data Parallel:** Divide the data into batches and run the same model on multiple nodes or devices, combining the parameters on a master node. Thus the worker nodes train the model on batches of data and send the parameter updates to the master node, also known as the parameter server.

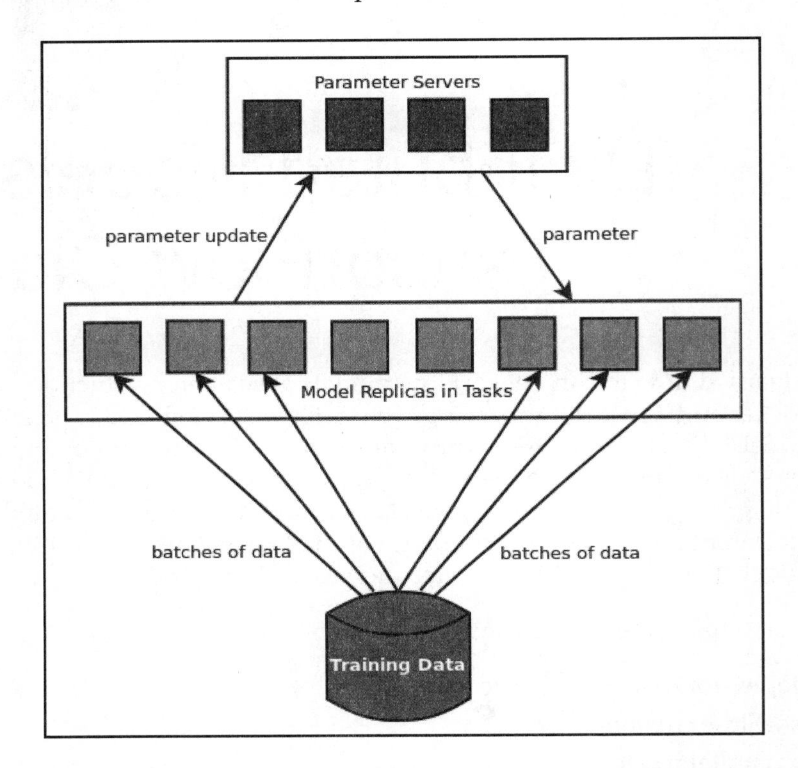

The preceding diagram shows the data parallel approach where the model replicas read the partitions of data in batches and send the parameter updates to the parameter servers, and parameter servers send the updated parameters back to the model replicas for the next batched computation of updates.

In TensorFlow, there are two ways to implement replicating the model on multiple nodes/devices under the data parallel strategy:

- **In-Graph Replication**: In this approach, there is a single client task that owns the model parameters and assigns the model calculations to multiple worker tasks.

- **Between-Graph Replication**: In this approach, each client task connects to its own worker in order to assign the model calculation, but all workers update the same shared model. In this model, TensorFlow automatically assigns one worker to be the chief worker so that the model parameters are initialized only once by the chief worker.

Within both these approaches, the parameters on the parameter server(s) can be updated in two different ways:

- **Synchronous Update**: In a synchronous update, the parameter server(s) wait to receive the updates from all the workers before updating the gradients. The parameter server aggregates the updates, for example by calculating the mean of all the aggregates and applying them to the parameters. After the update, the parameters are sent to all the workers simultaneously. The disadvantage of this method is that one slow worker may slow down the updates for everyone.
- **Asynchronous Update**: In an asynchronous update, the workers send the updates to parameter server(s) as they are ready, and then the parameter server applies the updates as it receives them and sends them back. The disadvantage of this method is that by the time the worker calculates the parameters and sends the updates back, the parameters could have been updated several times by other workers. This problem can be alleviated by several methods such as lowering the batch size or lowering the learning rate. It is a surprise that the asynchronous method even works, but in reality, they do work !!!

TensorFlow clusters

A TensorFlow (TF) *cluster* is one mechanism that implements the distributed strategies that we have just discussed. At the logical level, a TF cluster runs one or more *jobs*, and each *job* consists of one or more *tasks*. Thus a job is just a logical grouping of the tasks. At the process level, each task runs as a TF server. At the machine level, each physical machine or node can run more than one task by running more than one server, one server per task. The *client* creates the graph on different servers and starts the execution of the graph on one server by calling the remote session.

As an example, the following diagram depicts two clients connected to two jobs named `m1`:

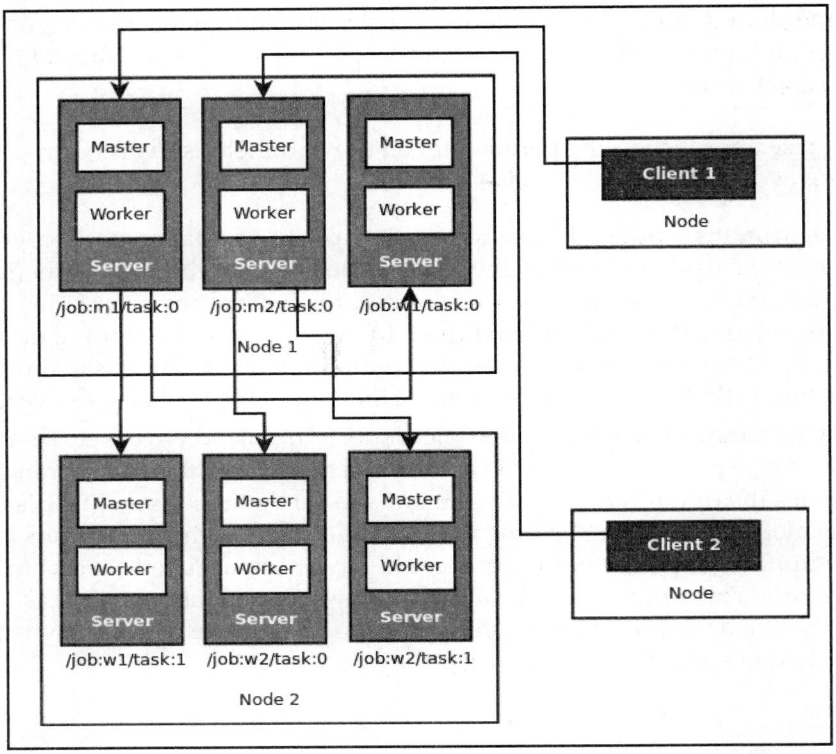

The two nodes are running three tasks each, and the job `w1` is spread across two nodes while the other jobs are contained within the nodes.

A TF server is implemented as two processes: master and worker. The master coordinates the computation with other tasks and the worker is the one actually running the computation. At a higher level, you do not have to worry about the internals of the TF server. For the purpose of our explanation and examples, we will only refer to TF tasks.

To create and train a model in data parallel fashion, use the following steps:

1. Define the cluster specifications
2. Create a server to host a task
3. Define the variable nodes to be assigned to parameter server tasks
4. Define the operation nodes to be replicated on all worker tasks

5. Create a remote session
6. Train the model in the remote session
7. Use the model for prediction

Defining cluster specification

In order to create a cluster, first, define a cluster specification. The cluster specification generally consists of two jobs: `ps` to create parameter server tasks and `worker` to create worker tasks. The `worker` and `ps` jobs contain the list of physical nodes where their respective tasks are running. As an example:

```
clusterSpec = tf.train.ClusterSpec({
  'ps': [
               'master0.neurasights.com:2222',   # /job:ps/task:0
               'master1.neurasights.com:2222'    # /job:ps/task:1
        ]
  'worker': [
               'worker0.neurasights.com:2222',   # /job:worker/task:0
               'worker1.neurasights.com:2222',   # /job:worker/task:1
               'worker0.neurasights.com:2223',   # /job:worker/task:2
               'worker1.neurasights.com:2223'    # /job:worker/task:3
            ]
          })
```

This specification creates two jobs, with two tasks in job `ps` spread across two physical nodes and four tasks in job `worker` spread across two physical nodes.

In our example code, we create all the tasks on a localhost, on different ports:

```
ps = [
        'localhost:9001',  # /job:ps/task:0
     ]
workers = [
        'localhost:9002',  # /job:worker/task:0
        'localhost:9003',  # /job:worker/task:1
        'localhost:9004',  # /job:worker/task:2
        ]
clusterSpec = tf.train.ClusterSpec({'ps': ps, 'worker': workers})
```

As you can see in the comments in the code, the tasks are identified with `/job:<job name>/task:<task index>`.

Create the server instances

Since the cluster contains one server instance per task, on every physical node, start the servers by passing them the cluster specification, their own job name and task index. The servers use the cluster specification to figure out what other nodes are involved in the computation.

```
server = tf.train.Server(clusterSpec, job_name="ps", task_index=0)
server = tf.train.Server(clusterSpec, job_name="worker", task_index=0)
server = tf.train.Server(clusterSpec, job_name="worker", task_index=1)
server = tf.train.Server(clusterSpec, job_name="worker", task_index=2)
```

In our example code, we have a single Python file that will run on all the physical machines, containing the following:

```
server = tf.train.Server(clusterSpec,
                         job_name=FLAGS.job_name,
                         task_index=FLAGS.task_index,
                         config=config
                         )
```

In this code, the `job_name` and the `task_index` are taken from the parameters passed at the command line. The package, `tf.flags` is a fancy parser that gives you access to the command-line arguments. The Python file is executed as follows, on every physical node (or in a separate terminal on the same node if you are using a localhost only):

```
# the model should be run in each physical node
# using the appropriate arguments
$ python3 model.py --job_name='ps' --task_index=0
$ python3 model.py --job_name='worker' --task_index=0
$ python3 model.py --job_name='worker' --task_index=1
$ python3 model.py --job_name='worker' --task_index=2
```

For even greater flexibility to run the code on any cluster, you can also pass the list of machines running parameter servers and workers through the command line: `-ps='localhost:9001' --worker='localhost:9002,localhost:9003,localhost:9004'`. You will need to parse them and set them appropriately in the cluster specifications dictionary.

To ensure that our parameter server only uses CPU and our worker tasks use GPU, we use the configuration object:

```
config = tf.ConfigProto()
config.allow_soft_placement = True

if FLAGS.job_name=='ps':
    #print(config.device_count['GPU'])
    config.device_count['GPU']=0
    server = tf.train.Server(clusterSpec,
                             job_name=FLAGS.job_name,
                             task_index=FLAGS.task_index,
                             config=config
                             )
    server.join()
    sys.exit('0')
elif FLAGS.job_name=='worker':
    config.gpu_options.per_process_gpu_memory_fraction = 0.2
    server = tf.train.Server(clusterSpec,
                             job_name=FLAGS.job_name,
                             task_index=FLAGS.task_index,
                             config=config
```

The parameter server is made to wait with `server.join()` while the worker tasks execute the training of the model and exit.

This is what our GPU looks like when all the four servers are running:

```
+-----------------------------------------------------------------------------+
| NVIDIA-SMI 384.90                 Driver Version: 384.90                    |
|-------------------------------+----------------------+----------------------+
| GPU  Name        Persistence-M| Bus-Id        Disp.A | Volatile Uncorr. ECC |
| Fan  Temp  Perf  Pwr:Usage/Cap|         Memory-Usage | GPU-Util  Compute M. |
|===============================+======================+======================|
|   0  Quadro P5000        Off  | 00000000:01:00.0 Off |                  N/A |
| N/A   58C    P0    34W /  N/A |  10372MiB / 16273MiB |     32%      Default |
+-------------------------------+----------------------+----------------------+

+-----------------------------------------------------------------------------+
| Processes:                                                       GPU Memory |
|  GPU       PID   Type   Process name                             Usage      |
|=============================================================================|
|    0      1505      C   python3                                       93MiB |
|    0     29841      C   python3                                     3423MiB |
|    0     29853      C   python3                                     3423MiB |
|    0     29862      C   python3                                     3423MiB |
+-----------------------------------------------------------------------------+
```

Define the parameter and operations across servers and devices

You can use the `tf.device()` function we used earlier in Chapter 1, to place the parameters on the `ps` tasks and the compute nodes of the graphs on the `worker` tasks.

Note that you can also place the graph nodes on specific devices by adding the device string to the task string as follows: `/job:<job name>/task:<task index>/device:<device type>:<device index>`.

For our demonstration example, we use the TensorFlow function `tf.train.replica_device_setter()` to place the variables and operations.

1. First, we define the worker device to be the current worker:

```
worker_device='/job:worker/task:{}'.format(FLAGS.task_index)
```

2. Next, define a device function using the `replica_device_setter`, passing the cluster specifications and current worker device. The `replica_device_setter` function figures out the parameter servers from the cluster specification, and if there are more than one parameter servers, then it distributes the parameters among them in a round robin fashion by default. The parameter placement strategy can be changed to a user-defined function or prebuilt strategies in the `tf.contrib` package.

```
device_func = tf.train.replica_device_setter(
    worker_device=worker_device,cluster=clusterSpec)
```

3. Finally, we create the graph inside the `tf.device(device_func)` block and train it. The creation and training of the graph is different for synchronous updates and asynchronous updates, hence we cover these in two separate subsections.

Define and train the graph for asynchronous updates

As discussed previously, and shown in the diagram here, in asynchronous updates all the worker tasks send the parameter updates when they are ready, and the parameter server updates the parameters and sends back the parameters. There is no synchronization or waiting or aggregation of parameter updates:

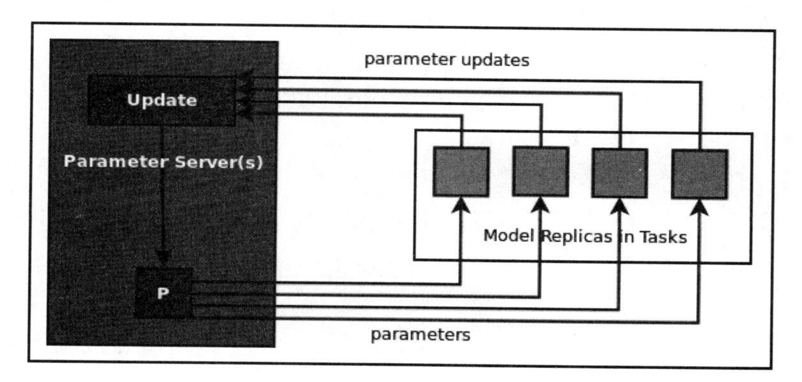

The full code for this example is in `ch-15_mnist_dist_async.py`. You are encouraged to modify and explore the code with your own datasets.

For asynchronous updates, the graph is created and trained with the following steps:

1. The definition of the graph is done within the `with` block:

```
with tf.device(device_func):
```

2. Create a global step variable using the inbuilt TensorFlow function:

```
global_step = tf.train.get_or_create_global_step()
```

3. This variable can also be defined as:

```
tf.Variable(0,name='global_step',trainable=False)
```

4. Define the datasets, parameters, and hyper-parameters as usual:

```
x_test = mnist.test.images
y_test = mnist.test.labels
n_outputs = 10  # 0-9 digits
n_inputs = 784  # total pixels
learning_rate = 0.01
n_epochs = 50
batch_size = 100
n_batches = int(mnist.train.num_examples/batch_size)
n_epochs_print=10
```

5. Define the placeholders, weights, biases, logits, cross-entropy, loss op, train op, accuracy as usual:

```
# input images
x_p = tf.placeholder(dtype=tf.float32,
                     name='x_p',
                     shape=[None, n_inputs])
# target output
y_p = tf.placeholder(dtype=tf.float32,
                     name='y_p',
                     shape=[None, n_outputs])
w = tf.Variable(tf.random_normal([n_inputs, n_outputs],
                                 name='w'
                                 )
                )
b = tf.Variable(tf.random_normal([n_outputs],
                                 name='b'
                                 )
                )
logits = tf.matmul(x_p,w) + b

entropy_op = tf.nn.softmax_cross_entropy_with_logits(labels=y_p,
                                                      logits=logits
                                                      )
loss_op = tf.reduce_mean(entropy_op)

optimizer = tf.train.GradientDescentOptimizer(learning_rate)
train_op = optimizer.minimize(loss_op,global_step=global_step)

correct_pred = tf.equal(tf.argmax(logits, 1), tf.argmax(y_p, 1))
accuracy_op = tf.reduce_mean(tf.cast(correct_pred, tf.float32))
```

These definitions will change when we learn how to build the synchronous update.

6. TensorFlow provides a supervisor class that helps in creating sessions for training and is very useful in a distributed training setting. Create a supervisor object as follows:

```
init_op = tf.global_variables_initializer
sv = tf.train.Supervisor(is_chief=is_chief,
                         init_op = init_op(),
                         global_step=global_step)
```

7. Use the supervisor object to create a session and run the training under this session block as usual:

```
with sv.prepare_or_wait_for_session(server.target) as mts:
    lstep = 0

    for epoch in range(n_epochs):
        for batch in range(n_batches):
            x_batch, y_batch = mnist.train.next_batch(batch_size)
            feed_dict={x_p:x_batch,y_p:y_batch}
            _,loss,gstep=mts.run([train_op,loss_op,global_step],
                                 feed_dict=feed_dict)
            lstep +=1
        if (epoch+1)%n_epochs_print==0:
            print('worker={},epoch={},global_step={}, \
                  local_step={},loss={}'.
                  format(FLAGS.task_index,epoch,gstep,lstep,loss))
    feed_dict={x_p:x_test,y_p:y_test}
    accuracy = mts.run(accuracy_op, feed_dict=feed_dict)
    print('worker={}, final accuracy = {}'
        .format(FLAGS.task_index,accuracy))
```

On starting the parameter server, we get the following output:

```
$ python3 ch-15_mnist_dist_async.py --job_name='ps' --task_index=0
I tensorflow/core/common_runtime/gpu/gpu_device.cc:1030] Found device 0
with properties:
    name: Quadro P5000 major: 6 minor: 1 memoryClockRate(GHz): 1.506
pciBusID: 0000:01:00.0
totalMemory: 15.89GiB freeMemory: 15.79GiB
I tensorflow/core/common_runtime/gpu/gpu_device.cc:1120] Creating
TensorFlow device (/device:GPU:0) -> (device: 0, name: Quadro P5000, pci
bus id: 0000:01:00.0, compute capability: 6.1)
E1213 16:50:14.023235178   27224 ev_epoll1_linux.c:1051]     grpc epoll fd:
23
```

```
I tensorflow/core/distributed_runtime/rpc/grpc_channel.cc:215] Initialize
GrpcChannelCache for job ps -> {0 -> localhost:9001}
I tensorflow/core/distributed_runtime/rpc/grpc_channel.cc:215] Initialize
GrpcChannelCache for job worker -> {0 -> localhost:9002, 1 ->
localhost:9003, 2 -> localhost:9004}
I tensorflow/core/distributed_runtime/rpc/grpc_server_lib.cc:324] Started
server with target: grpc://localhost:9001
```

On starting the worker tasks we get the following three outputs:

The output from worker 1:

```
$ python3 ch-15_mnist_dist_async.py --job_name='worker' --task_index=0
I tensorflow/core/common_runtime/gpu/gpu_device.cc:1030] Found device 0
with properties:
    name: Quadro P5000 major: 6 minor: 1 memoryClockRate(GHz): 1.506
pciBusID: 0000:01:00.0
totalMemory: 15.89GiB freeMemory: 9.16GiB
I tensorflow/core/common_runtime/gpu/gpu_device.cc:1120] Creating
TensorFlow device (/device:GPU:0) -> (device: 0, name: Quadro P5000, pci
bus id: 0000:01:00.0, compute capability: 6.1)
E1213 16:50:37.516609689   27507 ev_epoll1_linux.c:1051]     grpc epoll fd:
23
I tensorflow/core/distributed_runtime/rpc/grpc_channel.cc:215] Initialize
GrpcChannelCache for job ps -> {0 -> localhost:9001}
I tensorflow/core/distributed_runtime/rpc/grpc_channel.cc:215] Initialize
GrpcChannelCache for job worker -> {0 -> localhost:9002, 1 ->
localhost:9003, 2 -> localhost:9004}
I tensorflow/core/distributed_runtime/rpc/grpc_server_lib.cc:324] Started
server with target: grpc://localhost:9002
I tensorflow/core/distributed_runtime/master_session.cc:1004] Start master
session 1421824c3df413b5 with config: gpu_options {
per_process_gpu_memory_fraction: 0.2 } allow_soft_placement: true
worker=0,epoch=9,global_step=10896, local_step=5500, loss =
1.2575616836547852
worker=0,epoch=19,global_step=22453, local_step=11000, loss =
0.7158586382865906
worker=0,epoch=29,global_step=39019, local_step=16500, loss =
0.43712112307548523
worker=0,epoch=39,global_step=55513, local_step=22000, loss =
0.3935799300670624
worker=0,epoch=49,global_step=72002, local_step=27500, loss =
0.3877961337566376
worker=0, final accuracy = 0.8865000009536743
```

The output from worker 2:

```
$ python3 ch-15_mnist_dist_async.py --job_name='worker' --task_index=1
I tensorflow/core/common_runtime/gpu/gpu_device.cc:1030] Found device 0
with properties:
    name: Quadro P5000 major: 6 minor: 1 memoryClockRate(GHz): 1.506
pciBusID: 0000:01:00.0
totalMemory: 15.89GiB freeMemory: 12.43GiB
I tensorflow/core/common_runtime/gpu/gpu_device.cc:1120] Creating
TensorFlow device (/device:GPU:0) -> (device: 0, name: Quadro P5000, pci
bus id: 0000:01:00.0, compute capability: 6.1)
E1213 16:50:36.684334877   27461 ev_epoll1_linux.c:1051]     grpc epoll fd:
23
I tensorflow/core/distributed_runtime/rpc/grpc_channel.cc:215] Initialize
GrpcChannelCache for job ps -> {0 -> localhost:9001}
I tensorflow/core/distributed_runtime/rpc/grpc_channel.cc:215] Initialize
GrpcChannelCache for job worker -> {0 -> localhost:9002, 1 ->
localhost:9003, 2 -> localhost:9004}
I tensorflow/core/distributed_runtime/rpc/grpc_server_lib.cc:324] Started
server with target: grpc://localhost:9003
I tensorflow/core/distributed_runtime/master_session.cc:1004] Start master
session 2bd8a136213a1fce with config: gpu_options {
per_process_gpu_memory_fraction: 0.2 } allow_soft_placement: true
worker=1,epoch=9,global_step=11085, local_step=5500, loss =
0.6955764889717102
worker=1,epoch=19,global_step=22728, local_step=11000, loss =
0.5891970992088318
worker=1,epoch=29,global_step=39074, local_step=16500, loss =
0.4183048903942108
worker=1,epoch=39,global_step=55599, local_step=22000, loss =
0.32243454456329346
worker=1,epoch=49,global_step=72105, local_step=27500, loss =
0.5384714007377625
worker=1, final accuracy = 0.8866000175476074
```

The output from worker 3:

```
$ python3 ch-15_mnist_dist_async.py --job_name='worker' --task_index=2
I tensorflow/core/common_runtime/gpu/gpu_device.cc:1030] Found device 0
with properties:
    name: Quadro P5000 major: 6 minor: 1 memoryClockRate(GHz): 1.506
pciBusID: 0000:01:00.0
totalMemory: 15.89GiB freeMemory: 15.70GiB
I tensorflow/core/common_runtime/gpu/gpu_device.cc:1120] Creating
TensorFlow device (/device:GPU:0) -> (device: 0, name: Quadro P5000, pci
bus id: 0000:01:00.0, compute capability: 6.1)
E1213 16:50:35.568349791   27449 ev_epoll1_linux.c:1051]     grpc epoll fd:
23
```

```
I tensorflow/core/distributed_runtime/rpc/grpc_channel.cc:215] Initialize
GrpcChannelCache for job ps -> {0 -> localhost:9001}
I tensorflow/core/distributed_runtime/rpc/grpc_channel.cc:215] Initialize
GrpcChannelCache for job worker -> {0 -> localhost:9002, 1 ->
localhost:9003, 2 -> localhost:9004}
I tensorflow/core/distributed_runtime/rpc/grpc_server_lib.cc:324] Started
server with target: grpc://The full code for this example is in
ch-15_mnist_dist_sync.py. You are encouraged to modify and explore the code
with your own datasets.localhost:9004
I tensorflow/core/distributed_runtime/master_session.cc:1004] Start master
session cb0749c9f5fc163e with config: gpu_options {
per_process_gpu_memory_fraction: 0.2 } allow_soft_placement: true
I tensorflow/core/distributed_runtime/master_session.cc:1004] Start master
session 55bf9a2b9718a571 with config: gpu_options {
per_process_gpu_memory_fraction: 0.2 } allow_soft_placement: true
worker=2,epoch=9,global_step=37367, local_step=5500, loss =
0.8077645301818848
worker=2,epoch=19,global_step=53859, local_step=11000, loss =
0.26333487033843994
worker=2,epoch=29,global_step=70299, local_step=16500, loss =
0.6506651043891907
worker=2,epoch=39,global_step=76999, local_step=22000, loss =
0.20321622490882874
worker=2,epoch=49,global_step=82499, local_step=27500, loss =
0.4170967936515808
worker=2, final accuracy = 0.8894000053405762
```

We printed the global step and local step. The global step indicates the count of steps across all the worker tasks while the local step is a count within that worker task, that is why local tasks count up to 27,500 and are the same for every epoch for every worker, but since workers are doing the global steps at their own pace, the number of global steps has no symmetry or pattern across epochs or across workers. Also, we see that the final accuracy is different for each worker, since each worker executed the final accuracy at a different time, with different parameters available at that time.

Define and train the graph for synchronous updates

As discussed before, and depicted in the diagram here, in synchronous updates, the tasks send their updates to the parameter server(s), and ps tasks wait for all the updates to be received, aggregate them, and then update the parameters. The worker tasks wait for the updates before proceeding to the next iteration of computing parameter updates:

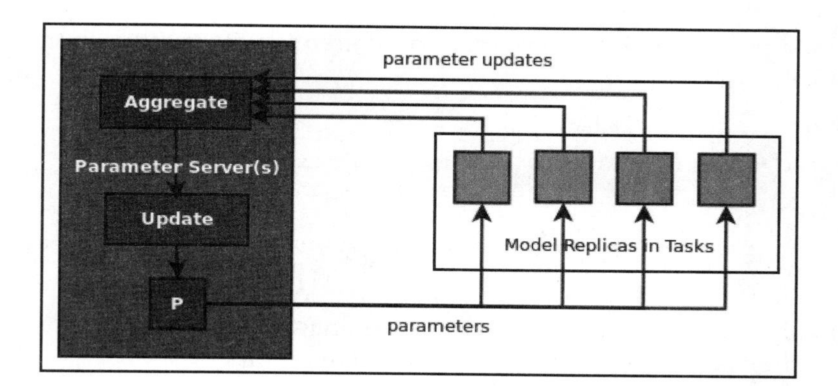

The full code for this example is in `ch-15_mnist_dist_sync.py`. You are encouraged to modify and explore the code with your own datasets.

For synchronous updates, the following modifications need to be made to the code:

1. The optimizer needs to be wrapped in SyncReplicaOptimizer. Thus, after defining the optimizer, add the following code:

```
# SYNC: next line added for making it sync update
optimizer = tf.train.SyncReplicasOptimizer(optimizer,
    replicas_to_aggregate=len(workers),
    total_num_replicas=len(workers),
    )
```

2. This should be followed by adding the training operation as before:

```
train_op = optimizer.minimize(loss_op,global_step=global_step)
```

3. Next, add the initialization function definitions, specific to the synchronous update method:

```
if is_chief:
    local_init_op = optimizer.chief_init_op()
else:
    local_init_op = optimizer.local_step_init_op()
chief_queue_runner = optimizer.get_chief_queue_runner()
init_token_op = optimizer.get_init_tokens_op()
```

4. The supervisor object is also created differently with two additional initialization functions:

```
# SYNC: sv is initialized differently for sync update
sv = tf.train.Supervisor(is_chief=is_chief,
    init_op = tf.global_variables_initializer(),
    local_init_op = local_init_op,
    ready_for_local_init_op = optimizer.ready_for_local_init_op,
    global_step=global_step)
```

5. Finally, within the session block for training, we initialize the sync variables and start the queue runners if it is the chief worker task:

```
# SYNC: if block added to make it sync update
if is_chief:
    mts.run(init_token_op)
    sv.start_queue_runners(mts, [chief_queue_runner])
```

The rest of the code remains the same as an asynchronous update.

The TensorFlow libraries and functions for supporting distributed training are under continuous development. Hence, be on the lookout for new functionality added or function signatures changed. At the timing of writing of this book, we used TensorFlow 1.4.

Summary

In this chapter, we learned how to distribute the training of our models across multiple machines and devices, using TensorFlow clusters. We also learned model parallel and data parallel strategies for the distributed execution of TensorFlow code.

The parameter updates can be shared with synchronous or asynchronous updates to parameter servers. We learned how to implement code for synchronous and asynchronous parameter updates. With the skills learned in this chapter, you will be able to build and train very large models with very large datasets.

In the next chapter, we shall learn how to deploy TensorFlow models on mobile and embedded devices running iOS and Android platforms.

16
TensorFlow Models on Mobile and Embedded Platforms

TensorFlow models can also be used in applications running on mobile and embedded platforms. TensorFlow Lite and TensorFlow Mobile are two flavors of TensorFlow for resource-constrained mobile devices. TensorFlow Lite supports a subset of the functionality compared to TensorFlow Mobile. TensorFlow Lite results in better performance due to smaller binary size with fewer dependencies.

To integrate TensorFlow into your application, first, train a model using the techniques we mention throughout the book and then save the model. The saved model can now be used to do the inference and prediction in the mobile application.

To learn how to use TensorFlow models on mobile devices, in this chapter we cover the following topics:

- TensorFlow on mobile platforms
- TF Mobile in Android apps
- TF Mobile demo on Android
- TF Mobile demo on iOS
- TensorFlow Lite
- TF Lite demo on Android
- TF Lite demo on iOS

TensorFlow on mobile platforms

TensorFlow can be integrated into mobile apps for many use cases that involve one or more of the following machine learning tasks:

- Speech recognition
- Image recognition
- Gesture recognition
- Optical character recognition
- Image or text classification
- Image, text, or speech synthesis
- Object identification

To run TensorFlow on mobile apps, we need two major ingredients:

- A trained and saved model that can be used for predictions
- A TensorFlow binary that can receive the inputs, apply the model, produce the predictions, and send the predictions as output

The high-level architecture looks like the following figure:

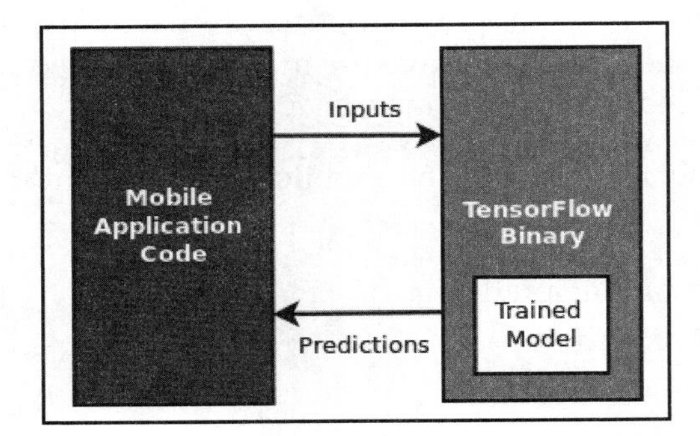

The mobile application code sends the inputs to the TensorFlow binary, which uses the trained model to compute predictions and send the predictions back.

TF Mobile in Android apps

The TensorFlow ecosystem enables it to be used in Android apps through the interface class `TensorFlowInferenceInterface`, and the TensorFlow Java API in the jar file `libandroid_tensorflow_inference_java.jar`. You can either use the jar file from the JCenter, download a precompiled jar from `ci.tensorflow.org`, or build it yourself.

The inference interface has been made available as a JCenter package and can be included in the Android project by adding the following code to the `build.gradle` file:

```
allprojects {
   repositories {
      jcenter()
   }
}
dependencies {
   compile 'org.tensorflow:tensorflow-android:+'
}
```

Instead of using the pre-built binaries from the JCenter, you can also build them yourself using Bazel or Cmake by following the instructions at this link: `https://github.com/tensorflow/tensorflow/blob/r1.4/tensorflow/contrib/android/README.md`.

Once the TF library is configured in your Android project, you can call the TF model with the following four steps:

1. Load the model:

```
TensorFlowInferenceInterface inferenceInterface =
     new TensorFlowInferenceInterface(assetManager, modelFilename);
```

2. Send the input data to the TensorFlow binary:

```
inferenceInterface.feed(inputName,
    floatValues, 1, inputSize, inputSize, 3);
```

3. Run the prediction or inference:

```
inferenceInterface.run(outputNames, logStats);
```

4. Receive the output from the TensorFlow binary:

```
inferenceInterface.fetch(outputName, outputs);
```

TF Mobile demo on Android

In this section, we shall learn about recreating the Android demo app provided by the TensorFlow team in their official repo. The Android demo will install the following four apps on your Android device:

- `TF Classify`: This is an object identification app that identifies the images in the input from the device camera and classifies them in one of the pre-defined classes. It does not learn new types of pictures but tries to classify them into one of the categories that it has already learned. The app is built using the inception model pre-trained by Google.
- `TF Detect`: This is an object detection app that detects multiple objects in the input from the device camera. It continues to identify the objects as you move the camera around in continuous picture feed mode.
- `TF Stylize`: This is a style transfer app that transfers one of the selected predefined styles to the input from the device camera.
- `TF Speech`: This is a speech recognition app that identifies your speech and if it matches one of the predefined commands in the app, then it highlights that specific command on the device screen.

The sample demo only works for Android devices with an API level greater than 21 and the device must have a modern camera that supports `FOCUS_MODE_CONTINUOUS_PICTURE`. If your device camera does not have this feature supported, then you have to add the path submitted to TensorFlow by the author: `https://github.com/tensorflow/tensorflow/pull/15489/files`.

The easiest way to build and deploy the demo app on your device is using Android Studio. To build it this way, follow these steps:

1. Install Android Studio. We installed Android Studio on Ubuntu 16.04 from the instructions at the following link: `https://developer.android.com/studio/install.html`
2. Check out the TensorFlow repository, and apply the patch mentioned in the previous tip. Let's assume you checked out the code in the `tensorflow` folder in your home directory.

3. Using Android Studio, open the Android project in the path `~/tensorflow/tensorflow/examples/Android`. Your screen will look similar to this:

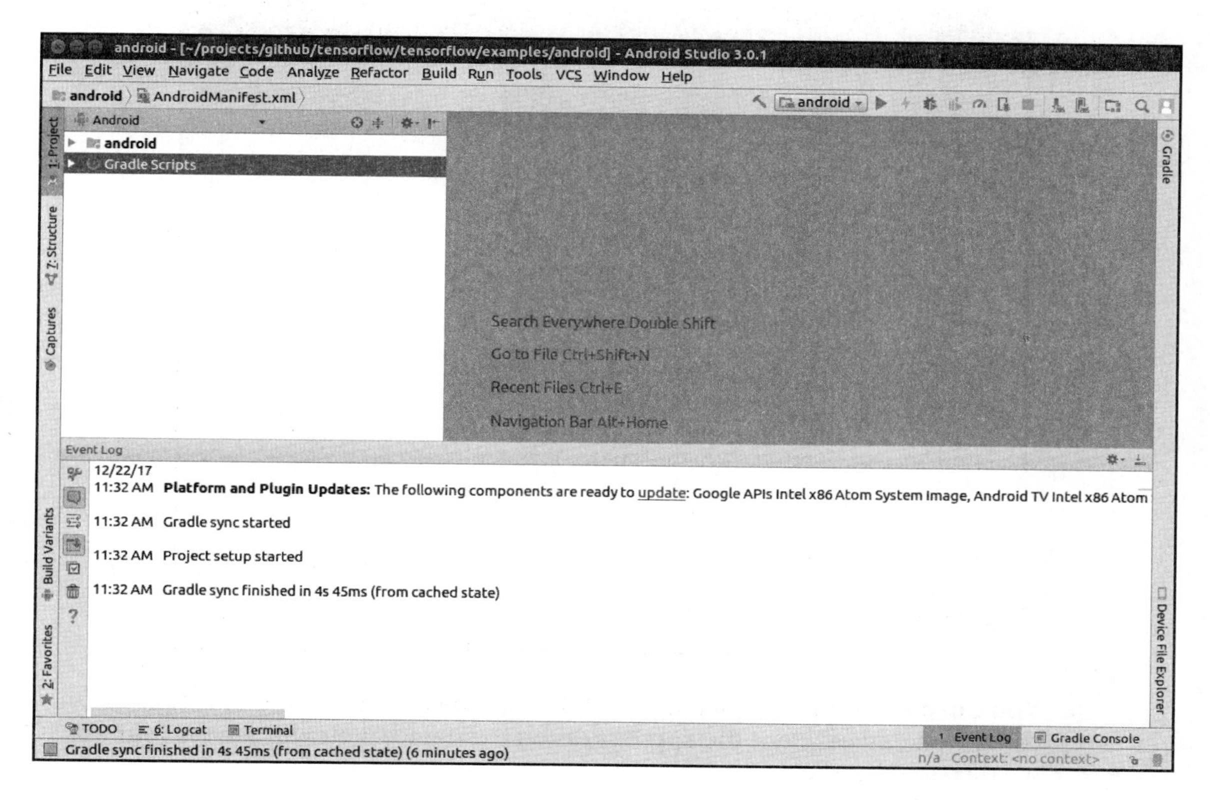

4. Expand the Gradle Scripts option from the left bar and then open the `build.gradle` file.
5. In the `build.gradle` file, locate the `def nativeBuildSystem` definition and set it to `'none'`. In the version of the code we checked out, this definition is at line 43:

```
def nativeBuildSystem = 'none'
```

6. Build the demo and run it on either a real or simulated device. We tested the app on these devices:

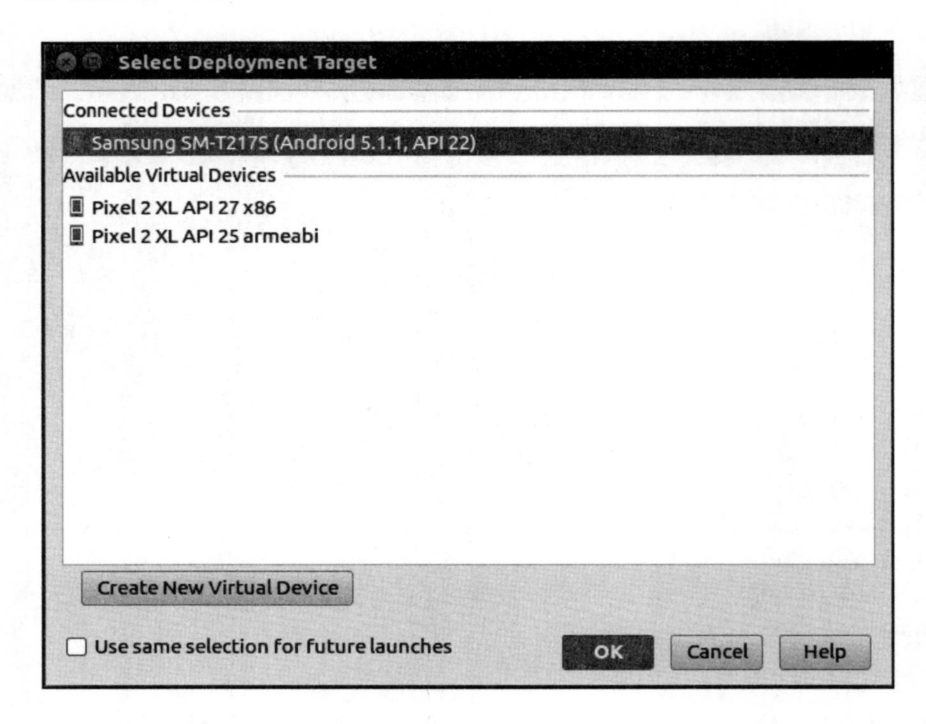

7. You can also build the apk and install the apk file on the virtual or actual connected device. Once the app installs on the device, you will see the four apps we discussed earlier:

TF Example Apps in Android Simulator

You can also build the whole demo app from the source using Bazel or Cmake by following the instructions at this link: `https://github.com/tensorflow/tensorflow/tree/r1.4/tensorflow/examples/android`

TF Mobile in iOS apps

TensorFlow enables support for iOS apps by following these steps:

1. Include TF Mobile in your app by adding a file named `Profile` in the root directory of your project. Add the following content to the `Profile`:

   ```
   target 'Name-Of-Your-Project'
          pod 'TensorFlow-experimental'
   ```

2. Run the `pod install` command to download and install the TensorFlow Experimental pod.
3. Run the `myproject.xcworkspace` command to open the workspace so you can add the prediction code to your application logic.

To create your own TensorFlow binaries for iOS projects, follow the instructions at this link: `https://github.com/tensorflow/tensorflow/tree/master/tensorflow/examples/ios`

Once the TF library is configured in your iOS project, you can call the TF model with the following four steps:

1. Load the model:

   ```
   PortableReadFileToProto(file_path, &tensorflow_graph);
   ```

2. Create a session:

   ```
   tensorflow::Status s = session->Create(tensorflow_graph);
   ```

3. Run the prediction or inference and get the outputs:

   ```
   std::string input_layer = "input";
   std::string output_layer = "output";
   std::vector<tensorflow::Tensor> outputs;
   tensorflow::Status run_status = session->Run(
       {{input_layer, image_tensor}},
       {output_layer}, {}, &outputs);
   ```

4. Fetch the output data:

   ```
   tensorflow::Tensor* output = &outputs[0];
   ```

TF Mobile demo on iOS

In order to build the demo on iOS, you need Xcode 7.3 or later. Follow these steps to build the iOS demo apps:

1. Check out the TensorFlow code in a `tensorflow` folder in your home directory.
2. Open a terminal window and execute the following commands from your home folder to download the Inception V1 model, extract the label and graph files, and move these files into the data folders inside the sample app code:

   ```
   $ mkdir -p ~/Downloads
   $ curl -o ~/Downloads/inception5h.zip \
   https://storage.googleapis.com/download.tensorflow.org/models/incep
   tion5h.zip \
     && unzip ~/Downloads/inception5h.zip -d ~/Downloads/inception5h
   $ cp ~/Downloads/inception5h/* \
     ~/tensorflow/tensorflow/examples/ios/benchmark/data/
   $ cp ~/Downloads/inception5h/* \
     ~/tensorflow/tensorflow/examples/ios/camera/data/
   $ cp ~/Downloads/inception5h/* \
     ~/tensorflow/tensorflow/examples/ios/simple/data/
   ```

3. Navigate to one of the sample folders and download the experimental pod:

   ```
   $ cd ~/tensorflow/tensorflow/examples/ios/camera
   $ pod install
   ```

4. Open the Xcode workspace:

   ```
   $ open tf_simple_example.xcworkspace
   ```

5. Run the sample app in the device simulator. The sample app will appear with a Run Model button. The camera app requires an Apple device to be connected, while the other two can run in a simulator too.

TensorFlow Lite

TF Lite is the new kid on the block and still in the developer view at the time of writing this book. TF Lite is a very small subset of TensorFlow Mobile and TensorFlow, so the binaries compiled with TF Lite are very small in size and deliver superior performance. Apart from reducing the size of binaries, TensorFlow employs various other techniques, such as:

- The kernels are optimized for various device and mobile architectures
- The values used in the computations are quantized
- The activation functions are pre-fused
- It leverages specialized machine learning software or hardware available on the device, such as theAndroid NN API

The workflow for using the models in TF Lite is as follows:

1. **Get the model:** You can train your own model or pick a pre-trained model available from different sources, and use the pre-trained as is or retrain it with your own data, or retrain after modifying some parts of the model. As long as you have a trained model in the file with an extension .pb or .pbtxt, you are good to proceed to the next step. We learned how to save the models in the previous chapters.
2. **Checkpoint the model**: The model file only contains the structure of the graph, so you need to save the checkpoint file. The checkpoint file contains the serialized variables of the model, such as weights and biases. We learned how to save a checkpoint in the previous chapters.
3. **Freeze the model**: The checkpoint and the model files are merged, also known as freezing the graph. TensorFlow provides the `freeze_graph` tool for this step, which can be executed as follows:

```
$ freeze_graph
     --input_graph=mymodel.pb
     --input_checkpoint=mycheckpoint.ckpt
     --input_binary=true
     --output_graph=frozen_model.pb
     --output_node_name=mymodel_nodes
```

4. **Convert the model**: The frozen model from step 3 needs to be converted to TF Lite format with the `toco` tool provided by TensorFlow:

```
$ toco
      --input_file=frozen_model.pb
      --input_format=TENSORFLOW_GRAPHDEF
      --output_format=TFLITE
      --input_type=FLOAT
      --input_arrays=input_nodes
      --output_arrays=mymodel_nodes
      --input_shapes=n,h,w,c
```

5. The `.tflite` model saved in step 4 can now be used inside an Android or iOS app that employs the TFLite binary for inference. The process of including the TFLite binary in your app is continuously evolving, so we recommend the reader follows the information at this link to include the TFLite binary in your Android or iOS app: `https://github.com/tensorflow/tensorflow/tree/master/tensorflow/contrib/lite/g3doc`

Generally, you would use the `graph_transforms:summarize_graph` tool to prune the model obtained in step 1. The pruned model will only have the paths that lead from input to output at the time of inference or prediction. Any other nodes and paths that are required only for training or for debugging purposes, such as saving checkpoints, are removed, thus making the size of the final model very small.

The official TensorFlow repository comes with a TF Lite demo that uses a pre-trained `mobilenet` to classify the input from the device camera in the 1001 categories. The demo app displays the probabilities of the top three categories.

TF Lite Demo on Android

To build a TF Lite demo on Android, follow these steps:

1. Install Android Studio. We installed Android Studio on Ubuntu 16.04 from the instructions at the following link: `https://developer.android.com/studio/install.html`
2. Check out the TensorFlow repository, and apply the patch mentioned in the previous tip. Let's assume you checked out the code in the `tensorflow` folder in your home directory.

3. Using Android Studio, open the Android project from the path `~/tensorflow/tensorflow/contrib/lite/java/demo`. If it complains about a missing SDK or Gradle components, please install those components and sync Gradle.
4. Build the project and run it on a virtual device with API > 21.

   We received the following warnings, but the build succeeded. You may want to resolve the warnings if the build fails:

```
Warning:The Jack toolchain is deprecated and will not
run. To enable support for Java 8 language features built
into the plugin, remove 'jackOptions { ... }' from your
build.gradle file, and add

android.compileOptions.sourceCompatibility 1.8
android.compileOptions.targetCompatibility 1.8

Future versions of the plugin will not support usage of
'jackOptions' in build.gradle.
To learn more, go to
https://d.android.com/r/tools/java-8-support-message.html

Warning:The specified Android SDK Build Tools version
(26.0.1) is ignored, as it is below the minimum supported
version (26.0.2) for Android Gradle Plugin 3.0.1.
Android SDK Build Tools 26.0.2 will be used.
To suppress this warning, remove "buildToolsVersion
'26.0.1'" from your build.gradle file, as each version of
the Android Gradle Plugin now has a default version of
the build tools.
```

You can also build the whole demo app from the source using Bazel with the instructions at the following link: `https://github.com/tensorflow/tensorflow/tree/master/tensorflow/contrib/lite`.

TF Lite demo on iOS

In order to build the demo on iOS, you need Xcode 7.3 or later. Follow these steps to build the iOS demo apps:

1. Check out the TensorFlow code in a `tensorflow` folder in your home directory.
2. Build the TF Lite binary for iOS from the instructions at this link: `https://github.com/tensorflow/tensorflow/tree/master/tensorflow/contrib/lite`.
3. Navigate to the sample folder and download the pod:

```
$ cd ~/tensorflow/tensorflow/contrib/lite/examples/ios/camera
$ pod install
```

4. Open the Xcode workspace:

```
$ open tflite_camera_example.xcworkspace
```

5. Run the sample app in the device simulator.

Summary

In this chapter, we learned about using TensorFlow models on mobile applications and devices. TensorFlow provides two ways to run on mobile devices: TF Mobile and TF Lite. We learned how to build TF Mobile and TF Lite apps for iOs and Android. We used TensorFlow demo apps in this chapter as an example. The reader is encouraged to explore the source code of these demo apps and use TF Mobile and TF Lite to empower their own mobile applications with machine learning models built using TensorFlow.

In next chapter, we shall learn about how to use TensorFlow in R statistical software with the R packages released by RStudio.

17
TensorFlow and Keras in R

R is an open source platform that includes an environment and a language for statistical computing. It also has a desktop and web-based IDE known as R Studio. More information about R is available at the following link: `https://www.r-project.org/`. R provides support for TensorFlow and Keras by providing the following R packages:

- `tensorflow` package offers support for the TF core API
- `tfestimators` package provides support for the TF estimators API
- `keras` package provides support for the Keras API
- `tfruns` package for the TensorBoard-style visualization of models and training sessions

In this chapter, we shall learn how to use TensorFlow in R and will cover the following topics:

- Installing TensorFlow and Keras packages in R
- TF core API in R
- TF estimator API in R
- Keras API in R
- TensorBoard in R
- `tfruns` package in R

Installing TensorFlow and Keras packages in R

To install the three R packages that support TensorFlow and Keras in R, execute the following commands in R.

1. First, install `devtools`:

```
install.packages("devtools")
```

2. Install the `tensorflow` and `tfestimators` package:

```
devtools::install_github("rstudio/tensorflow")
devtools::install_github("rstudio/tfestimators")
```

3. Load the `tensorflow` library and install the required functions:

```
library(tensorflow)
install_tensorflow()
```

4. By default, the install function creates a virtual environment and installs the `tensorflow` package in the virtual environment.

There are four installation methods available, which can be specified using the method parameter:

`auto`	**Automatically choose a default for the current platform**
`virtualenv`	Install into the virtual environment located at ~/.virtualenvs/r-tensorflow
`conda`	Install into the Anaconda Python environment named r-tensorflow
`system`	Install into the system Python environment

5. By default, the install function installs the CPU-only version of TensorFlow. To install the GPU version, use a version parameter:

`gpu`	**Installs tensorflow-gpu**
`nightly`	Installs the nightly CPU-only build
`nightly-gpu`	Installs the nightly GPU build
`n.n.n`	Installs a specific version, such as 1.3.0
`n.n.n-gpu`	Installs the GPU build of a specific version, such as 1.3.0

If you want the TensorFlow library to use a specific version of Python, use the following functions or set `TENSORFLOW_PYTHON` environment variables:

- `use_python('/usr/bin/python2')`
- `use_virtualenv('~/venv')`
- `use_condaenv('conda-env')`
- `Sys.setenv(TENSORFLOW_PYTHON='/usr/bin/python2')`

We installed TensorFLow in R on Ubuntu 16.04 using the following command:

```
install_tensorflow(version="gpu")
```

Note that the installation does not support Python 3 at the time of writing this book.

6. Install the keras package:

```
devtools::install_github("rstudio/keras")
```

7. Install Keras in the virtual environment:

```
library(keras)
install_keras()
```

8. To install the GPU version, use:

```
install_keras(tensorflow = "gpu")
```

9. Install the `tfruns` package:

```
devtools::install_github("rstudio/tfruns")
```

TF core API in R

We learned about the TensorFlow core API in Chapter 1. In R, this API is implemented with the `tensorflow` R package.

As an example, we provide a walkthrough of the MLP Model for classifying handwritten digits from the MNIST dataset at the following link: `https://tensorflow.rstudio.com/tensorflow/articles/examples/mnist_softmax.html`.

You can follow along with the code in the Jupyter R notebook `ch-17a_TFCore_in_R`.

1. First, load the library:

```
library(tensorflow)
```

2. Define the hyper-parameters:

```
batch_size <- 128
num_classes <- 10
steps <- 1000
```

3. Prepare the data:

```
datasets <- tf$contrib$learn$datasets
mnist <- datasets$mnist$read_data_sets("MNIST-data", one_hot =
TRUE)
```

The data is loaded from the TensorFlow dataset library and is already normalized to fall into the [0,1] range.

4. Define the model:

```
# Create the model
x <- tf$placeholder(tf$float32, shape(NULL, 784L))
W <- tf$Variable(tf$zeros(shape(784L, num_classes)))
b <- tf$Variable(tf$zeros(shape(num_classes)))
y <- tf$nn$softmax(tf$matmul(x, W) + b)

# Define loss and optimizer
y_ <- tf$placeholder(tf$float32, shape(NULL, num_classes))
cross_entropy <- tf$reduce_mean(-tf$reduce_sum(y_ * log(y),
reduction_indices=1L))
```

```
train_step <-
tf$train$GradientDescentOptimizer(0.5)$minimize(cross_entropy)
```

5. Train the model:

```
# Create session and initialize variables
sess <- tf$Session()
sess$run(tf$global_variables_initializer())

# Train
for (i in 1:steps) {
  batches <- mnist$train$next_batch(batch_size)
  batch_xs <- batches[[1]]
  batch_ys <- batches[[2]]
  sess$run(train_step,
           feed_dict = dict(x = batch_xs, y_ = batch_ys))
}
```

6. Evaluate the model:

```
correct_prediction <- tf$equal(tf$argmax(y, 1L), tf$argmax(y_, 1L))
accuracy <- tf$reduce_mean(tf$cast(correct_prediction, tf$float32))
score <-sess$run(accuracy,
         feed_dict = dict(x = mnist$test$images,
                          y_ = mnist$test$labels))

cat('Test accuracy:', score, '\n')
```

The output is as follows:

```
Test accuracy: 0.9185
```

Pretty cool!

Find more examples of TF Core in R at the following link: `https://tensorflow.rstudio.com/tensorflow/articles/examples/`
Further documentation on the `tensorflow` R package can be found at the following link: `https://tensorflow.rstudio.com/tensorflow/reference/`.

TF estimator API in R

We learned about the TensorFlow estimator API in Chapter 2. In R, this API is implemented with the `tfestimator` R package.

As an example, we provide a walkthrough of the MLP Model for classifying handwritten digits from the MNIST dataset at the following link: `https://tensorflow.rstudio.com/tfestimators/articles/examples/mnist.html`.

You can follow along with the code in the Jupyter R notebook `ch-17b_TFEstimator_in_R`.

1. First, load the libraries:

```
library(tensorflow)
library(tfestimators)
```

2. Define the hyper-parameters:

```
batch_size <- 128
n_classes <- 10
n_steps <- 100
```

3. Prepare the data:

```
# initialize data directory
data_dir <- "~/datasets/mnist"
dir.create(data_dir, recursive = TRUE, showWarnings = FALSE)

# download the MNIST data sets, and read them into R
sources <- list(
  train = list(
    x =
"https://storage.googleapis.com/cvdf-datasets/mnist/train-images-id
x3-ubyte.gz",
    y =
"https://storage.googleapis.com/cvdf-datasets/mnist/train-labels-id
x1-ubyte.gz"
  ),
  test = list(
    x =
"https://storage.googleapis.com/cvdf-datasets/mnist/t10k-images-idx
3-ubyte.gz",
    y =
```

```
"https://storage.googleapis.com/cvdf-datasets/mnist/t10k-labels-idx
1-ubyte.gz"
  )
)

# read an MNIST file (encoded in IDX format)
read_idx <- function(file) {
  # create binary connection to file
  conn <- gzfile(file, open = "rb")
  on.exit(close(conn), add = TRUE)
  # read the magic number as sequence of 4 bytes
  magic <- readBin(conn, what="raw", n=4, endian="big")
  ndims <- as.integer(magic[[4]])
  # read the dimensions (32-bit integers)
  dims <- readBin(conn,what="integer",n=ndims,endian="big")
  # read the rest in as a raw vector
  data <- readBin(conn,what="raw",n=prod(dims),endian="big")
  # convert to an integer vecto
  converted <- as.integer(data)
  # return plain vector for 1-dim array
  if (length(dims) == 1)
    return(converted)
  # wrap 3D data into matrix
  matrix(converted,nrow=dims[1],ncol=prod(dims[-1]),byrow=TRUE)
}

mnist <- rapply(sources,classes="character",how
="list",function(url) {
  # download + extract the file at the URL
  target <- file.path(data_dir, basename(url))
  if (!file.exists(target))
    download.file(url, target)
  # read the IDX file
  read_idx(target)
})

# convert training data intensities to 0-1 range
mnist$train$x <- mnist$train$x / 255
mnist$test$x <- mnist$test$x / 255
```

The data is read from the downloaded gzip files and then normalized to fall into the [0,1] range.

4. Define the model:

```
# construct a linear classifier
classifier <- linear_classifier(
  feature_columns = feature_columns(
    column_numeric("x", shape = shape(784L))
  ),
  n_classes = n_classes # 10 digits
)

# construct an input function generator
mnist_input_fn <- function(data, ...) {
  input_fn(
    data,
    response = "y",
    features = "x",
    batch_size = batch_size,
    ...
  )
}
```

5. Train the model:

```
train(classifier,input_fn=mnist_input_fn(mnist$train),steps=n_steps
)
```

6. Evaluate the model:

```
evaluate(classifier,input_fn=mnist_input_fn(mnist$test),steps=200)
```

The output is as follows:

```
Evaluation completed after 79 steps but 200 steps was specified
```

average_loss	loss	global_step	accuracy
0.35656	45.13418	100	0.9057

Pretty cool!!

Find more examples of TF Estimators in R at the following link: https://tensorflow.rstudio.com/tfestimators/articles/examples/

Further documentation on the tensorflow R package can be found at the following link: https://tensorflow.rstudio.com/tfestimators/reference/

Keras API in R

We learned about the Keras API in Chapter 3. In R, this API is implemented with the keras R package. The keras R package implements most of the functionality of the Keras Python interface, including both the sequential API and the functional API.

As an example, we provide a walkthrough of the MLP Model for classifying handwritten digits from the MNIST dataset at the following link: https://keras.rstudio.com/articles/examples/mnist_mlp.html.

You can follow along with the code in the Jupyter R notebook ch-17c_Keras_in_R.

1. First, load the library:

```
library(keras)
```

2. Define the hyper-parameters:

```
batch_size <- 128
num_classes <- 10
epochs <- 30
```

3. Prepare the data:

```
# The data, shuffled and split between train and test sets
c(c(x_train, y_train), c(x_test, y_test)) %<-% dataset_mnist()

x_train <- array_reshape(x_train, c(nrow(x_train), 784))
x_test <- array_reshape(x_test, c(nrow(x_test), 784))

# Transform RGB values into [0,1] range
x_train <- x_train / 255
x_test <- x_test / 255

cat(nrow(x_train), 'train samples\n')
cat(nrow(x_test), 'test samples\n')

# Convert class vectors to binary class matrices
y_train <- to_categorical(y_train, num_classes)
y_test <- to_categorical(y_test, num_classes)
```

The comments are self-explanatory: the data is loaded from the Keras dataset library and then transformed to a 2D arrray and normalized to fall into the [0,1] range.

4. Define the model:

```
model <- keras_model_sequential()
model %>%
  layer_dense(units=256,activation='relu',input_shape=c(784)) %>%
  layer_dropout(rate = 0.4) %>%
  layer_dense(units = 128, activation = 'relu') %>%
  layer_dropout(rate = 0.3) %>%
  layer_dense(units = 10, activation = 'softmax')

summary(model)

model %>% compile(
    loss = 'categorical_crossentropy',
    optimizer = optimizer_rmsprop(),
    metrics = c('accuracy')
)
```

5. A sequential model is defined and compiled. We get the model definition as follows:

```
Layer (type)                 Output Shape              Param #
================================================================
dense_26 (Dense)             (None, 256)               200960
________________________________________________________________
dropout_14 (Dropout)         (None, 256)               0
________________________________________________________________
dense_27 (Dense)             (None, 128)               32896
________________________________________________________________
dropout_15 (Dropout)         (None, 128)               0
________________________________________________________________
dense_28 (Dense)             (None, 10)                1290
================================================================
Total params: 235,146
Trainable params: 235,146
Non-trainable params: 0
```

6. Train the model:

```
history <- model %>% fit(
    x_train, y_train,
    batch_size = batch_size,
    epochs = epochs,
    verbose = 1,
    validation_split = 0.2
)

plot(history)
```

The output of the fit function is stored in the history object, which contains the loss and metrics values from the training epochs. The data in the history object is plotted and the result is as follows:

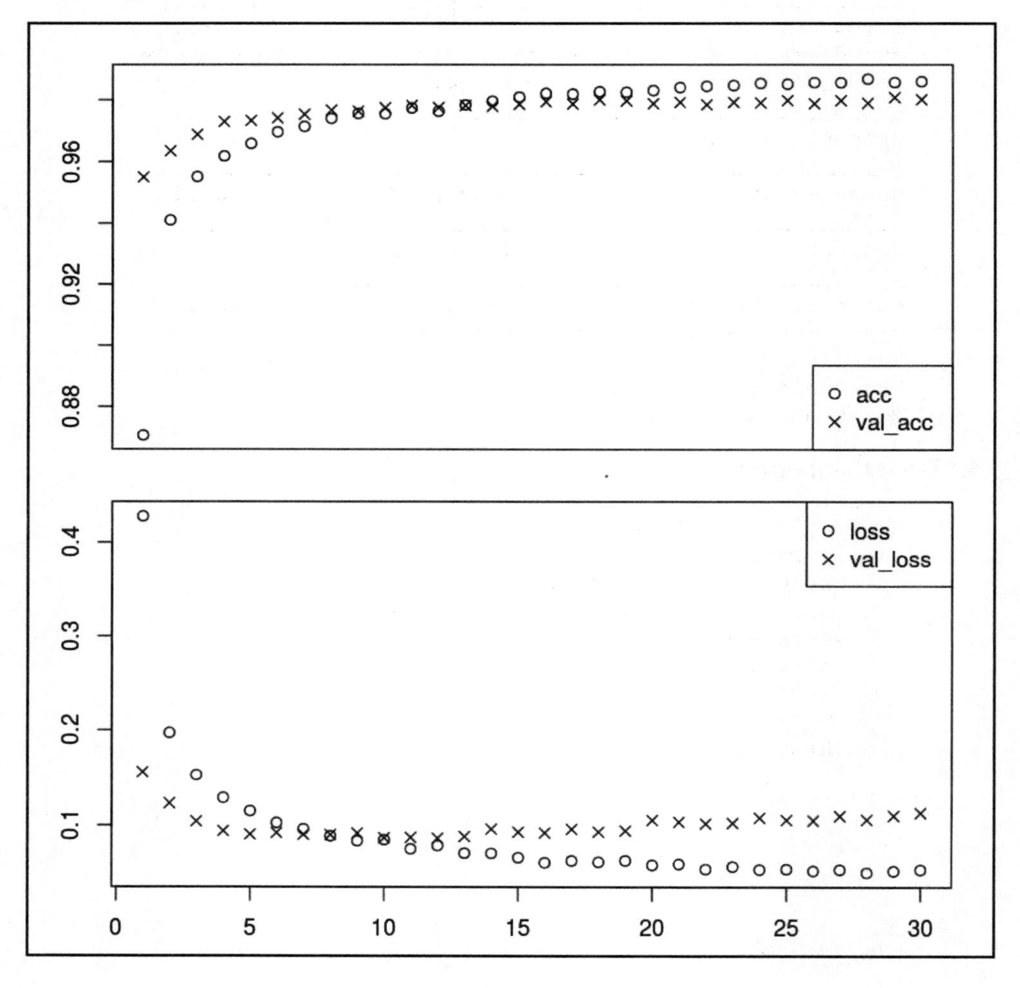

Training and Validation Accuracy (y-axis) in Epochs (x-axis)

7. Evaluate the model:

```
score <- model %>% evaluate(
    x_test, y_test,
    verbose = 0
)

# Output metrics
cat('Test loss:', score[[1]], '\n')
cat('Test accuracy:', score[[2]], '\n')
```

The output is as follows:

```
Test loss: 0.1128517
Test accuracy: 0.9816
```

Pretty cool!!

Find more examples of Keras in R at the following link: `https://keras.rstudio.com/articles/examples/index.html`
Further documentation on the Keras R package can be found at the following link: `https://keras.rstudio.com/reference/index.html`

TensorBoard in R

You can follow along with the code in the Jupyter R notebook `ch-17d_TensorBoard_in_R`.

You can view the TensorBoard with the `tensorboard()` function as follows:

```
tensorboard('logs')
```

Here, `'logs'` is the folder where the TensorBoard logs should be created.

The data will be shown as the epochs execute and the data is recorded. In R, collecting the data for TensorBoard depends on the package being used:

- If you are using the `tensorflow` package, then attach the `tf$summary$scalar` operations to the graph
- If you are using the `tfestimators` package, then TensorBoard data is automatically written to the `model_dir` parameter that is specified while creating the estimator
- If you are using the `keras` package, then you have to include the `callback_tensorboard()` function while training the model using the `fit()` function

We modify the training in the Keras example provided earlier as follows:

```
# Training the model --------
tensorboard("logs")

history <- model %>% fit(
    x_train, y_train,
    batch_size = batch_size,
    epochs = epochs,
    verbose = 1,
    validation_split = 0.2,
    callbacks = callback_tensorboard("logs")
)
```

When we execute the notebook, we get the following output for the training cell:

```
Started TensorBoard at http://127.0.0.1:4233
```

When we click on the link, we see the scalars plotted in TensorBoard:

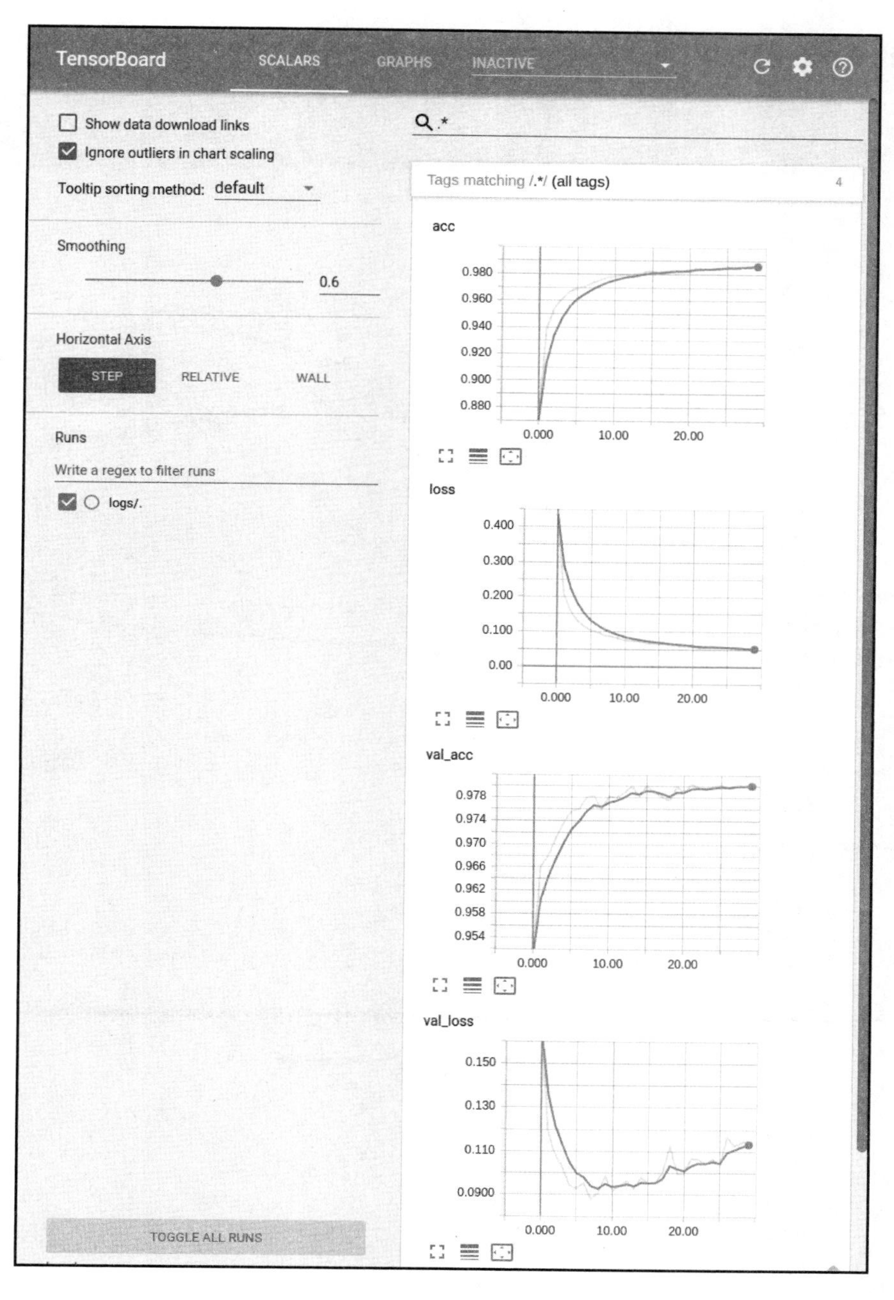

TensorBoad Visualization of Plots

On clicking the Graphs tab, we see the computation graph in TensorBoard:

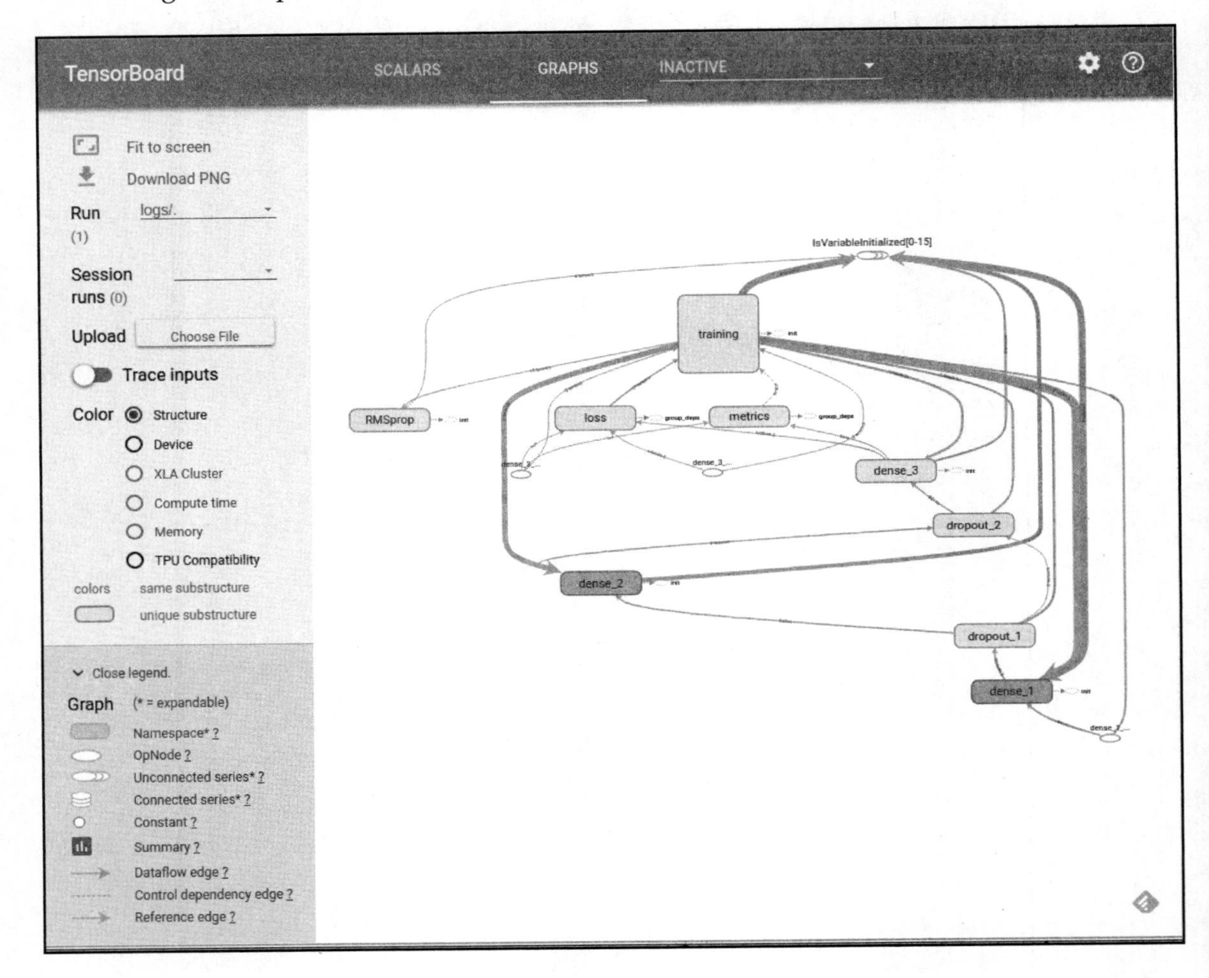

TensorBoard Visualization of Computation Graph

Further documentation on TensorBoard in R can be found at the following link: https://tensorflow.rstudio.com/tools/tensorboard.html.

The tfruns package in R

You can follow along with the code in the Jupyter R notebook ch-17d_TensorBoard_in_R.

The tfruns package is very useful tool provided in R that helps in tracking multiple runs for training the models. The run data is automatically captured by the tfruns package for models built using the keras and tfestimators packages in R. Using tfruns is pretty simple and easy. Just create your code in an R file and then execute the file using the training_run() function. For example, if you have a mnist_model.R file, then execute it using the training_run() function in the interactive R console as follows:

```
library(tfruns)
training_run('mnist_model.R')
```

Once the training is finished, the window displaying the summary of the run appears automatically. We get the following output in the window from the `mnist_mlp.R` we got from the `tfruns` GitHub repository (`https://github.com/rstudio/tfruns/blob/master/inst/examples/mnist_mlp/mnist_mlp.R`).

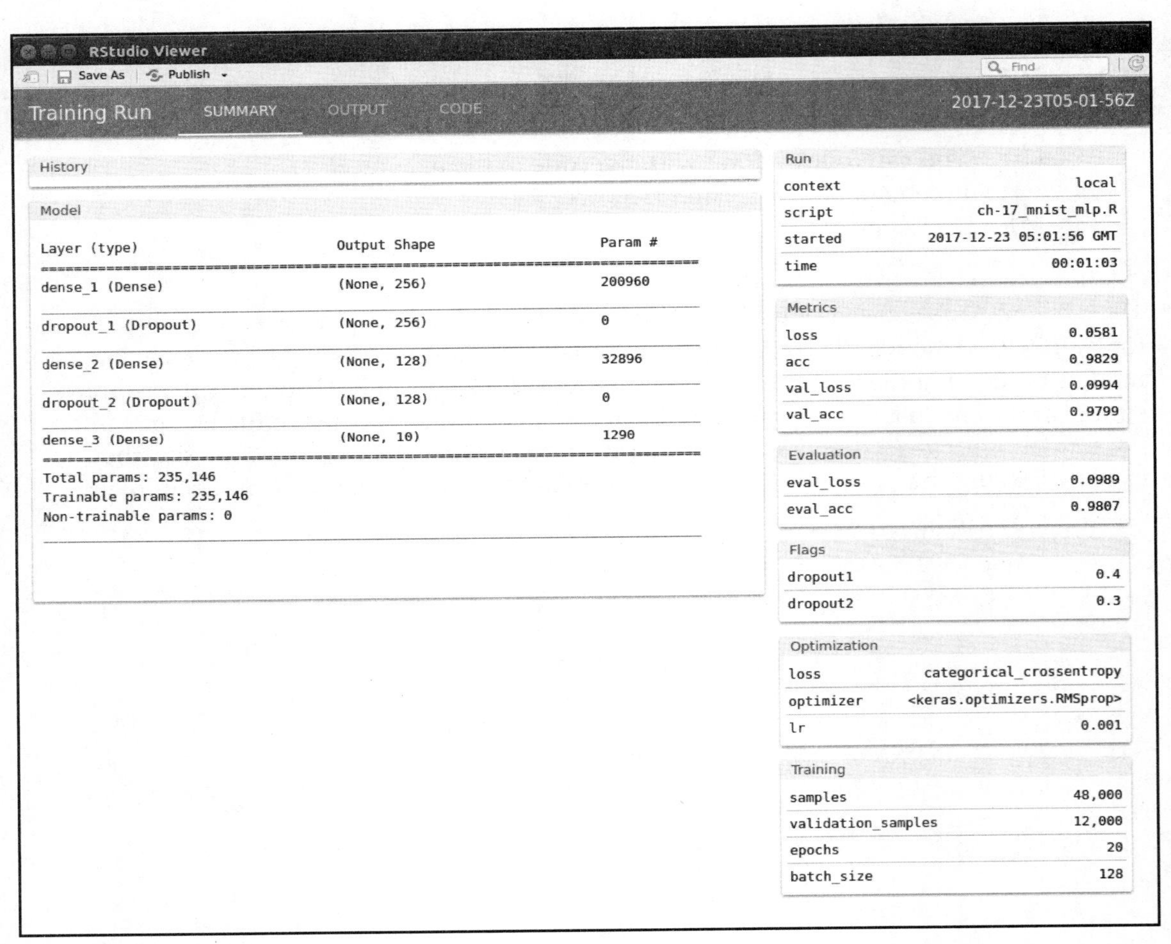

tfruns visualization of the model run

In the Viewer window, the output tab contains the plots:

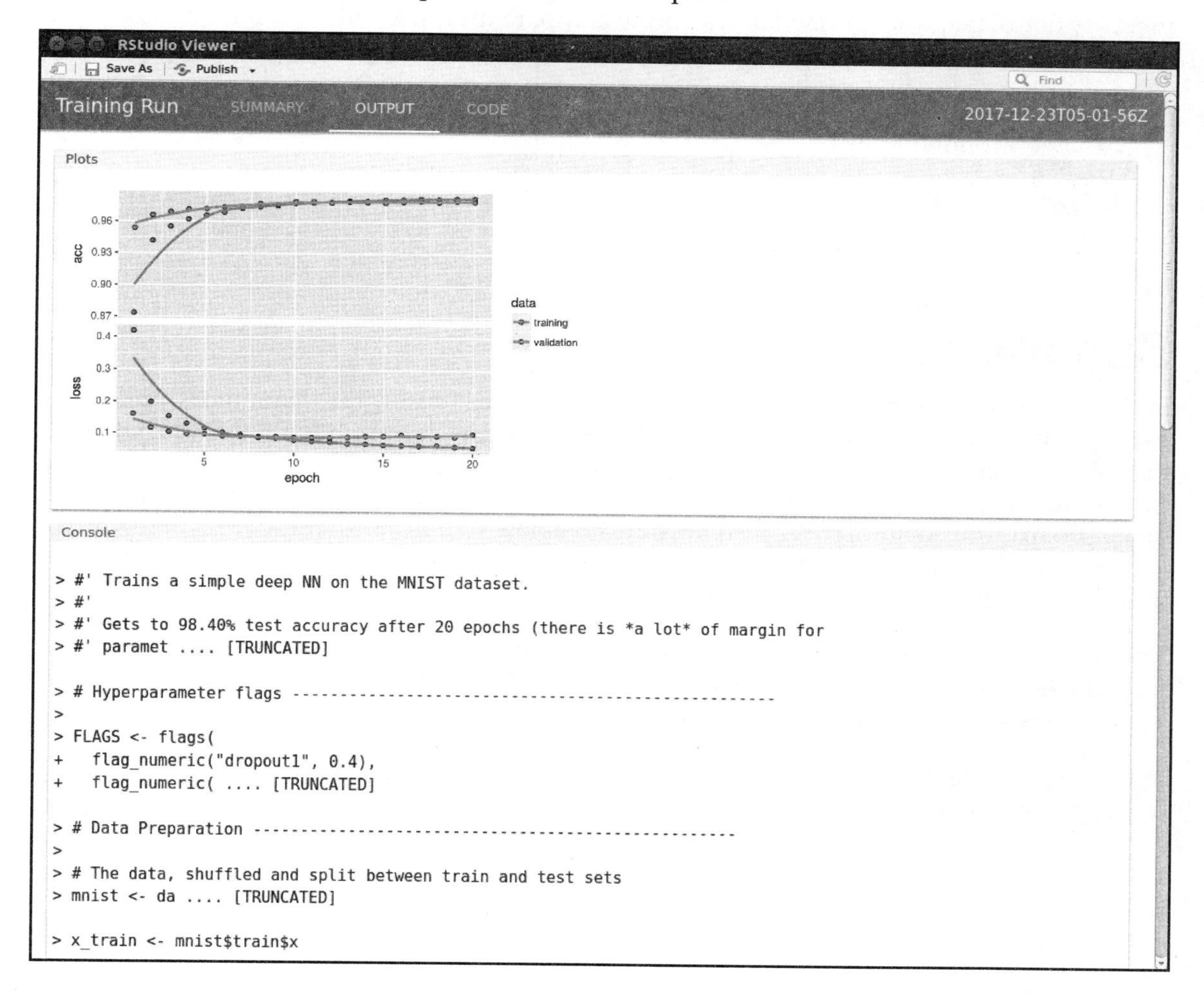

tfruns visualization of the accuracy and loss values

The `tfruns` package installs an addin to RStudio and can also be accessed from the `Addins` menu option. The package also allows you to compare multiple runs and publish run reports to RPubs or RStudio Connect. You can also choose to save the reports locally.

Further documentation on the `tfruns` package in R can be found at the following links:
`https://tensorflow.rstudio.com/tools/tfruns/reference/`
`https://tensorflow.rstudio.com/tools/tfruns/articles/overview.html`.

Summary

In this chapter, we learned how to use TensorFlow Core, TensorFlow Estimators, and Keras packages in R to build and train machine learning models. We provided a walkthrough of the MNIST examples from RStudio and provided links for further documentation of the TensorFlow and Keras R packages. We also learned how to use the visualization tool TensorBoard from within R. We also introduced a new tool from R Studio, `tfruns`, which allows you to create reports for multiple runs, analyze and compare them, and save them locally or publish them.

The ability to work directly in R is useful because plenty of production data science and machine learning code is written using R, and now you can integrate TensorFlow into the same codebase and run it within the R environment.

In the next chapter, we shall learn some techniques for debugging the code for building and training TensorFlow models.

18

Debugging TensorFlow Models

As we learned in this book, TensorFlow programs are used to build and train models that can be used for prediction in various kinds of tasks. When training the model, you build the computation graph, run the graph for training, and evaluate the graph for predictions. These tasks repeat until you are satisfied with the quality of the model, and then save the graph along with the learned parameters. In production, the graph is built or restored from a file and populated with the parameters.

Building deep learning models is a complex art and the TensorFlow API and its ecosystem are equally complex. When we build and train models in TensorFlow, sometimes we get different kinds of errors, or the models do not work as expected. As an example, how often do you see yourself getting stuck in one or more of the following situations:

- Getting NaN in loss and metrics output
- The loss or some other metric doesn't improve even after several iterations

In such situations, we would need to debug the code written using the TensorFlow API.

To fix the code so that it works, one could use the debugger or other methods and tools provided by the platform, such as the Python debugger (`pdb`) in Python and the GNU debugger (`gdb`) in Linux OS. The TensorFlow API also provides some additional support to fix the code when things go wrong.

In this chapter, we shall learn the additional tools and techniques available in TensorFlow to assist in debugging:

- Fetching tensor values with `tf.Session.run()`
- Printing tensor values with `tf.Print()`
- Asserting on conditions with `tf.Assert()`
- Debugging with the TensorFlow debugger (`tfdbg`)

Fetching tensor values with tf.Session.run()

You can fetch the tensor values you want to print with `tf.Session.run()`. The values are returned as a NumPy array and can be printed or logged with Python statements. This is the simplest and easiest approach, with the biggest drawback being that the computation graph executes all the dependent paths, starting from the fetched tensor, and if those paths include the training operations, then it advances one step or one epoch.

Therefore, most of the time you would not call `tf.Session.run()` to fetch tensors in the middle of the graph, but you would execute the whole graph and fetch all the tensors, the ones you need to debug along with the ones you do not need to debug.

The function `tf.Session.partial_run()` is also available for situations where you may want to execute part of the graph, but it is a highly experimental API and not ready for production use.

Printing tensor values with tf.Print()

Another option to print values for debugging purposes is to use `tf.Print()`. You can wrap a tensor in `tf.Print()` to print its values in the standard error console when the path containing the `tf.Print()` node is executed. The `tf.Print()` function has the following signature:

```
tf.Print(
    input_,
    data,
    message=None,
    first_n=None,
    summarize=None,
    name=None
   )
```

The arguments to this function are as follows:

- `input_` is a tensor that gets returned from the function without anything being done to it
- `data` is the list of tensors that get printed
- `message` is a string that gets printed as a prefix to the printed output
- `first_n` represents the number of steps to print the output; if this value is negative then the value is always printed whenever the path is executed

- `summarize` represents the number of elements to print from the tensor; by default, only three elements are printed

You can follow along with the code in the Jupyter notebook `ch-18_TensorFlow_Debugging`.

Let us modify the MNIST MLP model we created earlier to add the print statement:

```
model = tf.Print(input_=model,
                 data=[tf.argmax(model,1)],
                 message='y_hat=',
                 summarize=10,
                 first_n=5
                 )
```

When we run the code, we get the following in Jupyter's console:

```
I tensorflow/core/kernels/logging_ops.cc:79] y_hat=[0 0 0 7 0 0 0 0 0 0...]
I tensorflow/core/kernels/logging_ops.cc:79] y_hat=[0 7 7 1 8 7 2 7 7 0...]
I tensorflow/core/kernels/logging_ops.cc:79] y_hat=[4 8 0 6 1 8 1 0 7 0...]
I tensorflow/core/kernels/logging_ops.cc:79] y_hat=[0 0 1 0 0 0 0 5 7 5...]
I tensorflow/core/kernels/logging_ops.cc:79] y_hat=[9 2 2 8 8 6 6 1 7 7...]
```

The only disadvantage of using `tf.Print()` is that the function provides limited formatting functionality.

Asserting on conditions with tf.Assert()

Yet another way to debug TensorFlow models is to insert conditional asserts. The `tf.Assert()` function takes a condition, and if the condition is false, it then prints the lists of given tensors and throws `tf.errors.InvalidArgumentError`.

1. The `tf.Assert()` function has the following signature:

```
tf.Assert(
    condition,
    data,
    summarize=None,
    name=None
)
```

2. An assert operation does not fall in the path of the graph like the `tf.Print()` function. To make sure that the `tf.Assert()` operation gets executed, we need to add it to the dependencies. For example, let us define an assertion to check that all the inputs are positive:

```
assert_op = tf.Assert(tf.reduce_all(tf.greater_equal(x,0)),[x])
```

3. Add`assert_op` to the dependencies at the time of defining the model, as follows:

```
with tf.control_dependencies([assert_op]):
    # x is input layer
    layer = x
    # add hidden layers
    for i in range(num_layers):
        layer = tf.nn.relu(tf.matmul(layer, w[i]) + b[i])
    # add output layer
    layer = tf.matmul(layer, w[num_layers]) + b[num_layers]
```

4. To test this code, we introduce an impurity after epoch 5, as follows:

```
if epoch > 5:
    X_batch = np.copy(X_batch)
    X_batch[0,0]=-2
```

5. The code runs fine for five epochs and then throws the error:

```
epoch: 0000   loss = 6.975991
epoch: 0001   loss = 2.246228
epoch: 0002   loss = 1.924571
epoch: 0003   loss = 1.745509
epoch: 0004   loss = 1.616791
epoch: 0005   loss = 1.520804

-----------------------------------------------------------------
InvalidArgumentError          Traceback (most recent call last)
...
InvalidArgumentError: assertion failed: [[-2 0 0]...]
...
```

Apart from the `tf.Assert()` function, which can take any valid conditional expression, TensorFlow provides the following assertion operations that check for specific conditions and have a simple syntax:

- `assert_equal`
- `assert_greater`
- `assert_greater_equal`
- `assert_integer`
- `assert_less`
- `assert_less_equal`
- `assert_negative`
- `assert_none_equal`
- `assert_non_negative`
- `assert_non_positive`
- `assert_positive`
- `assert_proper_iterable`
- `assert_rank`
- `assert_rank_at_least`
- `assert_rank_in`
- `assert_same_float_dtype`
- `assert_scalar`
- `assert_type`
- `assert_variables_initialized`

As an example, the previously mentioned example assert operation can also be written as follows:

```
assert_op = tf.assert_greater_equal(x,0)
```

Debugging with the TensorFlow debugger (tfdbg)

The TensorFlow debugger (`tfdbg`) works the same way at a high level as other popular debuggers, such as `pdb` and `gdb`. To use a debugger, the process is generally as follows:

1. Set the breakpoints in the code at locations where you want to break and inspect the variables
2. Run the code in debug mode
3. When the code breaks at a breakpoint, inspect it and then move on to next step

Some debuggers also allow you to interactively watch the variables while the code is executing, not just at the breakpoint:

1. In order to use `tfdbg`, first import the required modules and wrap the session inside a debugger wrapper:

    ```
    from tensorflow.python import debug as tfd

    with tfd.LocalCLIDebugWrapperSession(tf.Session()) as tfs:
    ```

2. Next, attach a filter to the session object. Attaching a filter is the same as setting a breakpoint in other debuggers. For example, the following code attaches a `tfdbg.has_inf_or_nan` filter which breaks if any of the intermediate tensors have `nan` or `inf` values:

    ```
    tfs.add_tensor_filter('has_inf_or_nan_filter', tfd.has_inf_or_nan)
    ```

3. Now when the code executes the `tfs.run()`, the debugger will start a debugger interface in the console where you can run various debugger commands to watch the tensor values.
4. We have provided the code for trying out `tfdbg` in the `ch-18_mnist_tfdbg.py` file. We see the `tfdbg` console when we execute the code file with `python3`:

```
python3 ch-18_mnist_tfdbg.py
```

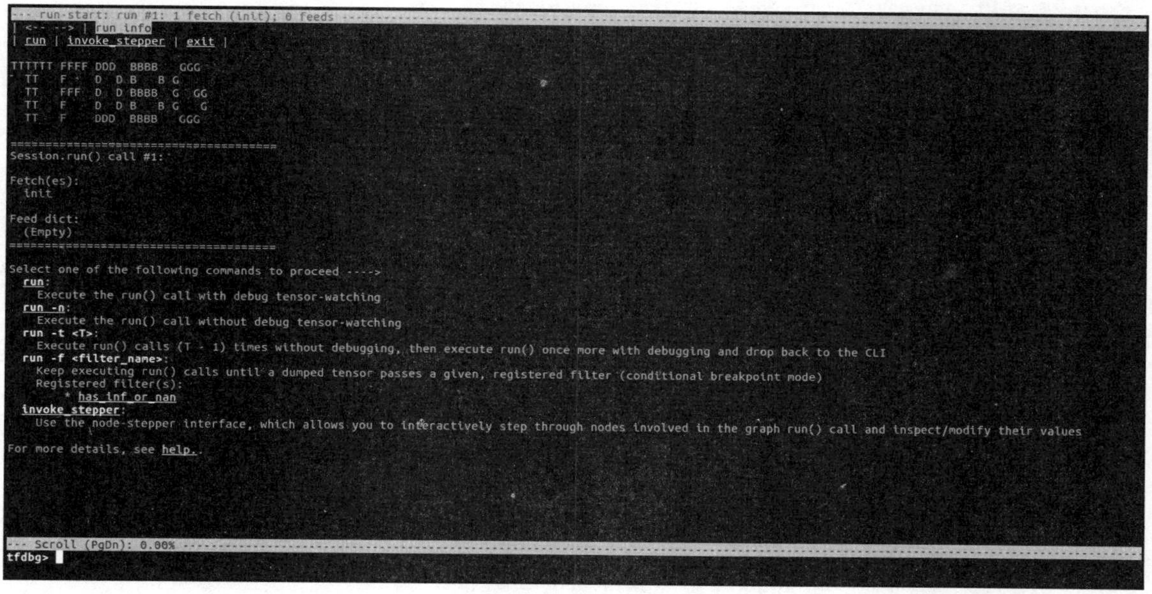

```
--- run-start: run #1: 1 fetch (init); 0 feeds ----------------------------
| <-- --> | run info
| run | invoke_stepper | exit |

TTTTTT FFFF DDD  BBBB   GGG
  TT   F    D  D B   B G
  TT   FFF  D  D BBBB  G  GG
  TT   F    D  D B   B G   G
  TT   F    DDD  BBBB   GGG

======================================
Session.run() call #1:

Fetch(es):
  init

Feed dict:
  (Empty)
======================================

Select one of the following commands to proceed ---->
  run:
    Execute the run() call with debug tensor-watching
  run -n:
    Execute the run() call without debug tensor-watching
  run -t <T>:
    Execute run() calls (T - 1) times without debugging, then execute run() once more with debugging and drop back to the CLI
  run -f <filter_name>:
    Keep executing run() calls until a dumped tensor passes a given, registered filter (conditional breakpoint mode)
    Registered filter(s):
        * has_inf_or_nan
  invoke_stepper:
    Use the node-stepper interface, which allows you to interactively step through nodes involved in the graph run() call and inspect/modify their values

For more details, see help..

--- Scroll (PgDn): 0.00% --------------------------------------------------
tfdbg>
```

5. Give the command `run -f has_inf_or_nan` at the `tfdbg>` prompt. The code breaks after the first epoch, because we populated the data with the `np.inf` value:

```
--- run-end: run #552: 2 fetches; 2 feeds ------------------------------------------
| <-- --> | lt -f has_inf_or_nan
| list_tensors | node_info | print_tensor | list_inputs | list_outputs | run_info | help |
54 dumped tensor(s) passing filter "has_inf_or_nan":

t (ms)    Size (B) Op type                        Tensor name
[2.590]   306.49k  Switch                         Assert/AssertGuard/Assert/Switch_1:1
[3.053]   6.43k    MatMul                         MatMul:0
[3.318]   6.42k    Add                            add:0
[3.537]   6.42k    Relu                           Relu:0
[3.771]   12.68k   MatMul                         MatMul_1:0
[4.153]   12.68k   Add                            add_1:0
[4.411]   12.68k   Relu                           Relu_1:0
[4.600]   4.09k    MatMul                         MatMul_2:0
[4.875]   4.08k    Add                            add_2:0
[5.408]   4.08k    Print                          Print:0
[6.072]   4.09k    Reshape                        Reshape:0
[6.403]   4.13k    SoftmaxCrossEntropyWithLogits  SoftmaxCrossEntropyWithLogits:1
[6.410]   624      SoftmaxCrossEntropyWithLogits  SoftmaxCrossEntropyWithLogits:0
[6.810]   584      Reshape                        Reshape_2:0
[7.009]   172      Mean                           Mean:0
[8.235]   4.17k    Mul                            gradients/SoftmaxCrossEntropyWithLogits_grad/mul:0
[8.367]   4.13k    Reshape                        gradients/Reshape_grad/Reshape:0
[8.586]   256      Sum                            gradients/add_2_grad/Sum_1:0
[8.590]   4.12k    Sum                            gradients/add_2_grad/Sum:0
[8.754]   4.13k    Reshape                        gradients/add_2_grad/Reshape:0
[8.905]   264      Reshape                        gradients/add_2_grad/Reshape_1:0
[9.023]   4.16k    Identity                       gradients/add_2_grad/tuple/control_dependency:0
[9.211]   298      Identity                       gradients/add_2_grad/tuple/control_dependency_1:0
[9.221]   12.72k   MatMul                         gradients/MatMul_2_grad/MatMul:0
[9.345]   1.48k    MatMul                         gradients/MatMul_2_grad/MatMul_1:0
[9.492]   12.76k   Identity                       gradients/MatMul_2_grad/tuple/control_dependency:0
[9.495]   1.51k    Identity                       gradients/MatMul_2_grad/tuple/control_dependency_1:0
[9.504]   302      ApplyGradientDescent           GradientDescent/update_b_out/ApplyGradientDescent:0
[9.668]   1.51k    ApplyGradientDescent           GradientDescent/update_w_out/ApplyGradientDescent:0
[9.719]   12.72k   ReluGrad                       gradients/Relu_1_grad/ReluGrad:0
[9.949]   12.71k   Sum                            gradients/add_1_grad/Sum:0
[9.956]   346      Sum                            gradients/add_1_grad/Sum_1:0
[10.101]  354      Reshape                        gradients/add_1_grad/Reshape_1:0
[10.121]  12.72k   Reshape                        gradients/add_1_grad/Reshape:0
[10.280]  12.75k   Identity                       gradients/add_1_grad/tuple/control_dependency:0
[10.280]  388      Identity                       gradients/add_1_grad/tuple/control_dependency_1:0
[10.564]  394      ApplyGradientDescent           GradientDescent/update_b_0001/ApplyGradientDescent:0
[10.612]  6.47k    MatMul                         gradients/MatMul_1_grad/MatMul:0
[10.618]  2.23k    MatMul                         gradients/MatMul_1_grad/MatMul_1:0
[10.812]  2.26k    Identity                       gradients/MatMul_1_grad/tuple/control_dependency_1:0
[10.822]  6.51k    Identity                       gradients/MatMul_1_grad/tuple/control_dependency:0
[11.028]  6.47k    ReluGrad                       gradients/Relu_grad/ReluGrad:0
[11.034]  2.26k    ApplyGradientDescent           GradientDescent/update_w_0001/ApplyGradientDescent:0
[11.247]  276      Sum                            gradients/add_grad/Sum_1:0
[11.254]  6.46k    Sum                            gradients/add_grad/Sum:0
[11.399]  6.47k    Reshape                        gradients/add_grad/Reshape:0
[11.406]  284      Reshape                        gradients/add_grad/Reshape_1:0
[11.563]  6.50k    Identity                       gradients/add_grad/tuple/control_dependency:0
[11.565]  318      Identity                       gradients/add_grad/tuple/control_dependency_1:0
--- Scroll (PgDn): 0.00% -----------------------------------------------------------
tfdbg>
```

6. Now you can use the `tfdbg` console or the clickable interface to inspect the values of various tensors. For example, we look at the values of one of the gradients:

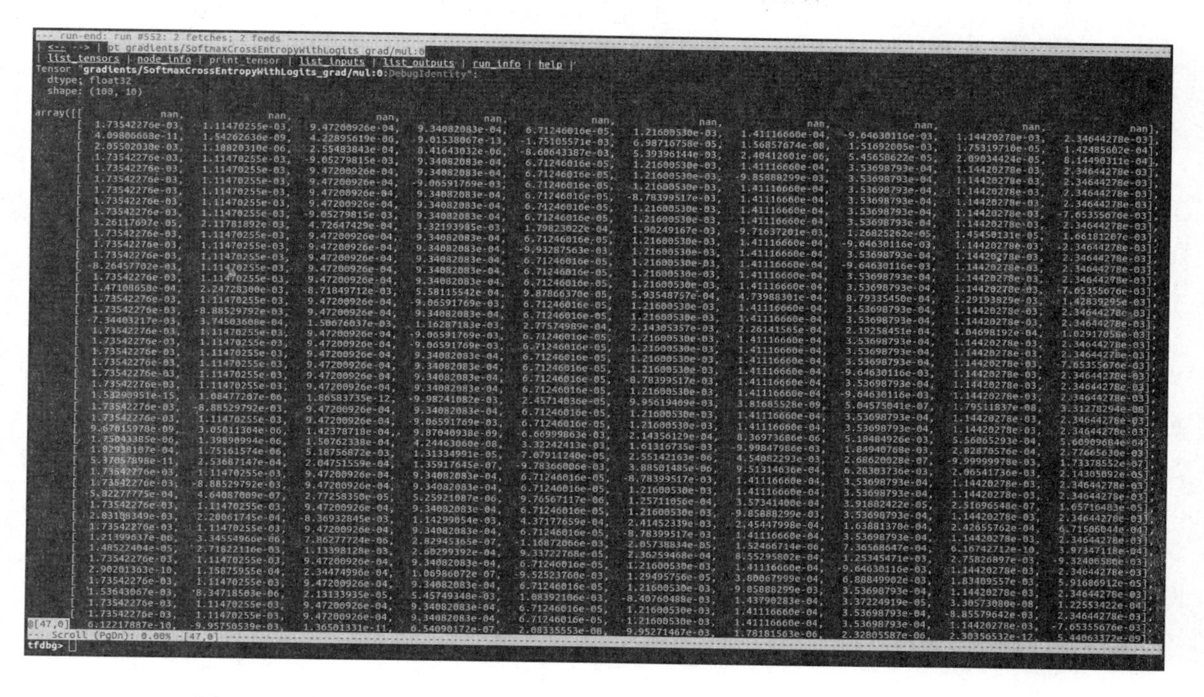

You can find more information about using the `tfdbg` console and inspecting the variables at the following link: `https://www.tensorflow.org/programmers_guide/debugger`.

Summary

In this chapter, we learned how to debug the code for building and training models in TensorFlow. We learned that we can fetch the tensors as NumPy arrays using `tf.Session.run()`. We can also print the values of tensors by adding `tf.Print()` operations in the computation graph. We also learned how to raise errors when certain conditions fail to hold during execution with `tf.Assert()` and other `tf.assert_*` operations. We closed the chapter with an introduction to the TensorFlow debugger (`tfdbg`) for setting breakpoints and watching the values of tensors like we would do for debugging the code in the Python debugger (`pdb`) or the GNU debugger (`gdb`).

This chapter brings our journey to a new milestone. We do not expect that the journey ends here, but we believe that the journey just got started and you will further expand and apply the knowledge and skills gained in this book.

We are keenly looking forward to hearing your experiences, feedback, and suggestions.

Tensor Processing Units

A **Tensor Processing Unit** (**TPU**) is an **application-specific integrated circuit** (**ASIC**) that implements hardware circuits optimized for the computation requirements of deep neural networks. A TPU is based on a **Complex Instruction Set Computer** (**CISC**) instruction set that implements high-level instructions for running complex tasks for training deep neural networks. The heart of the TPU architecture resides in the systolic arrays that optimize the matrix operations.

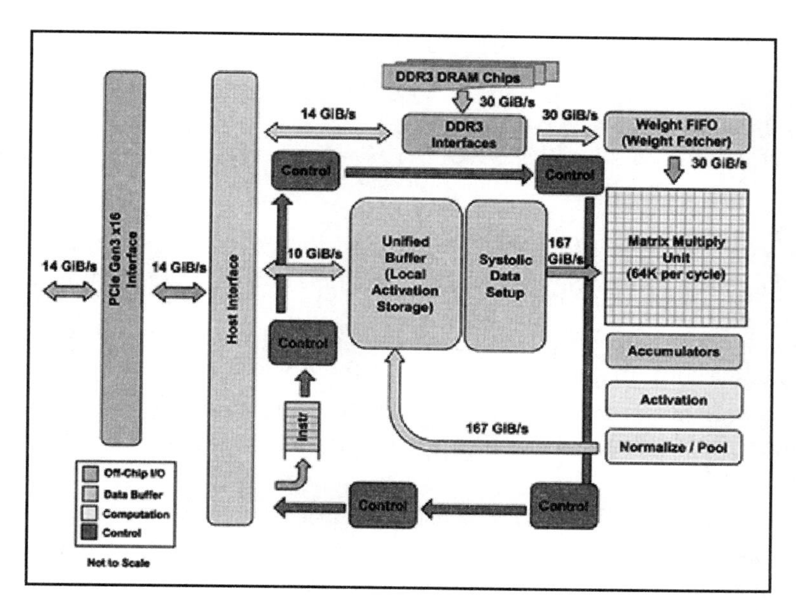

The Architecture of TPU

Image from: https://cloud.google.com/blog/big-data/2017/05/images/149454602921110/tpu-15.png

TensorFlow provides a compiler and software stack that translates the API calls from TensorFlow graphs into TPU instructions. The following block diagram depicts the architecture of TensorFlow models running on top of the TPU stack:

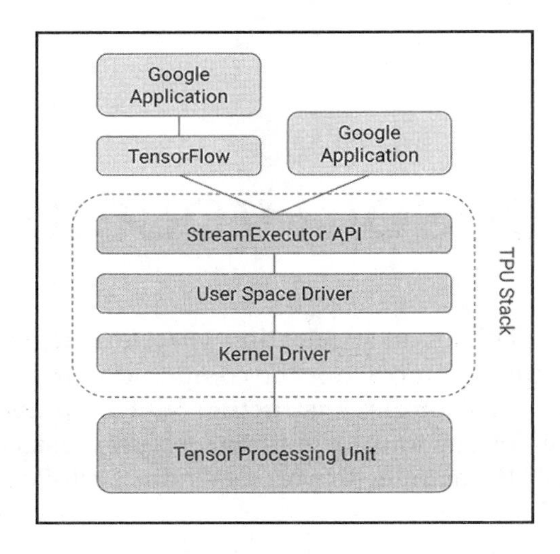

Image from: https://cloud.google.com/blog/big-data/2017/05/images/149454602921110/tpu-2.png

For more information on the TPU architecture, read the blog at the following link: `https://cloud.google.com/blog/big-data/2017/05/an-in-depth-look-at-googles-first-tensor-processing-unit-tpu`.

The TensorFlow API for the TPU resides in the `tf.contrib.tpu` module. For building the models on the TPU, the following three TPU-specific TensorFlow modules are used:

- `tpu_config`: The `tpu_config` module allows you to create the configuration object that contains information about the host that would be running the model.
- `tpu_estimator`: The `tpu_estimator` module encapsulates the estimator in the `TPUEstimator` class. To run the estimator on the TPU, we create an object of this class.
- `tpu_optimizer`: The `tpu_optimizer` module wraps the optimizer. For example, in the following sample code we wrap the SGD optimizer in the `CrossShardOptimizer` class from the `tpu_optimizer` module.

As an example, the following code builds the CNN model for the MNIST dataset on the TPU with the TF Estimator API:

The following code is adapted from `https://github.com/tensorflow/tpu-demos/blob/master/cloud_tpu/models/mnist/mnist.py`.

```
import tensorflow as tf

from tensorflow.contrib.tpu.python.tpu import tpu_config
from tensorflow.contrib.tpu.python.tpu import tpu_estimator
from tensorflow.contrib.tpu.python.tpu import tpu_optimizer

learning_rate = 0.01
batch_size = 128

def metric_fn(labels, logits):
    predictions = tf.argmax(logits, 1)
    return {
        "accuracy": tf.metrics.precision(
            labels=labels, predictions=predictions),
    }

def model_fn(features, labels, mode):
    if mode == tf.estimator.ModeKeys.PREDICT:
        raise RuntimeError("mode {} is not supported yet".format(mode))

    input_layer = tf.reshape(features, [-1, 28, 28, 1])
    conv1 = tf.layers.conv2d(
        inputs=input_layer,
        filters=32,
        kernel_size=[5, 5],
        padding="same",
        activation=tf.nn.relu)
    pool1 = tf.layers.max_pooling2d(inputs=conv1, pool_size=[2, 2],
                                    strides=2)
    conv2 = tf.layers.conv2d(
        inputs=pool1,
        filters=64,
        kernel_size=[5, 5],
        padding="same",
        activation=tf.nn.relu)
    pool2 = tf.layers.max_pooling2d(inputs=conv2, pool_size=[2, 2],
                                    strides=2)
    pool2_flat = tf.reshape(pool2, [-1, 7 * 7 * 64])
    dense = tf.layers.dense(inputs=pool2_flat, units=128,
```

```
                                  activation=tf.nn.relu)
    dropout = tf.layers.dropout(
        inputs=dense, rate=0.4,
        training=mode == tf.estimator.ModeKeys.TRAIN)
    logits = tf.layers.dense(inputs=dropout, units=10)
    onehot_labels = tf.one_hot(indices=tf.cast(labels, tf.int32), depth=10)

    loss = tf.losses.softmax_cross_entropy(
        onehot_labels=onehot_labels, logits=logits)

    if mode == tf.estimator.ModeKeys.EVAL:
        return tpu_estimator.TPUEstimatorSpec(
            mode=mode,
            loss=loss,
            eval_metrics=(metric_fn, [labels, logits]))

    # Train.
    decaying_learning_rate = tf.train.exponential_decay(learning_rate,
                                                      tf.train.get_global_step(),
                                                      100000,0.96)

    optimizer = tpu_optimizer.CrossShardOptimizer(
            tf.train.GradientDescentOptimizer(
                learning_rate=decaying_learning_rate))

    train_op = optimizer.minimize(loss,
            global_step=tf.train.get_global_step())
    return tpu_estimator.TPUEstimatorSpec(mode=mode,
            loss=loss, train_op=train_op)

def get_input_fn(filename):
    def input_fn(params):
        batch_size = params["batch_size"]

        def parser(serialized_example):
            features = tf.parse_single_example(
                serialized_example,
                features={
                    "image_raw": tf.FixedLenFeature([], tf.string),
                    "label": tf.FixedLenFeature([], tf.int64),
                })
            image = tf.decode_raw(features["image_raw"], tf.uint8)
            image.set_shape([28 * 28])
            image = tf.cast(image, tf.float32) * (1. / 255) - 0.5
            label = tf.cast(features["label"], tf.int32)
            return image, label

        dataset = tf.data.TFRecordDataset(
```

```
            filename, buffer_size=FLAGS.dataset_reader_buffer_size)
        dataset = dataset.map(parser).cache().repeat()
        dataset = dataset.apply(
            tf.contrib.data.batch_and_drop_remainder(batch_size))
        images, labels = dataset.make_one_shot_iterator().get_next()
        return images, labels
    return input_fn

# TPU config

master = 'local' #URL of the TPU instance
model_dir = '/home/armando/models/mnist'
n_iterations = 50  # number of iterations per TPU training loop
n_shards = 8   # number of TPU chips

run_config = tpu_config.RunConfig(
        master=master,
        evaluation_master=master,
        model_dir=model_dir,
        session_config=tf.ConfigProto(
            allow_soft_placement=True,
            log_device_placement=True
        ),
        tpu_config=tpu_config.TPUConfig(n_iterations,
                                        n_shards
        )
    )

estimator = tpu_estimator.TPUEstimator(
    model_fn=model_fn,
    use_tpu=True,
    train_batch_size=batch_size,
    eval_batch_size=batch_size,
    config=run_config)

train_file = '/home/armando/datasets/mnist/train' # input data file
train_steps = 1000 # number of steps to train for

estimator.train(input_fn=get_input_fn(train_file),
                max_steps=train_steps
                )

eval_file = '/home/armando/datasets/mnist/test' # test data file
eval_steps = 10

estimator.evaluate(input_fn=get_input_fn(eval_file),
                   steps=eval_steps
                   )
```

More examples of building models on a TPU can be found at the following link: `https://github.com/tensorflow/tpu-demos`.

Other Books You May Enjoy

If you enjoyed this book, you may be interested in these other books by Packt:

TensorFlow 1.x Deep Learning Cookbook
Antonio Gulli, Amita Kapoor

ISBN: 9781788293594

- Install TensorFlow and use it for CPU and GPU operations
- Implement DNNs and apply them to solve different AI-driven problems.
- Leverage different data sets such as MNIST, CIFAR-10, and Youtube8m with TensorFlow and learn how to access and use them in your code.
- Use TensorBoard to understand neural network architectures, optimize the learning process, and peek inside the neural network black box.
- Use different regression techniques for prediction and classification problems
- Build single and multilayer perceptrons in TensorFlow
- Implement CNN and RNN in TensorFlow, and use it to solve real-world use cases.
- Learn how restricted Boltzmann Machines can be used to recommend movies.
- Understand the implementation of Autoencoders and deep belief networks, and use them for emotion detection.
- Master the different reinforcement learning methods to implement game playing agents.
- GANs and their implementation using TensorFlow.

Machine Learning with TensorFlow 1.x
Quan Hua, Shams Ul Azeem, Saif Ahmed

ISBN: 9781786462961

- Explore how to use different machine learning models to ask different questions of your data
- Learn how to build deep neural networks using TensorFlow 1.x
- Cover key tasks such as clustering, sentiment analysis, and regression analysis using TensorFlow 1.x
- Find out how to write clean and elegant Python code that will optimize the strength of your algorithms
- Discover how to embed your machine learning model in a web application for increased accessibility
- Learn how to use multiple GPUs for faster training using AWS

Leave a review - let other readers know what you think

Please share your thoughts on this book with others by leaving a review on the site that you bought it from. If you purchased the book from Amazon, please leave us an honest review on this book's Amazon page. This is vital so that other potential readers can see and use your unbiased opinion to make purchasing decisions, we can understand what our customers think about our products, and our authors can see your feedback on the title that they have worked with Packt to create. It will only take a few minutes of your time, but is valuable to other potential customers, our authors, and Packt. Thank you!

Index

E

F

G

H

I

K

L

M

N

O

P

Q

R

S

T

U

V

W